T0324208

CHROMATOGRAPHY

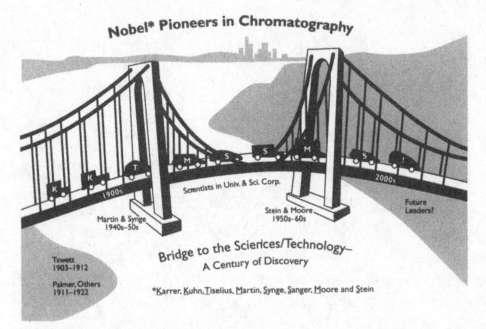

The evolution of chromatography: The bridge to the sciences and technology. Some of the early scientists who invented, rediscovered, and/or advanced chromatography include M. S. Tswett, L. S. Palmer, R. Kuhn, A. W. K. Tiselius, A. J. P. Martin, R. L. M. Synge, F. Sanger, S. Moore, and W. B. Stein, and the awardees in the earlier book, *Chromatography: A Century of Discovery (1900–2000)—the Bridge to the Sciences/Technology*, Vol. 64, Elsevier, Amsterdam, 2001; see Chapters 1, 2, 5, and 6 therein and the present Chapters 3, 4, 5, and 11. What are the common features of their discoveries?

CHROMATOGRAPHY

A Science of Discovery

Edited by

Robert L. Wixom
Charles W. Gehrke
University of Missouri

Associate Editors

Deborah L. Chance
Thomas P. Mawhinney
University of Missouri, Columbia

A JOHN WILEY & SONS, INC., PUBLICATION

Published by John Wiley & Sons, Inc., Hoboken, New Jersey.
Published simultaneously in Canada

For general information on our other products and services or for technical support, please contact our Customer Care Department within the United States at (800) 762-2974, outside the United States at (317) 572-3993 or fax (317) 572-4002.

Wiley also publishes it books in variety of electronic formats. Some content that appears in print may not be available in electronic formats. For more information about Wiley products, visit our web site at www.wiley.com.

Library of Congress Cataloging-in-publication Data:

Chromatography : a science of discovery / edited by Robert L. Wixom, Charles W. Gehrke.
   p. cm.
  Includes bibliographical references and index.
  ISBN 978-0-470-28345-5 (cloth)
  1. Chromatographic analysis–History–20th century–21st century. I. Wixom, Robert L. II. Gehrke, Charles W.
  QD79.C4C48376 2010
  543′.8–dc22
                       ·2009020801

CONTENTS

DEDICATION

To Professor Emeritus Charles W. Gehrke (deceased February 10, 2009), *who employed his clarity, resourcefulness, and organizational skills in the demanding task as senior editor of this book; and who will be remembered by faculty, students, and administration for his friendliness, persuasiveness, and integrity (see Editors/Authors section of this book).*

To Professor Emeritus Robert L. Wixom (deceased July 8, 2009), *who was determined to share with the new scientists of today, the rich history and fundamentals of thought and action which brought us to the chromatography of today and point us toward the future;*

and who will be remembered academically for his quest for truth and willingness to take the extra time and go the extra mile so that one might learn something new (see Editors/ Authors section of this book).

To Professor Ernst Bayer (deceased 2001), *who made significant contributions as co-editor of the earlier book, Chromatography: A Century of Discovery (1900– 2000)—The Bridge to the Sciences/Technology, Vol. 64, Elsevier, Amsterdam, 2001 (see his biography therein), and who has made outstanding multiple research contributions in chromatography.*

To our scientific colleagues, *who have contributed significantly to the advances of chromatography and then described their research in this volume; and who have contributed scientific thoughts, ideas for experiments, and means for communication.*

To our relevant research institutions, *whether university, professional societies, or corporate/government agencies that have supported related research endeavors.*

To our respective family members, *who have been both patient and helpful as we moved through the writing process.*

Our thanks are also extended to the editors of John Wiley & Sons, Inc.

PREFACE

The frontispiece illustration depicts the aim of this book: to capture the coherent message of our earlier chromatography book [1] and the present one—namely, to bridge, or to provide the connection of, the earlier chromatography (1900–2000) with the chromatography of the present book (2000–2008) and today's science. Those multifaceted messages are based on connecting the *pioneers of chromatography*—Mikail S. Tswett, Leroy S. Palmer, and several others [1], the early *builders of chromatography*—R. Kuhn, A. W. K. Tiselius, A. J. P. Martin, R. L. M. Synge, F. Sanger, S. Moore, and W. B. Stein, and the *awardees* (see [1] and Chapters 3 and 5 of this book). The *bridge* or the connections to later and wider facets of chromatography leads to examples of *the science of discovery*.

Now, a hundred years after M. S. Tswett, his robust infant of chromatography has grown to be a major subject with dozens of references for each chapter in the 2001 chromatography book [1], >50 pages in the appendixes of the 2001 online supplement [2], and this book. This expanding amplitude of chromatography and the scientific literature makes journal and/or book reading slower and perhaps tedious in the drive to be thorough. Fortunately, the recent electronic revolution has delivered the tools to resolve that dilemma—namely, journals online, abstract journals online, and search engines. Consequently, the appendixes in the online version of *Chromatography* [2] are not being updated for this volume. To summarize, turn on the power for the Internet.

Chromatography has observed the classical flow of many sciences—to question, to undertake related experiments, to probe the subject, to build on recent observations and hypotheses, to interpret the experiments, and when sufficient evidence has accumulated, to develop theories.

Some unique aspects of this book follow: Chapter 1 introduces the evidence for chromatography to be more than a method of separation, but is now a discipline of science. While some view chromatography as comprising multiple areas, Chapter 2 presents the case for chromatography as a unified science. After introducing the early "pioneers" and "builders" of chromatography in Chapter 3, an additional 25 Nobel Prize awardees (1937–1999) and 19 in (2000–2007) relied on chromatography as a method in their overall research. This review led to a discussion of "sharp turns," "breakthroughs," "scientific revolutions," and "paradigm shifts." The multiple branches of chromatography known in the 1960s have continued to grow over the subsequent decades to now be recognized as "trails of research" (Chapter 4). Our key Chapter 5 presents the 2000–2007 awardees and key contributors along with a description of their research.

In Chapter 6, several invited scientists describe recent advances in column technology and its validation. Chapters 7, 8, and 9 examine the contributions of chromatography in subject areas: agricultural, space, and biological/medical sciences; pharmaceutical science; and the environment, natural products, and chemical analysis and synthesis. Chapter 10 provides a comprehensive compilation of references on the history of chromatography. Chapter 11, first describes three institutions that have contributed over the decades to chromatography: the Pittsburgh Conference on Analytical Chemistry and Applied Spectrometry (PITTCON); the Chemical Heritage Foundation, and the American Chemical Society. Next is a section on recognizing and using chromatographic separations as molecular interaction amplifiers. Perspectives for the future are given by awardees and significant contributors to the field of chromatography. Finally, the book concludes with how chromatography has contributed to the advances in systems biology, genomics, and proteomics along with the initial reliance on other methods and theory in order to advance these new areas of science. The continuous theme throughout this book portrays chromatography as "the science of discovery."

This book is recommended for students in the sciences and research chromatographers at all levels of experience: professional scientists; research investigators in academia, government, and industry; science libraries in academia, government, industry, and professional societies; and historians and philosophers of science and educators and their advanced students. This book builds on its 2001 predecessor [1] to identify the additional more recent (2000–2008) major advances in chromatography and the discoveries that will influence many sciences in this ongoing twenty-first century. Science has the goal of discovery and hence "the science of discovery" is implicit in many of the following chapters.

This book describes chromatography as the "bridge"—as a central science—a key foundation built on the twentieth century for major advances and discoveries yet to come across many sciences of the twenty-first century.

REFERENCES

1. C. W. Gehrke, R. L. Wixom, and E. Bayer, (Eds.), *Chromatography: A Century of Discovery (1900–2000)—The Bridge to the Sciences/Technology*, Vol. 64, Elsevier, Amsterdam, 2001.
2. C. W. Gehrke, R. L. Wixom, and E. Bayer, (Eds.), *Chromatography: A Century of Discovery (1900–2000)—The Bridge to the Sciences/Technology*, Chromatography—A New Discipline of Science, Internet Chapters, Appendixes, and Indexes, 2001. [See Internet at Chem. Web Preprint Server (http://www.chemweb.com/preprint)].

ACKNOWLEDGMENTS

We, the Editors, have had many helpful conversations with and advice from V. G. Berezkin, P. R. Brown, T. L. Chester, V. A. Davankov, W. G. Jennings, J. Janak, and P. Sandra and the awardees. We have had the benefit of many helpful discussions with other chromatographers, particularly at the national meeting of PITTCON and the American Chemical Society (ACS). Many of these scientists have contributions in the text. We have appreciated the valuable comments and insights of Dr. Leslie S. Ettre.

We have appreciated the many contributions of valuable librarians at the University of Missouri: Kate Anderson, Rachel Brekhaus, Brenda Graves-Blevins, Janice Dysart, Rebecca S. Graves, E. Diane Johnson, Amanda McConnell, Rachel Scheff, and Caryn Scoville. The Editors have received helpful input from the librarians of the Chemical Heritage Foundation initiated by the American Chemical Society and other sponsors (Philadelphia, PA) and *Chemical Abstracts* (published by ACS, Columbus, OH). Preparation of copy for this book is due in large part to the excellent secretarial skills plus accuracy and patience of Rosemary Crane, Tina Jenkins, and Cynthia Santos at the University of Missouri, Columbia.

The editors have warmly appreciated the graphic artwork by Sammae Heard, MU graphic artist, and the pen and ink drawings by Corrine Barbour, MU graduate art student.

The research for and the preparation, writing, and editing of this book were supported by the University of Missouri, Columbia (USA):

- Chancellor, Brady Deaton, Vice Provost for Research, Jim Coleman and their Office of Research
- School of Medicine and Dean William Crist
- College of Agriculture, Food and Natural Resources, and Dean Thomas Payne
- Department of Biochemistry and Chair Gerald Hazelbauer
- Experiment Station Chemical Laboratories (Agriculture), and Director Thomas P. Mawhinney
- Analytical Biochemistry Laboratory (ABC Labs), Columbia, Missouri, USA and CEO Byron Hill

Postscript. Unfortunately, the editors/authors of this book were unable to see its publication as they passed away before the final copy was completed. Thinking ahead, as good scientists must, Dr. Wixom enlisted the assistance in final editing of this manuscript of two University of Missouri colleagues and fellow chromatographers: Deborah L. Chance, Research Assistant Professor in the Departments of Molecular Microbiology & Immunology and Child Health, and Thomas P. Mawhinney, Professor in the Departments of Biochemistry and Child Health and current Director of the Missouri Agricultural Experiment Station Chemical Laboratories.

We thank Dr. Wixom and Dr. Gehrke for their grand contribution to the field in the assembling of this historical and prospective document, and the staff of John Wiley & Sons for their assistance in seeing that this work is published.

DEBORAH L. CHANCE and THOMAS P. MAWHINNEY
Associate Editors

EDITORS/AUTHORS

Charles William Gehrke was born on July 18, 1917 in New York City. He studied at The Ohio State University, receiving a B.A. in 1939, a B.Sc. in Education in 1941, and an M.S. in Bacteriology in 1941. From 1941 to 1945, he was Professor and Chairman of the Department of Chemistry at Missouri Valley College, Marshall, Missouri, teaching chemistry and physics to World War II Navy midshipmen (from destroyers, battleships, and aircraft carriers in the South Pacific) for officer training. These young men returned to the war as deck and flight officers. In 1946, he returned as instructor in agricultural biochemistry to The Ohio State University, receiving his Ph.D. in 1947. In 1949, he joined the College of Agriculture at the University of Missouri–Columbia (UMC), retiring in Fall 1987 from positions as Professor of Biochemistry, manager of the Agricultural Experiment Station Chemical Laboratories, College of Agriculture, Food and Natural Resources, and Director of the University Interdisciplinary Chromatography Mass-Spectrometry facility. His duties also included those of State Chemist for the Missouri Fertilizer and Limestone Control laws. He was Scientific Coordinator at the Cancer Research Center in Columbia from 1989 to 1997.

Gehrke is the author of over 260 scientific publications in analytical chemistry and biochemistry. His research interests include the development of quantitative, high-resolution gas and liquid chromatographic methods for amino acids, purines, pyrimidines, major and modified nucleosides in RNA, DNA, and methylated "CAP" structures in mRNA; fatty acids; biological markers in the detection of cancer; characterization and interaction of proteins, chromatography of biologically important molecules, structural characterization of carcinogen–RNA/DNA adducts; and automation of analytical methods for nitrogen, phosphorus, and potassium in fertilizers. He developed automated spectrophotometric methods of lysine, methionine, and cystine.

He has lectured on gas–liquid chromatography of amino acids in Japan, in China, and at many universities and institutes in the United States and Europe. In the 1970s, Gehrke analyzed all the lunar samples returned by Apollo flights 11, 12, and 14–17 for amino acids and extractable organic compounds as a co-investigator with Cyril Ponnamperuma, University of Maryland, and with a consortium of scientists at the National Aeronautics and Space Administration (NASA), Ames Research Center, California, and the University of Maryland, College Park, Maryland.

Awards and Honors. In 1971, Dr. Gehrke received the annual Association of Official Analytical Chemists' (AOAC) Harvey W. Wiley Award in Analytical Chemistry. He was recipient of the Senior Faculty Member Award, UMC College of Agriculture, in 1973. Invited by the Soviet Academy of Sciences, he gave a summary presentation on organic substances in lunar fines at the August 1974 Oparin International Symposium on the "Origin of Life." In 1975, he was selected as a member of the American Chemical Society Charter Review Board for Chemical Abstracts. Sponsored by five Central American governments, he taught the chromatographic analysis of amino acids at the Central American Research Institute for Industry in Guatemala in 1975.

Gehrke was elected to Who's Who in Missouri Education and was a recipient of the UMC Faculty–Alumni Gold Medal Award in 1975 and the Kenneth A. Spencer Award from the Kansas City Section of the American Chemical Society for meritorious achievement in agricultural and food chemistry from 1979 to 1980. He received the Tswett Chromatography Memorial Medal from the Scientific Council on Chromatography, Academy of Sciences of the USSR, Moscow in 1978 and the Sigma XI Senior Research Award by the UMC Chapter in 1980. In 1986, Gehrke was given the American Chemical Society Midwest Chemist Award. He was an invited speaker on "Modified Nucleosides and Cancer" in Freiburg, German Federal Republic, 1982, and gave presentations as an invited scientist throughout Japan, the People's Republic of China, Taiwan, the Philippines, and Hong Kong (in 1982 and 1987). He was elected to the Board of Directors and Editorial Board of the AOAC, 1979–1980; was President-Elect of the Association of Official Analytical Chemists International Organization, 1982–1983; and was honored by election as their Centennial President from 1983 to 1984. He developed an article, "Libraries of Instruments," to describe interdisciplinary research programs on strengthening research in American Universities.

Gehrke was founder, board member, and former Chairman of the Board of Directors (1968–1992) of the Analytical Biochemistry Laboratories, Inc., a private corporation, in Columbia, MO, of 300 scientists, engineers, biologists, and chemists specializing in chromatographic instrumentation, and addressing worldwide problems on environmental and pharmaceutical issues to the corporate sector. He was a member of the board of SPIRAL Corporation, Dijon, France.

Over 60 masters and doctoral students have received their advanced degrees in analytical biochemistry under his direction. In addition to his extensive contributions to amino acid analysis by gas chromatography, Gehrke and colleagues have pioneered in the development of sensitive, high-resolution, quantitative high-performance liquid chromatographic methods for over 100 major and modified nucleosides in RNA,

DNA, transfer RNA (tRNA), and messenger RNA (mRNA), and then applied these methods in collaborative research with scientists in molecular biology across the world (1970s–1990s). At the 1982 International Symposium on Cancer Markers, Freiburg, German Federal Republic, E. Borek, Professor of Biochemistry, Columbia University, stated that "Professor Gehrke's chromatographic methods are being used successfully by more than half of the scientists in attendance at these meetings."

His involvement in chromatography began in the early 1960s with investigations on improved gas chromatographic (GC) methods for fatty acid analysis. Gehrke is widely known for developing a comprehensive quantitative GC method for the analysis of amino acids in biological samples and ultramicroscale methods for life molecules in moon samples in the 1970s. This method was used and advanced in the analysis of lunar samples when he was co-investigator with NASA. In the 1970s, his major interests shifted toward the development of quantitative high-performance liquid chromatographic (HPLC) methods for the analysis of various important substances in biological samples, especially the modified nucleosides in tRNA as biomarkers in cancer research.

Major Research Contributions and Publications. Dr. Gehrke

- Developed eight methods adopted as official methods by AOAC International (formerly Association of Analytical Chemists), sampling Ca, Mg, K, P, and N.
- Was the first to develop and automate AOAC Official Chemical Methods for fertilizers (1950s–1980s).
- Was the first to discover quantitative GLC of total protein amino acids (1960s–1970s), 45 publications.
- Was the first to develop quantitative HPLC of total nucleosides in tRNA, mRNA, rRNAs, and DNAs (1970s–1990s), 31 publications.
- Was the first to use HPLC-MS nucleoside chromatography in molecular biology (1987–1994), 23 publications.
- Was the first to use GLC and HPLC methods for metabolites in body fluids as potential biological markers (1971–1994), 54 publications.
- Was the first to use GLC in analysis of Apollo 11, 12, and 14–17 moon samples at ultrahigh sensitivity levels (1969–1974), 10 publications.
- Was the first to propose a lunar/Mars-based analytical laboratory (1989–1990 and 1995).

Dr. Gehrke was the author, co-author, or editor of the following books:

- 1979—author of a chapter in L. S. Ettre and A. Zlatkis (Eds.), *75 Years of Chromatography—a Historical Dialogue*, pp. 75–86 (Elsevier Science Publishers BV, Amsterdam, The Netherlands).
- 1987—C. W. Gehrke (CWG), K. C. Kuo, and R. L. Zumwalt (Eds.), *Amino Acid Analysis by Gas Chromatography*, in three volumes (CRC Press, Boca Raton, FL), 19 chapters by 29 authors (5 chapters by CWG).

- 1990—C. W. Gehrke and K. Kuo (Authors/Eds.), *Chromatography and Modification of Nucleosides*, a three-volume, 1206-page treatise published by Elsevier in the *Journal of Chromatography* Library Series addressing the topics (1) analytical methods for major and modified nucleosides, (2) biochemical roles and function of modification, (3) modified nucleosides in cancer and normal metabolism, and (4) a comprehensive database of structural information on tRNAs and nucleosides by HPLC, GC, MS, NMR, UV, and FT-IR combined techniques.

- 1993–C. Ponnamperuma and C. W. Gehrke (Eds.), *Proceedings of the Ninth College Park Colloquium—A Lunar-Based Chemical Analysis Laboratory* (A. Deepak Publishing, Hampton, VA).

- 1997—C. W. Gehrke, Mitchell K. Hobish, Robert W. Zumwalt, Michel Prost, and Jean Degrés, *A Lunar-Based Analytical Laboratory*, C. Ponnamperuma memorial volume (A. Deepak Publishing, Hampton, VA).

- 1954–1995—nine additional chapters and reviews in other scientific journals and books.

- 260 research papers on analytical biochemistry and chromatography (1950–2000).

In 1989 and 1993, C. W. Gehrke and C. Ponnamperuma of the University of Maryland were named co-principal investigators on a proposal to address the scientific technical concerns of placing a chemical laboratory on the moon that would be automated, miniaturized, and computer robotic-operated and would support NASA programs in the study of five aspects of the exploration of space: (1) astronaut health, (2) closed-environment life support, (3) lunar resources, (4) exobiology, and (5) planetology.

Awards. Gehrke received the American Chemical Society National Award in Separations Science and Technology in 1999 and the American Chemical Society National Award in Chromatography in 2000.

His latest published book with Dr. Robert L. Wixom and Dr. Ernst Bayer, co-editors, was *Chromatography: A Century of Discovery (1900–2000)*, published by Elsevier Sciences, B.V., Vol. 64, in 2001 (709 pages). The present book on chromatography builds on that 2001 book.

Postscript. After 91 years of life and 62 years as a vigorous scientist, Dr. Gehrke passed away on February 10, 2009. Some recent comments heard on his character are: "A great man has fallen, but left many seeds for his students and colleagues who will carry on," "enthusiastic supporter of life sciences," "never took shortcuts," "dedicated teacher," "an entrepreneur who was always moving on," and "a devoted husband, father, and grandfather" (addition by Robert L. Wixom, co-editor).

Robert L. Wixom, co-editor of this book, was born on July 6, 1924 in Philadelphia. In 1947, he graduated with a B.Sc. in Chemistry from Earlham College, Richmond, IN. His graduate studies and thesis were at the University of Illinois under the guidance of Professor William C. Rose, and he received his Ph.D. in Biochemistry in 1952.

Wixom held teaching/research faculty appointments in the Department of Biochemistry, School of Medicine, University of Arkansas (1952–1964) and the Department of Biochemistry, School of Medicine/College of Agriculture, UMC (1964–1992). He took year-long sabbatical/research leaves at Oxford University (1961–1962), the University of Wisconsin, Madison, WI (1970–1971), the Massachusetts Institute of Technology (MIT), Cambridge, MA (1978–1979), and the Fox Chase Institute for Cancer Research, Philadelphia (1985–1986). His 40 years of research (45 peer-reviewed papers, two reviews) and graduate teaching focused mainly on amino acid and protein metabolism. He taught intermediate and advanced biochemistry to medical students, graduate students in diverse departments, and undergraduate students with a variety of majors. Wixom guided the Advanced Biochemistry Laboratory course at UMC for 20 years, which covered several experiments in chromatography, and for 15 years taught a course on biochemical information retrieval. He has received three teaching awards. He served as a Departmental Representative to the Graduate Faculty Senate (1980–1993) and its Chair (1989–1992); this included a key role in three major new university programs. He officially retired in 1992 as Professor Emeritus of Biochemistry, but continued many similar activities.

Reflecting other earlier interests, Wixom was the co-initiator of the UMC Environmental Affairs Council and served as their first chair for 3 years (1991–1994). He initiated and served as senior editor of the 1996 book, *Environmental Challenges for Higher Education: Integration of Sustainability into Academic Programs*. The preceding experiences served as the educational background for his role as co-editor of the 2001 book, *Chromatography: A Century of Discovery (1900–2000)—The Bridge to the Sciences/Technology* and now its sequel, *Chromatography: A Science of Discovery*, John Wiley & Sons, Inc., 2010.

Postscript. After celebrating 85 years of life, much of it as a teacher and seeker of knowledge in life as well as in the laboratory, Dr. Wixom passed away on July 8, 2009. "Distinguished as a scientist, educator and outdoorsman," "energetic," "courageous," "passionate," "dedicated to service," and "persistent," were among the many comments about this "always the teacher," "family man," Bob Wixom (addition by Deborah L. Chance and Thomas P. Mawhinney, associate editors and University colleagues of the editors).

CONTRIBUTORS

Armstrong, Daniel W. Professor and Chair, Department of Chemistry and Biochemistry, University of Texas at Arlington, Arlington, TX 76019-0065, USA

Bayer, Ernst Professor Emeritus, Research Center for Nucleic Acid and Peptide Chemistry, Institüt für Organische Chemie, Universität Tübingen, 72076 Tübingen, Auf der Morgenstelle 18, Germany

Berezkin, Viktor G. Doctor of Sciences (Chemistry), Professor, A. V. Topchiev Institute of Petrochemical Synthesis, Russian Academy of Sciences, Lenin Av. 29, Moscow, 11991, Russia

Berger, Terry Department of Chemistry, AccelaPure Corporation, Princeton University, 9435 Downing St., Englewood, FL 34224, USA

Brown, Phyllis R. Professor Emeriti, University of Rhode Island, Kingston, RI 02881, USA

Burger, Barend (Ben) Victor Professor Emeritus, Department of Chemistry and Polymer Science, Stellenbosch University, Private Bag XI, 7602 Matieland, South Africa

Butters, Terry D. Department of Biochemistry, Oxford University, Glycobiology Institute, South Parks Road, Oxford OX1 3QU, UK

Chester, Thomas L. Adjunct Research Professor, Department of Chemistry, University of Cincinnati, PO Box 210172, Cincinnati, OH 45221-0172, USA

Cussler, Edward L. Chemical Engineering and Materials Science, University of Minnesota, Minneapolis, MN 55455, USA

Davankov, Vadim Nesmeyanov-Institute of Element-Organic Compounds, Moscow, 119991, Russia

Dong, Michael W. Senior Scientist in Analytical Chemistry, Genetech, Small Molecule Pharmaceutical Sciences, South San Francisco, CA, USA

Engelhardt, Heinz Professor Emeritus, Fachrichtung 12-6 Instrummentelle Analytik/Umweltanalytik, Univesität Des Saarlandes Im Stadtwald; Bau 12, D-66123 Saarbrücken, Germany

Gehrke, Charles W. Professor Emeritus, Department of Biochemistry and the Agricultural Experiment Station Chemical Laboratories, College of Agriculture, Food and Natural Resources, University of Missouri, Columbia, MO 65212 (founder of ABC Laboratories), USA

Guo, Yong Principal Scientist and team leader in Analytical Development at Johnson & Johnson Pharmaceutical & Research Development LLC, 1000 Route 202, Rariton, NJ 08869, USA

Haddad, Paul R. Professor & Federation Fellow, Director, Australian Center for Research on Separation Science; Director, Pfizer Analytical Research Center, University of Tasmania, Hobart Tas 7001, Australia

Hage, David S. Charles Bessey Professor of Analytical and Bioanlytical Chemistry, Department of Chemistry, 704 Lincoln Hall, Lincoln, NE 68588, USA

Hancock, William S. Professor of Chemistry and Chemical Biology and Bradstreet Chair in Bioanalytical Chemistry, Northeastern University, Boston, MA 02115, USA

Hennion, Marie-Claire Professor of Analytical Sciences, Ecole Supérieure de Physique et de Chimie de Paris (ESPCI), Laboratoire Environnement et Chimie Analytique, 10 rue Vauquelin, Paris 75005, France

Hopper, Terry N. Director of GMP Analytical Services, ABC Laboratories, Columbia, MO 65202, USA

Jacoby, Mitch Senior Editor, Chemical & Engineering News, Chicago Bureau

Janak, Jaroslav, Professor Emeritus of Analytical Chemistry, Institute of Analytical Chemistry, Academy of Sciences of the Czech Republic, 97 Veveri St., Brno 60200, Czech Republic

Jennings, Walter ("Walt") G. Professor Emeritus, University of California, Davis and co-founder, J&W Scientific, Inc., 91 Blue Ravine Road, Folsom, CA 95630; also co-founder, Air Toxics, Ltd.

Johnson, Lyle Environmental Manager (retired), Analytical Biochemistry Laboratories (ABC), Columbia, MO 65202, USA

Kirkland, J. Jack Little Falls Analytical Division, Newport Site, Hewlett-Packard Co., 538 First State Boulevard, Newport, DE 19804, USA

Koch, Del Principal Scientist in Chemical Services, Analytical Biochemistry Laboratories (ABC), Columbia, MO 65202, USA

Lowe, Christopher R. Institute of Biotechnology, Cambridge University, Tennis Court Road, Cambridge, CB2 1QT, UK

Majors, Ronald E. Agilent Technologies, Inc., 2850 Centerville Road, Wilmington, DE 19808; also *Column Watch and Sample Prep Perspectives* editor, LC·GC

Maloney, Todd D. Analytical Sciences Research and Development, Lilly Research Laboratories, Eli Lilly and Company, Lilly Corporate Center, Indianapolis, IN 46285, USA

Marriott, Philip Professor, Department of Applied Chemistry, Royal Melbourne Institute of Technology (RMIT), GPO Box 2476V, Melbourne, Victoria 3001, Australia

Neville, David C. A. Research Scientist Oxford Glycobiology Institute, South Parks Road, Oxford OX1 3QU, UK

Noland, Pat Vice President of Pharmaceutical Analysis, ABC Laboratories, Columbia, MO 65202, USA

Okamoto, Yoshio EcoTopia Science Institute, Nagoya University, Furo-cho, Chikusa-ku, Nagoya 464-8603, Japan

Pawliszyn, Janusz Professor, Department of Chemistry, University of Waterloo, Waterloo, Ontario N2K 3G1, Canada

Pirkle, William H. Professor Emeritus, Department of Chemistry, University of Illinois, Urbana, IL 61801, USA

Porath, Jerker O. Professor Emeritus, Center for Surface Biotechnology, Uppsala Biomedical Center, Uppsala University, Box 577, SE-751 23 Uppsala, Sweden

Saelzer, Roberto Professor, Department of Pharmacy, Dirección de Docencia, Universidad de Concepción, Casilla 160-C, Correo 3, Edmundo Larenas 64, Concepción, Chile

Sandra, Pat Laboratory of Organic Chemistry, University of Ghent, Krijgslaan 281 S4, 9000 Ghent, Belgium

Schomburg, Gerhard Professor, Department of Chromatography and Electrophoresis, Max-Planck-Institut für Kohlenforschung, Stiftstrasse 39, D-45470 Mülheim an der Ruhr, Germany

Schurig, Volker Institute of Organic Chemistry, University of Tübingen, Auf der Morgenstelle 18, D-72076, Tübingen, Germany

Sjövall, Jan B. Professor, Department of Medical Biochemistry and Biophysics, Karolinska Institute, S171 77 Stockholm, Sweden

Stalcup, Apryll Department of Chemistry, University of Cincinnati, PO Box 210172, Cincinnati, OH 45221-0172, USA

Svec, Frantisek Material Sciences Division, Facilities Director, The Molecular Foundry of the E. O. Lawrence Berkeley National Laboratory, University of California at Berkeley, Berkeley, CA 94720, USA

Tritton, Thomas President of Chemical Heritage Foundation, Philadelphia, PA

Uden, Peter C. Professor Emeritus, Department of Chemistry, University of Massachusetts, Amherst, MA 01003, USA

Unger, Klaus K. Professor, Institüt für Anorganische Chemie und Analytishce Chemie, Johanne Gutenburg-Universität, Duesber-Weg 24, 55099 Mainz, Germany

Van Breemen, Richard B. Department of Medicinal Chemistry and Pharmacognosy, UIC/NIH Center for Botanical Dietary Supplements Research, University of Illinois, College of Pharmacy, 833 South Wood Street, Chicago, IL 60612, USA

Varine, John Department of Chemistry, University of Pennsylvania, and President, PITTCON 2008

Vega, Mario Professor, Facultad de Farmacia, University of Concepción, Casilla 237, Correo 3, Concepcion, Concepción, Chile

Welch, Christopher J. Process Research Laboratory on Automation & Robotics, Merck & Co., Inc., PO Box 2000, RY800-C362, Rahway, NJ 07065, USA

Wixom, Robert L. Professor Emeritus, Department of Biochemistry, University of Missouri, Columbia, MO 65212, USA

Wren, Stephen Principal Scientist, Pharmaceutical and Analytical R&D Group, AstraZeneca, Silk Court, Hulley Road, Silk Road Business Park, Macclesfield, Cheshire, UK

Xu, Raymond N. Associate Research Investigator, Abbott Laboratories, Department R46W, Building AP13A, 100 Abbott Park Road, Abbott Park, IL 60064, USA

Yeung, Edward Iowa State University, Ames Laboratory—USDOE and Department of Chemistry, Iowa State University of Science and Technology, College of Liberal Arts and Sciences, Ames, IA 50011, USA

ASSOCIATE EDITORS POSTSCRIPT

Chance, Deborah L. Research Assistant Professor, Departments of Molecular Microbiology & Immunology, and Child Health, University of Missouri, Columbia, MO 65212, USA

Mawhinney, Thomas P. Director, Missouri Agricultural Experiment Station Chemical Laboratories, Professor, Departments of Biochemistry, and Child Health, University of Missouri, Columbia, MO 65211, USA

1

CHROMATOGRAPHY—A NEW DISCIPLINE OF SCIENCE

Robert L. Wixom

Department of Biochemistry, University of Missouri, Columbia

Charles W. Gehrke

Department of Biochemistry and the Agricultural Experiment Station Chemical Laboratories, College of Agriculture, Food and Natural Resources, University of Missouri, Columbia

Viktor G. Berezkin

A. V. Topchiev Institute of Petrochemical Synthesis, Russian Academy of Sciences, Moscow

Jaroslav Janak

Institute of Analytical Chemistry, Academy of Sciences of the Czech Republic, Brno

An essential condition for any fruitful research is the possession of suitable methods. Any scientific progress is progress in the method.

—M. S. Tswett [3]

Science has been defined in different ways at different times. Nowadays it typically involves forming an idea of the way something works, and then making careful measurements, experiments or observations to test the hypothesis. If the evidence keeps agreeing, the hypothesis grows more believable. If even one observation contradicts it,

Chromatography: A Science of Discovery. Edited by Robert L. Wixom and Charles W. Gehrke
Copyright © 2010 John Wiley & Sons, Inc.

the entire hypothesis is falsified and the search begins again. Gradual refinements of hypotheses leads to the development of a theory . . . . A theory is a set of hypotheses that have stood the test of time, so far at least have not been contradicted by evidence and have become extremely trustworthy . . . .

—C. Suplee, *Milestones of Science*, 2000

CHAPTER OUTLINE

1.A. INTRODUCTION

Chromatography has developed over the past century [1,2] and has major input into many areas of modern science [1,2]. The main original work of M. S. Tswett was published in the book *Chromatographic Adsorption Analysis* [3]. For a review of the beginnings of chromatography, see the description of the "pioneer of chromatography," Mikhail Semenovich Tswett, born of an Italian mother and Russian father in Italy (1872), educated in Switzerland in botany, and his later research on plant pigments—all of which led to his fundamental development of adsorption chromatography [1,3]. K. I. Sakodynskii made also a great contribution to the research and description of M. S. Tswett's life [4,5]. A special reference is made to the book *Michael Tswett—the Creator of Chromatography*, published by the Russian Academy of Sciences, Scientific Council on Adsorption and Chromatography (2003) [6]. The book was released on the occasion of the 100th anniversary of the discovery of chromatography and is consulted by a wide range of readers interested in the history of science and culture. The book is a study of the famous Russian researcher Michael Tswett (1872–1919)—the creator of the method of chromatography, which is now widely used both in science and technology. Finally, it was the merit of Eugenia M. Senchenkova, an associate of the S. I. Vavilov Institute of the History of Science and Technology of the Academy of Science in Moscow, who compiled the life story of Tswett into a book [6].

Later contributions in adsorption chromatography were made by the *pioneers*: Leroy S. Palmer (1887–1944) at the University of Missouri, Gottfried Kränzlin (1882), Theodor Lippmaa (1892–1944), and Charles Dhére (1876–1955) [1, Chaps. 1, 2]. Later three Nobel Prizes in Chemistry were awarded to five Nobel Laureates in Chemistry for their research in chromatography; these *discoverers* of chromatography in science were Arne W. Tiselius, Archer J. P. Martin, Richard L. M. Synge, Stanford Moore, and William H. Stein (see Chapter 3). The subsequent advances by Richard Kuhn and several other Nobel awardees are also described in Chapter 3.

Most readers are familiar with the subsequent development of partition chromatography (liquid–liquid partition chromatography and gas–liquid partition chromatography), paper chromatography, thin-layer chromatography, and ion-exchange chromatography [1, Chap. 1]. This detailed knowledge led to the sketch of historical relationships (shown in Fig. 1.1).

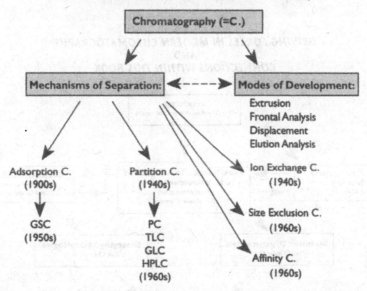

HISTORICAL FLOW OF SEMINAL CONCEPTS IN CHROMATOGRAPHY (1900s – 1960s)

Figure 1.1. Outline of the historical flow of scientific thought in chromatography (1900–1960s). The later developments in the 1970s–1980s led to UHPLC, HTLC, and associated hyphenated techniques with chromatography including MS/MS, NMR, IR and others (see Chapters 5–11). Also, it is understood that extrusion is not a chromatographic process or mode of development, but an older method of removing a developed chromatographic column. This figure will serve as a base outline for subsequent sections of this chapter. Partition chromatography and its sequential development occurred during the 1940s–1960s period. [*Note:* Additional comments and references may be found later in Ref. 1, Chap. 1, and later in this present chapter and book.]

Over subsequent years, the subject of chromatography has become prolific with respect to the number and variety of papers published. A brief reference to the earlier seminal concepts (see Section 1.H) in the current report will assist in building the needed bridge of communication.

1.B. LITERATURE ON CHROMATOGRAPHY

Books, to be understandable, need to be written in a linear pattern—paragraph by paragraph, chapter by chapter. However, the complexity of history, whether for chromatography or other subjects, requires the depiction of overall relationships, such as shown in Fig. 1.2.

Such close coherence or limited association persists for multiple investigations, many institutions, many professional societies [1, Chap. 3], and the many journals of original papers, abstract journals, review journals, books, and trade journals. For a detailed guide to the now very extensive field of chromatography literature, see Section 1.G in Ref. 1, the Appendixes in Ref. 7, and Ref. 8.

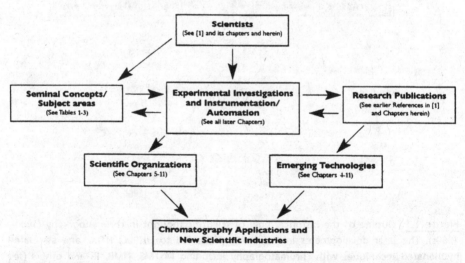

DRIVING FORCES IN MODERN CHROMATOGRAPHY
AND
CONNECTIONS WITHIN THIS BOOK

Figure 1.2. Driving forces in modern chromatography. This flowchart summarizes the known relationships of chromatography (or science in general) and will be expressed in greater detail in the stated chapters. The arrows highlight the connections, or the flow of thought, experiments, and the needed process of communication that leads to the emerging applications in new scientific industries.

1.C. WHAT IS CHROMATOGRAPHY?—A DEFINITION

Figure 1.1 suggests that chromatography is a collection of methods. Yes, there is a considerable overlap and transfer of procedures, equipment, and instruments. Hence some scientists consider information on chromatography as a "branch of science" as described in Table 1.1 (definitions 1 and 2). Does that description suffice? No, as it does not answer the question of definition. A clear definition of the scientific content of any vigorously developing scientific field is an important condition for revealing its principal features, major results of its development and structural evolution, including delineation of its boundaries [9]. Solving this problem is complicated by the fact that chromatography, like most other scientific disciplines, is continuously evolving. The most widespread definitions of chromatography, unfortunately, are not adequate. Therefore, it is the principal task of this section to elaborate a new definition of contemporary chromatography.

Chromatography was realized for the first time as an analytical "technological" process over a hundred years ago, but only in the more recent decades, investigators have noticed that many natural processes are, in fact, chromatographic. However, up to now, there is no commonly accepted logically valid definition of chromatography, although, as Socrates noted, "Precise logical definitions of concepts are the most essential conditions of true knowledge" [9]. Figuratively speaking, a definition is the shortest and, simultaneously, the most comprehensive characteristic of a given concept. Therefore, the best answer to the question "What is chromatography?" is primarily its definition [9].

On the basis of recommendations for constructing definitions that had been long ago developed in logic, Berezkin analyzed the two best-known collective, but differing, definitions of chromatography (Table 1.1), namely, the definitions elaborated by the International Union of Pure and Applied Chemistry (IUPAC) [10] and the Scientific Council on Chromatography, Russian Academy of Sciences [11]. Indeed, the first definition considers chromatography as a "method", the second one, as "science, process and method, simultaneously" [9].

It should be noted that the IUPAC definition practically repeats a definition of chromatography due to a well-known Dutch scientist, A. Keulemans [12], while the "triple" SCChrom definition is very close to a definition suggested by a well-known Russian scientist, M. S. Vigdergauz [13].

Earlier and even at the present time, one can hear an opinion that the elaboration of a more strict and precise definition of chromatography is not that important as a topical task for the development of the field. One can hardly agree with such a position, though supported by some respected chromatographers. It is hard to agree because of the lack of a precise (and commonly recognized) answer to the question "What is chromatography?" This lack of definition is undoubtedly a drawback in the development of this scientific discipline. The necessity has long been evident to provide a clear and logically based answer to this question, specifically, to elaborate a strict and sufficiently justified definition of chromatography.

TABLE 1.1. Definitions of Chromatography

Definitions	Source of Definition
1. Chromatography is a physical method of separation in which the components to be separated are distributed between two phases, one of which is stationary (stationary phase) while the other (the mobile phase) moves in a definite direction [10].[a]	IUPAC, 1993, International Commission [10]
2. Chromatography is a: (a) *Science* of intermolecular interactions and transport molecules or particles in a system of mutually immiscible phases moving relative to each other. (b) *Process* of multiple differentiated repeated distribution of chemical compounds (or particles), as a result of molecular interactions, between mutually immiscible phases (one of which is stationary) moving relative to each other leading to formation of concentration zones of individual components of original mixtures of such substances or particles. (c) *Method* of separation of mixtures of substances or particles based on differences in velocities of their movement in a system of mutually immiscible phases moving relative to each other [11].[b]	Scientific Council on Chromatography (Russian Acad. Sci.), 1997, *National Commission* [11], V. A. Davankov (Chair)
3. Chromatography is a scientific discipline (scientific field) that investigates formation, change, and movement of concentration zones of compounds (particles) of a studied sample in a flow of mobile phase, moving under conditions of interphase exchange relative to another (stationary) phase with sorption and/or sieve properties. A variant in separation is the use of selective influence on components of the analyzed mixture by one or a number of force fields [9].	V. G. Berezkin [9]
4. For a modified definition of chromatography by the Editors, see the text in Section 1.D.	The Editors

Source: V. G. Berezkin [9].

[a]Hereinafter, we will refer to this as the "IUPAC definition."

[b]Hereinafter, we will refer to this as the "SCChrom definition."

Thus, a general definition of chromatography was formulated [9, p. 56]:

A. *Chromatography* is a scientific discipline (scientific field) that investigates formation, change, and movement of concentration zones of analyte chemical compounds of a studied sample in a flow of mobile phase with respect to solid or liquid stationary phases or particles.

B. With selective influence (contact) of one, or a number of, sorbent(s) on components of the analyzed mixture; or

C. Under selective influence of one or a number of force fields on components of the analyzed mixture.

1.D. EVALUATIONS OF DEFINITIONS

We, as the Editors, suggest that this definition can be described as: *Chromatography is a scientific discipline (field of science) studying the formation, change, movement and*

TABLE 1.2. Principles and Methods of Chromatography[a]

Mobility–Equilibria	Phase	Chromatography
→	Liquid	"Tswett" Paper Thin-layer
	Gas	Gas
	Supercritical medium	Supercritical fluid
	Electrical flow	Electro
⇄	Sorbent liquid	Hypersorption Countercurrent Denuder
↓↑	Adsorption	Liquid–solid (gel) Gas–solid Supercritical fluid–solid
	Chemisorption	Ion exchange Affinity Complex-forming
	Absorption (partition)	Liquid–liquid Gas–liquid
	Restricted diffusion	Size-exclusion
	Physical field	Field-flow fractionation

Equilibrium (Isotherm)	Techniques	Channel	Hyphenated Mode
Linear	Zone	Column	One-dimensional
Nonlinear	Frontal	Slot	Multidimensional
	Displacement	Flat bed	Combined with spectral methods

Source: J. Janak.

[a]The arrows in this table reflect the movement of the mobile phase that contains the analyte in the stated chromatographic process.

TABLE 1.3. Integration of Seminal Concepts with Chromatography Leaders

Code	Seminal Concepts and Research Areas[a]	Chapter or Section Locations in the 2001 Book [1][b]	Chapter or Section Locations in this 2010 Book [14][c]
a	Theoretical contributions to C	6ABCE	5,6
b1	Early adsorption C	1B,6BE	1,3B
b2	Early partition C	1C,5,6BE	1,3B
b3	Paper C	1D,5	1,3B
b4	Thin-layer C	1D,5	1,3B,5
c	Ion-exchange C and ion C	1E,3CD,5	1,3B,5,7
d	Gas chromatography/capillary gas C	1–3,3CD,5,6BCDE	4C,5,6
e	Supports stationary- and bonded-phase C	5,6AB	3,6,9
f	Detectors in C	5,6ADE	4J
g	Size-exclusion C	5,6AE	4H,7,8
h	High-performance liquid C	4E,5,6ABCDE	4I,5,8
i	Affinity C/bioaffinity C/biosensors	5,6A	4E,5,11D
j	Petroleum C	1F,5,6BD	4A,5,11
k	Instrumentation in C	5,6ADE	4J
l	Electrophoresis/capillary electrophoresis	5,6ACDE	3B
m	Monolithic C		5,7
n	Synthetic and biological membrane separation and other techniques	5	5
o	Supercritical-fluid C/extraction	5,6A,6ADE	2,5
p	Hyphenated/coupled/tandem techniques in C	5,6ADE	4K,5–9,11C
q	Chiral C and ligand exchange C	5,6A	4F,5,6
r	Ultra-high-pressure liquid C	5	4I,6
s	Microfluidic C	5	
t	Biomedical science C	5,6ACD	5,7
u	Nutritional science C	4B,5,6ACD	7
v	Pharmacological science C		5,8
w	Environmental science C	4B,5	5,9
x	Space science C	4D	7
y	Genomic research C		5,11F
z	Proteomics research C		11F

[a]The symbol C in this column denotes *chromatography*.

[b]The numbers and letters in this column (column 3) refer to the chapter number in Ref. 1 and its subsections.

[c]References to the chapters in the present book are found in column 4.

separation of multiple concentration zones of chemical compounds (analytes) (or particles) of the studied sample in a flow of mobile phase relative to selective influence of one or a number of solid/liquid stationary phases or sorbents.

Separation may also be achieved with the influence of one or a number of force fields on components of the "centrifugal analyzed mixture" as in centrifuge sedimentation or in zone electrophoresis. The use of an external force field is not really chromatography, but a one-phase separation method that does not require movement of the mobile phase. Also, there is no stationary phase.

In Table 1.2, J. Janak presents possible variants of chromatography and discusses the theoretical aspects of the various chromatographics, as well as some thoughts on the process itself. The over 100 chromatography awardees and contributors in our 2001 book [1] and in this 2010 book [14] present the principles and applications of the chromatographic process.

In support of chromatography as a *discipline of science*, the Editors list the following 10 key attributes of chromatography in Section 1.G and describe seminal concepts of chromatography in Table 1.3 showing its widespread usage in science. Also, see Chapters 3 and 4 in this book on "paradigm shifts in chromatography" and the "trails of research".

1.E. PATHWAYS OF MODERN CHROMATOGRAPHY

Chromatography presents one of the greatest methodical phenomenon of the twentieth century with an extremely fruitful output for the future. It has not only matured by a growing theoretical background but also significantly advanced the methodical level of chemical research and control. In this way, it has opened new horizons and broken through the limits of manipulations and sensitivity determinations for many experiments. The results of chromatography have influenced knowledge in many basic scientific disciplines as chemistry, biology, and medicine, and applied scientific tasks as environmental, food, drug, space, and similar problems sciences and technologies. It has solved many problems of industrial production, and, last but not least, it has led to the establishment of a new important industrial branch of scientific instruments.

Possible variants of chromatography and analogous techniques are classified in terms of flow and equilibria directions, phase systems, and format of experiments in Table 1.2.

The development of chromatography cannot be comprehended as an isolated process. Building on Tswett's classical experiment, a series of column and flat-bed variants were performed contemporaneously with other improved separation methods, mainly electrophoresis. They "fertilized" each other (e.g., capillary gas chromatography → capillary zone electrophoresis) and also hybridized (electrochromatography).

All of the classical versions were performed on different adsorptive materials. Although they generated broad attention and represented a great deal of research and practice, they had the character of empirical improvements by trial and error.

The invention of the *partition* principle by A. J. P. Martin (1942) grew from his experimental work with fractional distillation and thinking about vapor–liquid equilibria. This invention, together with the concept of *theoretical plates* (TPs), had a major influence on the development of chromatography. The two-dimensional surface of any adsorbent used up to this time had been substituted by a three-dimensional space in the use of a liquid phase. This caused a far-reaching influence on linearization of the sorption isotherm of separated substances with extremely positive symmetrization of chromatographic curves on one hand and a broad spectrum of sorption media with tunable sorption properties on the other hand.

In the early 1950s, gas chromatography proved to be a real analytical method (1952). This variant of chromatography opened a rigorous theoretical treatment due to a more ideal behavior of the analyte in a gaseous state. The Dutch chemical engineers J. J. van Deemter, F. J. Zuiderweg, and A. Klinkenberg (1956) engaged in industrial gas flow processes and formulated the *rate theory*, which was accepted later as the so-called van Deemter equation describing the relationships among different types of diffusion and mass transfer phenomena and linear gas flow. This contribution was the second great impulse to further development in column technology by increasing the resolution power of columns from 10^2 TP of classical versions to 10^3 TP. Simultaneously, detection means have been improved in sensitivity limits from volume, molecular weight, and thermal conductivity measurements to flame ionization, mass spectrometry, and electron capture ionization means [1,8].

A further step was made by the American physicist, M. J. E. Golay. He applied his theoretical work on telegraph transport function to the gas flow in an open tube of small diameter, and thus introduced the capillary gas chromatography in practice [1]. This step increased the column resolution power to 10^6 TP.

These achievements illustrate how far from the optimum conditions the classical variants have been (phase ratios from, say, $2 : 1$ up to $1 : 10^2$).

In principle, this knowledge caused the rebirth of liquid chromatography in the 1970s. The Scottish physical chemist J. H. Knox (1983) had expanded the van Deemter equation to liquid–liquid system respecting three decimals in the diffusion coefficients values between gas and liquid and different mass transfer rate on gas–liquid and liquid–liquid interphases [1]. This resulted in greatly increased use and development of such columns and led to the beginnings of high-performance liquid chromatography (HPLC) and, later, high-performance thin-layer chromatography (HPTLC).

The theoretical background of chromatography had been profoundly influenced by the American theoretical chemist J. C. Giddings in the 1960s. His mathematical treatment of the dynamics of chromatography was later recognized as his unified theory of separation science (1991). His experimental work on high-density gas chromatography exposed the solvatization effect of compressed gas. This idea opened the use of supercritical fluids as the mobile phase by German analytical chemist, E. Klesper (1962). Supercritical-Fluid (SF) extraction by supercritical carbon dioxide and later by overpressured hot water was shown to be an extremely useful analytical means for trace analysis. Another goal by Giddings was the idea of dynamic sedimentation, known as *field-flow fractionation*, resulting in separation of particles, cells, and viruses. (It may be interesting to know that this idea was born during rafting on wild water by young sportsman, J. C. Giddings.)

1.F. SOME THOUGHTS ON THE CHROMATOGRAPHIC PROCESS

In addition to the components just described, modern chromatography has other key features:

- Miniaturization is clearly a trend in column technology characterized by study of the optimal surface area, pore size, and their homogeneity. However, there is a gap—a field for study—between the surface area and pore size of molecular sieves and nanoparticles and/or present monolithic columns.

- Methodical and technical means of chromatography are able to open new approaches in knowledge of many further natural processes. I believe some of such new cases can be identified or be found in nature—formation of mineral water composition in sedimentary rocks is a good example. Really, ion-exchange chromatography has been identified and experimentally verified in free nature. It is a dominating process of water infiltrated into Mesozoic sediments (trias) is migrating through tertiary shells (sarmat) having sea-imprinted elements as Na and Mg. This situation is typical for Carpatian Mountains bow. The calcium bicarbonate and sulfate waters are changed continuously to natrium rich and magnesium enriched types without any change in anion composition. It is not a "sci-fi" idea, but a "science of discovery" result, valid with high probability elsewhere as well. Understanding the geochemistry of mineral water of many spas or other health resorts (not only in Slovakia) is a great help in hydrogeological boring for such natural sources.

- Following this idea, there is a good analogy between field-flow fractionation and flow of water in riverbeds or of blood in body channel systems. Phase equilibria form there, and mass transfer effects exist (as in liquid or gas chromatography) on kidney membrane and lung tissue surfaces. Both organs can be the object as well as the subject of scientific experimentation with a diagnostic value by chromatographic means. In particular, sorption on, and emission from, skin can be interesting, because skin diseases are objects of an empirical and insufficiently known area of medicine.

- Many interphases (liquid-imprinted solid, gas–liquid, etc.) have been studied in chromatography, but the area of separation by or on a membrane does not seem to have been sufficiently researched at this time, although the membrane is a crucial part of any living cell. Transfer of chromatographic knowledge seems to be a hopeful task.

1.G. CHROMATOGRAPHY AS A SCIENTIFIC DISCIPLINE—ATTRIBUTES

Time marches on. A century has passed since the introduction of chromatography by Mikhail Tswett [1,2]. Since chromatography has grown far beyond a collection of methods, an overall review of the *attributes of modern chromatography* in the twenty-first century follows here, based on our earlier book [1] and updated:

- An organized path of study—see Refs. 1 and 7, as well as references cited therein and in this book.
- A broad and professional focus of research publications—original journals, review journals; see chapter references in Ref. 1 and appendixes in Ref. 7
- A considerable body of books, treatises, and handbooks; see the numerous chapter references in Ref. 1 and Apps. 4–7 in Ref. 7.

- A theoretical base that supports the methods and leads to further applications [1].
- A sense of direction and consistency within the subject area [1, Sec. 7.B].
- A comprehensive group of interacting professional societies with frequent meetings, seminars, conferences, and usually, but not always, an award for distinguished contributions [1, Chaps. 2, 3] (see also Chapter 11, below, this present volume).
- A strong core of excellent leaders in these societies and educators in major universities to actively seek new directions and continuous renewal and to impart new knowledge to students at several levels [1, Chaps. 5, 6] (see also Chapter 5, this volume).
- A set of detection instruments with accuracy, sensitivity, and selectively to meet the intellectual and laboratory challenges [7, Sec. S.10; 14] (see Chapter 6, this volume).
- A source of research funds: government agencies, research institutes, scientific industries, or private foundations [1].
- A broad outreach to other areas of science, industry, and society [1,14].

To summarize, scientific societies and science itself evolve, merge, and mature. Further amplification of these characteristics will be presented in the subsequent chapters of this book [14]. Clearly, *chromatography has the 10 key characteristics listed above and has become a major scientific discipline.*

1.H. RELATION OF SEMINAL CONCEPTS IN CHROMATOGRAPHY AND THE AWARDEES AND CONTRIBUTORS

Consistent with the seven boxes shown in Fig. 1.2, Chapters 2–11 emphasize the contributions of the awardees and contributing scientists, their description of their research accomplishments (e.g., research publications), and the pertinent overall seminal concepts. Hence, in Table 1.3, the Editors have devised a scheme for characterizing these *seminal concepts*, expressed as lowercase superscript letters "a" to "z." These letters will appear in many subsequent sections, particularly in Chapters 3–5, which present the awardees and contributors in alphabetic order; notation of these seminal concepts will facilitate the integration of subject areas. These features are related to the "Science of discovery," discussed later especially in Chapter 11.

1.I. SUMMARY

Chromatography has grown over the past century to be the central separation science; it has become the "bridge" (or the common denominator) for analytical methods. The principles and methods of chromatography are listed in Table 1.2 and the seminal concepts, in Table 1.3. Instead of measurement of only *one* or several components in a sample, chromatography facilitates the separation, detection, identification, and quantitative measurement with selective detectors of usually *all* the components in a sample. Its characteristics of sensitivity, selectivity, versatility, and quantitative features on micro, macro, and preparative scales have led to its rapid expansion. The driving forces of chromatography include the persistence and creativity of scientists, their experimental investigations, their interrelated seminal concepts, their research journals and other publications, and the relevant scientific organizations. The aims of this book are to summarize the past achievements, to delineate the new chromatographic discoveries by recent awardees and contributors during 2000–2008, and to thereby demonstrate the key features of modern chromatography. Comments on these areas in the subsequent chapters will further amplify the meaning of the phrase "a science of discovery."

REFERENCES

1. C. W. Gehrke, R. L. Wixom, and E. Bayer (Eds.), *Chromatography: A Century of Discovery (1900–2000)*, Vol. 64, Elsevier, Amsterdam, 2001.

2. L. S. Ettre (2008) Chapters in the *Evolution of Chromatography*, Imperial College Press, London; see App. 2 for his milestone papers on LC/GC.

3. V. G. Berezkin (Compiler), *M. S. Tswett, Chromatographic Adsorption Analysis*, selected works, Transl. Ed. Mary R. Masson, Ellis Horwood, New York, 1990.

4. K. I. Sakodynskii and K. V. Chmutov, *Chromatographia*, **5**, 471 (1972).

5. K. I. Sakodynskii, *Mikhail Tswett, Life and Work*, Viappiani, Milan, 1982.

6. E. M. Senchenkova (2003), *Mikhail Tswett—the Creator of Chromatography*, Russian Academy of Sciences, Scientific Council on Adsorption and Chromatography, Russia; Engl. transl. by M. A. Mayoroya and edited by V. A. Davankov and L. S. Ettre, 2003.

7. C. W. Gehrke, R. L. Wixom, and E. Bayer (Eds.), *Chromatography: A New Discipline of Science (1900–2000)*, Apps. 3–7; For a supplement, see online at Chem. Web Preprint Server (http://www.chemweb.com/preprint/).

8. C. Horvath (Ed.), *High-Performance Liquid Chromatography—Advances and Perspectives*, Vol. 2, Academic Press, New York, 1980.

9. V. G. Berezkin, *What is Chromatography? A New Approach Defining Chromatography*, 1st ed., Nauka (Science), Moscow, 2003; later, published by The Foundation: International Organization for the Promotion of Microvolume Separations (IOPMSnyw, Kenneypark 20, B-8500, Kortrigk, Belgium, Engl. transl.), 2004.

10. International Union of Pure and Applied Chemistry (IUPAC), Nomenclature for chromatography (recommendation), *Pure Appl. Chem.* **65**(4), 819 (1993).

11. V. A. Davankov (Chair), *Chromatography—Basic Terms—Terminology*, Commission of the Scientific Council on Chromatography, Russian Academy of Sciences National Commission, Issue 14, Moscow, 1997.

12. A. I. M. Keulemans, *Gas Chromatography*, Reinhold, New York, 1959.

13. M. S. Vigdergauz, in *Uspekhi Gazovoi Khromatographil (Advances in Gas Chromatography)*, Iss 4, Part 1, Kazan Branch USSR Akad. Sci. and D. I. Mendelev, 1975, p. 2 (in Russian).

14. R. L. Wixom and C. W. Gehrke (Eds.), *Chromatography: A Science of Discovery*, John Wiley & Sons, Inc., New York, 2010 (the present volume).

2

CHROMATOGRAPHY—A UNIFIED SCIENCE

Thomas L. Chester

Department of Chemistry, University of Cincinnati

CHAPTER OUTLINE

Figure 2.1. Thomas L. Chester.

Thomas L. Chester (Fig. 2.1) received his B.S. degree in Chemistry from the Florida State University in 1971. He then moved to Charleston, South Carolina, where he worked for the Verona Division of the Baychem Corporation (now Bayer) at their plant in Bushy Park. In Fall 1972, Tom enrolled in the graduate program at the University of Florida, where he earned the Ph.D. degree in 1976 under the direction of J. D. Winefordner. He then joined the Procter & Gamble Company, Cincinnati, Ohio, where he rose to Research Fellow in the Research & Development Department. He retired from P&G in 2007 and is now Adjunct Research Professor at the University of Cincinnati. Dr. Chester currently serves on the Editorial Advisory Boards of the *Journal of Chromatography A* and the *Journal of Liquid Chromatography*. He previously served on the A-page advisory panel for Analytical Chemistry. He was chair of the American Chemical Society (ACS) Subdivision of Chromatography and Separations Chemistry. He co-founded and served as President of Supercritical Conferences, the organization that produced the International Symposia on Supercritical Fluid Chromatography and Extraction, and served as Treasurer of the TriState Supercritical Fluids Discussion Group located in Cincinnati. Dr. Chester has authored over 70 publications and co-edited an ACS book, *Unified Chromatography*, 2001. His more recent research interests include chromatography modeling and optimization.

The Cincinnati Section of the American Chemical Society named Dr. Chester the 1993 Chemist of the Year. He was the recipient of the Keene P. Dimick Award in 1994 and the Chicago Chromatography Discussion Group Merit Award in 2007.

2.A. INTRODUCTION

In its broadest definition, *unified chromatography* is the simultaneous use of all parameters to accomplish a separation in the best manner possible. This definition broadens earlier concepts [1–7], but recognizes and builds on our history, opens the door to a practice of chromatography quite different from what we have done so far, and provides at least a glimpse at where we might be headed in the future. Our challenge in further developing chromatography is to abandon any perceived but unreal restrictions on what we can do, and then expand the scope of separations while keeping balance between practical utility and actual needs in the workplace. Let us explore some of our present limits and barriers while contemplating new possibilities.

2.B. MOBILE PHASES

From the early days of liquid chromatography (LC) and paper chromatography, mobile phases were fluids that could be easily handled. Chromatographers chose to use fluids that are well-behaved liquids at ambient temperature and pressure. They could not be so volatile that the compositions of mixed mobile phases would change in the course of generating a chromatogram. They could not be so viscous that mass transfer and capillary action were slow, or that inconveniently high pressure was required to generate flow through a packed column. There is a relatively small and well-known list of liquids that meet these requirements at ambient conditions, and the chief restriction preventing wider choices was the default condition of ambient temperature and pressure.

If we allow ourselves the ability to change the temperature and pressure (specifically, the outlet pressure in column LC or the gas pressure above a planar separation), many additional fluids become plausible [7]. For example, fluids that are normally gases at ambient conditions, such as CO_2, butane, and propane, are well-behaved, low-viscosity liquids at ambient temperature if the pressure is elevated sufficiently. These are relatively weak solvents but are fine for high-speed LC separation of soluble solutes at relatively low temperatures. In addition, including one of these fluids as a component in a mixed mobile phase with a more traditional solvent will greatly lower the viscosity and the pressure differential required to achieve flow through a column [8]. Diffusion rates will also be increased when the mobile-phase viscosity is lowered, thereby improving mass transfer and lowering analysis times. The use of a really weak mobile-phase component along with a strong component in a binary mixture does not seriously compromise the overall mobile-phase strength.

Among alcohols, methanol is the most often used in LC. Strength as a reversed-phase modifier increases as we progress to ethanol, propanol, etc., but viscosity also rises. Increasing the temperature greatly reduces the mobile-phase viscosity, lowers the pressure required to achieve the necessary flow, improves diffusion rates, and makes the use of "higher-viscosity" fluids practical. As with the earlier example of "gases" used as liquids by raising the pressure, "liquids" can be used well above their normal boiling points simply by elevating the pressure sufficiently to prevent boiling, again leading to new possibilities of chromatographic selectivity and speed. So, we can remove the restrictions defining "good" liquids at ambient temperature by allowing a higher temperature, and we can remove the restriction of operating under the normal boiling point by raising the pressure. Removing these restrictions vastly increases the mobile-phase components that we can consider using and also adds or expands selectivity tuning via temperature (and pressure when the mobile phase is compressible).

With liquid chromatography, solute–mobile-phase interactions are significant, and solute distribution into the mobile phase is increased well above what would be provided by the solute vapor pressure alone. Let us imagine beginning with a liquid chromatography setup and then somehow gradually lowering the strengths of the solute–mobile-phase interactions while increasing diffusion rates and lowering viscosity. This can be done discontinuously by progressing through a series of ever-smaller mobile-phase molecules, each with weaker intermolecular interactions. We can also do this continuously by reducing the mobile-phase density, thereby increasing the distance between molecules in the mobile phase and lowering intermolecular interaction strengths. Continuously changing the density can be accomplished in either pure fluids or in type I binary mixtures by first raising the pressure and temperature in concert to go around the critical point rather than going straight through the boiling transition to reach the vapor region. This is illustrated in Figure 2.2. Once around the critical point, pressure can be reduced to continuously lower the density.

Gas chromatography (GC) is the limiting case that is reached as the solute–mobile-phase interactions near zero [6,7]. In GC, the mobile phase is inert to the solutes and has no role in relative retention or selectivity; a solute's own vapor pressure (in the presence of stationary phase) controls distribution. Because the intermolecular interactions are nonexistent in GC mobile phases, which are inert carriers, the choice of fluid has no effect on relative retention or selectivity. Therefore, hydrogen and helium are the preferred mobile phases because they have the fastest diffusion rates of any gases.

When we consider changing the mobile phase properties continuously, GC becomes the limiting case when there are no intermolecular interactions in the mobile phase, and traditional LC becomes the limiting case when the mobile phase is at its practical limit of compression [6,7]. Clearly, the utilization of solute–mobile-phase interactions is essential in chromatography beyond GC. When we begin thinking in a unified sense, it becomes equally clear that these interactions can be manipulated not only by the mobile phase composition but also by temperature and pressure. This allows us to expand the choices for the mobile phase components. So, instead of categorizing chromatography according to the phase behavior of the mobile phase (GC, LC, etc.), the liberating perspective is to consider broader possibilities within a unified mobile phase continuum of temperature, pressure, and composition.

Before leaving the topic of mobile phases, we should also consider flow and transport. Pressure-driven flow has been used most often in chromatography, but flow can also

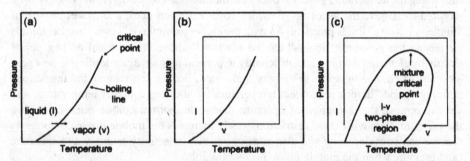

Figure 2.2. (a) A generic pressure–temperature phase diagram for a pure fluid showing only the liquid and vapor regions. The boiling line ends at the critical point where the distinction between liquid and vapor disappears. The horizontal arrow shows an isobaric path from liquid to vapor (i.e., boiling at atmospheric pressure). (b) A path demonstrating the change from liquid to vapor without going through a phase transition. (c) Similar phase-transition-free paths from liquid to vapor, and vice versa, are possible for type I binary mixtures; type I means that the two mixture components are miscible as liquids. (This diagram is at constant composition of the mixture; that is, it is an isopleth. The tie lines in the two-phase region are normal to the page. Mixture critical points are at the apex of isotherms, which would run normal to the page in a pressure–composition plane, but in general are not at the apex of isopleths.)

be electrically driven as in electrochromatography and electrophoresis. The likelihood of another flow-generating process seems remote, but should not be excluded from unified thinking. Neither should we arbitrarily exclude the possibility of combining several flow-generating or transport processes.

2.C. STATIONARY PHASES

Normal-phase LC, reversed-phase LC, ion exchange, size exclusion, hydrophilic interaction chromatography (HILIC), and other methods were initially developed and practiced separately. However, there are clearly opportunities in combining some of these mechanisms.

For example, imagine an oral-dose healthcare liquid product containing two actives requiring assay in a single separation—the first active ionized and very hydrophilic; the second active neutral and somewhat hydrophobic—and a soluble polymer added to the product for aesthetics. In a carelessly developed reversed-phase separation, the ionic active will likely elute very early if it is retained at all, and will not be well separated from other hydrophilic excipients in the product (e.g., buffer ions and other salts, hydrophilic colorants, hydrophilic flavorants). At the same time, the polymer may be strongly retained and may permanently foul the column after a dozen or so injections.

A solution to this involves choosing a stationary-phase support with pores small enough to exclude the polymer and elute it quickly to prevent fouling, plus the use of a pairing ion to increase retention of the ionic solute. So, already we are thinking somewhat beyond a reversed-phase separation by selectively adding size-exclusion and ion-pairing mechanisms to affect retention. But instead of adding a pairing ion to the mobile phase, consider a reversed phase that also has bonded ion-exchange character (e.g., see Ref. 9). This allows the mobile phase to remain free of nonvolatile additives, thereby preserving options for detection, such as mass spectrometric detection. Add to this the possibility of controlling the organic modifier independently from the pH, adding independent gradients of both modifier and pH, adding temperature and pressure control and the use of nontraditional fluids, and you start to see other possibilities. Perhaps needless to say, with such a wealth of options available, the complexity of decisionmaking goes up. However, with so much control of the separation, the time required for method development can be reduced when performed by a knowledgeable chromatography professional, and the resulting method can be tailored to provide the needed separation without necessarily resorting to a large plate number.

Let us also not forget about the stationary-phase physical form. Years ago the choice was a packed column or open tubular column. With packed columns, we had irregular or spherical particles. As practitioners, we seem quite content using totally porous spherical packings today, but is there any other format that makes sense? Solid-core spherical packings (also known as *pellicular packings*, *porous-shell packings*, *fused-core packings*, etc.) are more efficient than totally porous spherical packings and require less pressure [10]. Nonporous spherical packings are the limiting case as the porous layer is decreased, and as porosity is reduced, sample loading capacity is also diminished. A practical compromise between solute capacity and efficiency must be made.

Spherical packings require no orientation during the packing process, but would oriented particles of another shape provide some significant advantage? An extreme example is oriented-fiber packed columns [11]—and what are the shape possibilities for the fibers? Another extreme example is the no-particle porous stationary phase, or monolith [12].

So, with stationary phases we can change chemistry, porosity, size, shape, and other parameters to independently vary the specific retention mechanisms of solutes and tailor the selectivity of a separation to best suit our needs.

2.D. SOLUTE DERIVATIZATION

Gas chromatography practitioners have a long history of derivatizing solutes to accomplish some combination of increasing solute vapor pressure, improving thermal stability, or adding an easily detectable functional group. Liquid chromatography practitioners have also practiced solute derivatization, usually to add a detection feature or to form diastereomers from a solute racemate. We can think beyond this by considering derivatization to improve solute solubility in a particular mobile phase [13], to eliminate or hinder troublesome functional groups, and to specifically improve selectivity between a solute of interest and a neighboring peak in the chromatogram. A more recent trend among LC practitioners has been to minimize sample preparation as much as possible, but derivatization can be worthwhile if the benefit is large, particularly if the derivatization can be done quickly and in a single step. It should not be forgotten as a possibility in a unified approach.

2.E. OPTIMIZATION

In our history, we have tended to focus on one property of a separation and explore it alone as much as possible. We see this in the recent enthusiasm for sub-2-μm (<2-μm) particles in LC. A deliberate choice to reduce analysis times by using smaller particles is the driver in this case, but other secondary developments, such as higher-pressure pumps and reducing extra-column broadening, are necessary to support this effort. However, these supporting developments are done only in response to the driver.

The focus on a one-dimensional driver, like particle size, leads to large improvements in analysis times when methods are translated from larger particles [14], but unified thinking leads us to considering all adjustable parameters at the same time and can lead to an even better outcome. In the workplace, our goal in developing a separation is, or should be, to find the unique combination of parameter values that provides the required separation while minimizing overall cost. This usually means minimizing analysis time while putting realistic limits on pressure and solvent consumption and perhaps on the amount of stationary phase that we are willing to use. Requirements can also include robustness, ease of making the mobile phase, life of the column, or etc., if there is a way to explicitly state the needs and objectively measure the performance.

Because the underlying parameters controlling retention and resolution are highly interrelated in chromatography, a multivariate process is required for optimization. In this context, optimization does not mean improvement and does not focus on only one variable, but instead means that the best possible separation is developed within practical limits, such as maximum pressure, maximum mobile-phase consumption, available column and stationary phase formats, available mobile-phase components, temperature, pressure, and etc. When a method is optimized for resolution, it cannot be further improved for a given set of needs and constraints.

Needs and constraints will vary from one problem to the next, and some compromises will be necessary for practicality. We must realize that separations are complicated, that the parameters are strongly interrelated, that multivariate optimization is vastly superior to one-dimensional univariate optimization whenever more than one parameter is important, and that knowledge, flexibility, and rapid decisionmaking are the keys to success.

2.F. CONCLUSION

All chromatography practice is a subset of unified chromatography, so we are already practicing pieces of this bigger picture. We already have a good idea of how the usual parameters, when varied individually, affect retention and selectivity. However, we have little knowledge of interactions of these parameters. We have little or no practical experience varying parameters other than the obvious ones in the course of a separation: stationary phase choice, displacing ion and concentration in ion exchange, modifier identity and concentration in HPLC, temperature in GC, and so on.

Selectivity control is the key to rapid method development and fast analysis times—a practical separation is not possible regardless of the number of theoretical plates available if the selectivity is insufficient. More control over selectivity and speed is possible by utilizing more parameters and the interactions between parameters when they occur. Seeking new ways of adjusting selectivity of a separation, either before injection or during the separation, will add flexibility and control. We must identify perceived barriers to broader practice, test the reality of these barriers, and determine what is actually possible as we expand our capabilities in chromatography.

REFERENCES

1. R. E. Boehm and D. E. Martire, A unified theory of retention and selectivity in liquid chromatography. 1. Liquid–solid (adsorption) chromatography, *J. Phys. Chem.* **84**, 3620–3630 (1980).

2. D. E. Martire and R. E. Boehm, Unified theory of retention and selectivity in liquid chromatography. 2. Reversed-phase liquid chromatography with chemically bonded phases, *J. Phys. Chem.* **87**, 1045–1062 (1983).

3. D. Ishii and T. Takeuchi, Unified fluid chromatography, *J. Chromatogr. Sci.* **27**, 71–74 (1989).

4. D. E. Martire, Generalized treatment of spatial and temporal column parameters, applicable to gas, liquid and supercritical fluid chromatography: I. Theory, *J. Chromatogr.* **461**, 165–176 (1989).

5. J. C. Giddings, *Unified Separation Science*, Wiley, New York, 1991.

6. T. L. Chester, Chromatography from the mobile-phase perspective, *Anal. Chem.* **69**, 165A–169A (1997).

7. J. F. Parcher and T. L. Chester (Eds.), *Unified Chromatography*, ACS Symp. Series 748, American Chemical Society, 1999.

8. S. V. Olesik, Enhanced-fluidity liquid mixtures. Fundamental properties and chromatography, in *Advances in Chromatography*, 2008, CRC Press, Vol. 46, pp. 423–449.

9. Y. Sun, B. Cabovsak, C. E. Evans, T. H. Ridgway, and A. M. Stalcup, Retention characteristics of a new butylimidazolium-based stationary phase, *Anal. Bioanal. Chem.* **382**, 728–734 (2005).

10. J. J. DeStefano, T. J. Langlois, and J. J. Kirkland, Characteristics of superficially-porous silica particles for fast HPLC: Some performance comparisons with sub-2-Fm particles, *J. Chromatogr. Sci.* **46**, 254–260 (2008).

11. R. D. Stanelle, C. M. Straut, and R. K. J. Marcus, Nylon-6 Capillary-channeled polymer fibers as a stationary phase for the mixed-mode ion exchange/reversed-phase chromatography separation of proteins, *Chromatogr. Sci.* **45**, 415–421 (2007).

12. H. Zou, X. Huang, M. Ye, and Q. J. Luo, Monolithic stationary phases for liquid chromatography and capillary electrochromatography, *Chromatography A* **954**, 5–32 (2002).

13. L. A. Cole, J. G. Dorsey, and T. L. Chester, Investigation of derivatizing agents for polar solutes in supercritical fluid chromatography, *Analyst* **116**, 1287–1291 (1991).

14. F.-T. Ferse, D. Sievers, and M. Swartz, Methodenumstellung von HPLC auf UPLC, *LaborPraxis* **29**, 42–46 (2005).

3

PARADIGM SHIFTS IN CHROMATOGRAPHY: NOBEL AWARDEES

Robert L. Wixom

Department of Biochemistry, University of Missouri, Columbia

If science is the constellation of facts, theories and methods collected in current texts, then scientists are the men who, successfully or not, have striven to contribute on another element to that particular constellation. Scientific development becomes the piecemeal process by which these items have been added, singly or in combination, to the ever growing stockpile that constitutes scientific techniques and knowledge.

Three classes of problems—determination of significant facts, matching of facts with theory and articulation of theory—exhaust the literature of normal science, both empirical and theoretical. However, there are extraordinary problems, and it may well be their resolution that makes the scientific problems as a whole so particularly worthwhile. These limitations of accretion in *normal science* lead to the recognition of *paradigm shifts* in science.

—Thomas S. Kuhn, *The Structure of Scientific Revolutions*, 3rd ed., 1996, pp. 1, 34.

CHAPTER OUTLINE

3.A. INTRODUCTION

Knowledge is derived from both an individual's initiative and a society's programs; to examine one aspect without the other may lead to misleadings. To walk with balance between these poles requires diligence, perception, and judgment and yet still may not meet the views of all critics. The following endeavor is to show the interaction between individual scientists and society, changes from those who have preceded us to the present generation of scientists, and between scientific concepts (or areas) and the gifted scientists to be mentioned.

This chapter focuses on the major transformations in chromatography during the twentieth and early twenty-first centuries, but from a historical viewpoint—namely that of T. S. Kuhn, the author of the book *The Structure of Scientific Revolutions* (1962, 1970, 3rd ed. 1996). However, we must first introduce some of the scientific evidence for these changes and then their possible interpretation.

Our earlier book [1] presented some of the early historical development of chromatography, namely, the *pioneers* (M. S. Tswett, L. S. Palmer, C. Dhéré, and others). Their work came to the attention of R. Kuhn and several brilliant investigators who advanced chromatography and who received the Nobel Prize. Thus, this chapter begins with five Nobel Laureates, who may be considered *the builders* of chromatography, and is followed by other sections, leading to the overall nature of scientific advances (e.g., linear advances over time, paradigm shifts, or some combination thereof).

[*Note*: Many references that appear in this chapter will be abbreviated to conserve space. Some frequently repeated book references will be cited in an abbreviated style similar to that used for journals. The abbreviations used are given by title and are extracted from the full references listed below and are shown in open (unparenthesized) italics.]

Nobel Foundation, *Nobel Lectures—Chemistry* (Vol. 1, 1901–1921; Vol. 2, 1922–1941; Vol. 3, 1942–1962), Elsevier, Amsterdam, 1964–1966.

Nobel Foundation (T. Frängsmyr, S. Forsén, and/or B. Malström), *Nobel Lectures—Chemistry* (Vol. 4, 1963–1970; Vol. 5, 1971–1980; Vol. 6, 1981–1990; Vol. 7, 1991–1995; Vol. 8, 1996–2000; Vol. 9, 2001–2005), World Scientific, Singapore, 1992–2005.

L. K. James and J. L. Sturchio (Eds.), *Nobel Laureates in Chemistry* (1901–1992), American Chemical Society, Washington, DC and Chemical Heritage Foundation, Philadelphia, 1993.

Nobel Foundation (multiple editors), *Nobel Lectures—Physiol*(ogy or) *Med*(icine) (Vol. 1, 1901–1921; Vol. 2, 1922–1941; Vol. 3, 1942–1962; Vol. 4, 1963–1970), Elsevier, Amsterdam.

Nobel Foundation (T. Frängsmyr et al., Eds.), *Nobel Lectures—Physiol*(ogy or) *Med*(icine) (Vol. 5, 1971–1980; Vol. 6, 1981–1990; Vol. 7, 1991–1995; Vol. 8, 1996–2000; Vol. 9, 2001–2005), World Scientific, Singapore.

D. M. Fox et al. (Eds.), *Nobel Laureates in Med*(icine or) *Physiol*(ogy—A Biographical Dictionary), Garland Publishing, New York, 1990.

National Academy of Sciences (USA), *Biographical Memoirs—National Academy of Sciences of the USA*; hereafter abbreviated *Biogr. Mem. Natl. Acad. Sci. USA* (Vol. 1, 1877–Vol. 89, 2008).

Royal Society (London), *Biographical Memoirs of Fellows of the Royal Society* (London); hereafter abbreviated *Biogr. Mem. Fellows R. Soc.* (Vol. 1, 1955– Vol. 53, 2007).

M. Florkin and E. H. Stotz (Eds.), *Comprehensive Biochemistry* (Vol. 1, 1962– Vol. 46, 2007), Elsevier, Amsterdam.

A. Neuberger and L. L. M. Van Deenen (Eds.), *New Comprehensive Biochemistry* (Vol. 1, 1981–Vol. 43, 2008), Elsevier/North Holland Biomedical Press, Amsterdam.

L. S. Sherby and W. Odelbarg (Eds.), (The) *Who's Who of Nobel Prize Winners* (1901–2000), 4th ed., Onyx Press, Phoenix, AZ, 2002.

L. S. Ettre and A. Zlatkis (Eds.), *75 Years of Chromatography—a Historical Dialogue*, Elsevier, Amsterdam, 1979.

W. C. Gibson and J. Lederberg (Eds.), *The Excitement and Fascination of Science*, Annual Reviews (Vol. 1, 1965; Vol. 2, 1978, Vol. 3, 1990; Vol. 4, 1995), Palo Alto, CA.

P. E. Salpekar, (The) *Encyclopedia of Nobel Laureates—Chemistry* (1901–2005), 2 vols., Dominant R. H. Heynes, Publishers, 2007.

D. Rogers, *Nobel Laureate: Contributions to 20th Century Chemistry*, Royal Society of Chemistry, Cambridge, UK, 2006.

The reader will frequently encounter the phrase "see Chapter 1," and followed by a lowercase code letter; these expressions refer to the subject classification scheme ranging from "a" to "z" shown in Table 1.3 and discussed in Chapter 1 text, especially Section 1.B, which *introduces 27 seminal concepts or subject areas*. By using these "a"–"z" letters for the chromatographers in later chapters, plus the Author/Scientist Index, the reader will find those individuals who have conducted research in similar areas. This approach was adopted to facilitate the integration of different chapters and sections of our book.

3.B. NOBEL AWARDEES WHO ADVANCED CHROMATOGRAPHY

Leslie S. Ettre has identified five Nobel Laureates who have made major advances in chromatography [1, Table 2.1 in Chap. 2]. A brief scientific biography and selective references of each of these *builders of chromatography* follows.

ARNE W. K. TISELIUS
(1902–1971)

Uppsala University, Uppsala, Sweden

Nobel Prize in Chemistry, 1948

Figure 3.1. Arne W. K. Tiselius.

Arne W. K. Tiselius (Fig. 3.1), born in 1902, matriculated at the University of Uppsala in 1921 and received his first degree, Fil. Mag. (Master of Arts), in 1925. In 1925, he began his research career as a pupil of Theodor Svedberg at the Institute of Physical Chemistry in Uppsala. His first work was on ultracentrifugation, but soon, on the suggestion of Svedberg, he began to study the migration of particles in electric fields, namely, electrophoresis. It led to a Ph.D. thesis entitled *The Moving Boundary Method for Studying the Electrophoresis of Proteins* in 1930. In it he described an apparatus (a U-shaped tube) in an electrical field for observation of boundary movement by UV light absorption. The thesis was well received, and he was appointed Docent, indicating a possible future academic career. In order to qualify for a professorship in inorganic chemistry, Tiselius spent a few years with research on zeolites, particularly their ability to selectively adsorb substances. It was a forerunner to the work on molecular sieves taking place at his Institute 25 years later. During 1934–1935, Tiselius spent a year at Princeton University in the United States, where he studied adsorption phenomena. The time he spent there was to be of decisive importance for his future research, due in part to the contact network he established with prominent American researchers. He was encouraged to resume the work on electrophoresis and to start a new research line, which he called "adsorption analysis."

In 1937, a personal Research Professorship for Tiselius was made possible by a donation to the University of Uppsala. In 1946, Biochemistry was formally established as an independent department (later to become the Institute of Biochemistry under Tiselius' leadership). In the following years the boundary electrophoresis method was improved in many respects; for example, schlieren optics were used to follow the boundary movements. Many important discoveries were made, the foremost of which undoubtedly was the observation that blood serum contains four major groups of proteins, namely, albumin and α-, β-, and γ-globulins, as he called them. This classical, or free-flow, electrophoresis was in practice, but not in concept, superseded in later years by electrophoresis with migration in a solid medium—first paper, then starch or agar gels, and later synthetic polymers.

In the late 1930s, Tiselius began his research of adsorption analysis on the chromatography of sugars, amino acids, peptides, and other colorless substances on activated carbon. He designed and used an apparatus to observe the eluate from columns by a refractometric technique. The work led among other things to the definitions of frontal, elution, and displacement analysis [1].

"For his achievements in electrophoresis and adsorption analysis, especially for the discoveries of the complex nature of the serum proteins," Tiselius was awarded the Nobel Prize in Chemistry 1948 [1,2]. It added to his already great prestige and enabled him to obtain a new laboratory with the most modern equipment in 1952. It also helped him contact good graduate students and excellent guest researchers.

After the Nobel Prize, Tiselius had much less time to devote to his own research, but continued to work mainly through his students. The electrophoresis and chromatography research lines were continued and proliferated. Many important discoveries originating in his research laboratory appeared in the years to come. Tiselius' teaching and role in research guidance drew in many excellent graduate students: H. Svensson—electrophoresis (1946), S. Claesson—adsorption analysis (1946), J. O. Porath†—chromatography on ionic cellulose derivatives (1957), P. A. Albertsson—partition of cell particles and macromolecules (1960), P. Flodin†—dextran gels (1962), S. Hjertén†—free-zone electrophoresis (1967), and several others [1]; those indicated with a dagger (†) became leaders in the separation sciences on their own merits and later received their own awards in chromatography [1, Chap. 5].

However, Tiselius had long wanted to incorporate more biochemistry into the repertoire of the Institute. Thus, for example, before the 1954 double-helix paper, he believed that the secret of life was fundamentally a chemical problem and that he wanted to contribute to its solution. In this ambition, the Nobel Prize and the new laboratory greatly facilitated the inclusion of studies on enzymes, hormones, chloroplasts, seed proteins, muscle and blood serum proteins, cell particles, etc.

No account of the work of Arne Tiselius would be complete without reference to his achievements "on the other side of the counter," as he himself expressed it. He was a born leader and had an impressive personality. At meetings with whomever it may be, he listened to what the others had to say while smoking his pipe and then made a conclusion that everyone could agree on. Understandably, he was engaged in many activities concerning science in Sweden and internationally. Already in 1945, the Swedish government made him a member of a committee for improving the conditions of scientific research. It was followed by the position as chairman of several research foundations and of IUPAC, from 1951 to 1954. His most prestigious positions were as chairman of the Nobel Committee for Chemistry and later for the Nobel Foundation [2–6].

See Section 1.H, Table 1.3, a, b, c, g, l.

Gothenburg May 1st, 1999
By Per Floden to go with Tiselius. See Section 1.H.

ARCHER J. P. MARTIN
(1910–2002)
National Institute for Medical Research, London, Great Britain
Nobel Prize in Chemistry, 1952

Figure 3.2. Archer J. P. Martin.

Archer J. P. Martin (Fig. 3.2) received his graduate education with an emphasis in chemical engineering at Cambridge University, UK (Ph.D., 1936). His interest in distillation columns led to countercurrent separations and plate theory, including developing an apparatus of 45 tubes [each 5 ft long × 0.5 in. outer diameter (o.d.)]. Since he was then in the Dunn Nutritional Laboratory, these methods were applied, at the time, to the relatively new vitamin E. However, a 1933 visit by A. Winterstein from Richard Kuhn's laboratory led to a Cambridge demonstration of adsorption chromatography (carotene separation on a chalk column). Martin noted quite early the mathematical similarity of the two separation methods.

Martin next met Richard L. M. Synge at Cambridge University; Synge was working on the separation of amino acid derivatives while on a scholarship from the International Wool Secretariat. Together they designed a countercurrent apparatus with 39 theoretical plates to measure accurately the monoaminomonocarboxylic amino acids in wool [7]. Since the room was filled with chloroform vapor and one run took a week, they watched it in 4-h shifts to make the needed adjustments. Martin called it a "fiendish piece of apparatus" [8]. These complexities and difficulties led them to think of a design for only one mobile phase (a chloroform/ethanol mix) and a stationary phase of water on ordinary silica gel (available as a drying agent from a balance case, then ground, sieved, and moistened with water).

This silica-gel column, plus a mobile phase of water-saturated butanol and methyl orange as an indicator, provided the desired separation and detection of acetyl amino acids [8,9] and led to some theoretical considerations on zone movement, zone broadening, and partition coefficients. Such was the modest birth of partition chromatography in columns in 1941!

This advance led to the following simplifications: change of the support phase from silica to cellulose columns, then moved from columns to sheets, and exchanged from methyl orange to ninhydrin for detection of the free amino acids [10]. Thus paper chromatography (PC) arrived on the scientific scene in 1944. The method was also applicable to peptides, and by others for carbohydrates, anthocyanins and metal ions [8]. Soon circular disks were displaced by paper strips, then strips by paper sheets, and followed by two-dimensional paper chromatography.

Martin, after 2 years at the Lister Institute of Preventive Medicine, London, moved to the National Institute for Medical Research, where he was joined by A. T. James. Several laboratory adversities led them to test a gas (nitrogen) as the mobile phase to separate short-chain, volatile organic acids. Thus gas–liquid chromatography (GLC) arrived; its early beginnings were based on use of a glass column [0.25 in. inner diameter (i.d.) × 15 in. length], packed with Celite particles coated with a silicone oil as the stationary phase; the effluent acids were hand-titrated and plotted against time (e.g., a stopwatch held by James). By adding a steam-heated jacket, they separated the longer-chain fatty acids. This tedious manual work led to the later automated titration machines [11].

Martin noted that the amino acid analysis of the 1930s required 500 g of protein and needed 6 months of work to measure only the neutral amino acids [8]. By contrast, separation on the chromatography columns needed only several milligrams of protein; the later paper chromatograms, only micrograms of sample. These achievements in the 1940s–1950s served as the intellectual basis for the later quantitative methods of ion-exchange chromatography for amino acids (by Moore and Stein; see data presented later in this section, from the 1950s); and quantitative gas–liquid chromatography of amino acid derivatives (by Gehrke et al., see Ref. 1, Chap. 4C for data from the 1960s). The many subsequent developments confirm the early wisdom of the Nobel Prize Committee to recognize A. J. P. Martin and R. L. M. Synge in 1952 "for their invention of partition chromatography" [12,13]. Martin writes that there was some "magic," along with "science," in his specifics; the Editors would also add "experience," "intuition," "perseverance," and "creativity." He also received the M. S. Tswett Award (1974), the American Chemical Society National Award in Chromatography (1978), and other awards. One of his colleagues wrote that Martin "belongs to a small group of scientists who changed the face of chemistry and biochemistry in their own lifetimes" [14,15] and who saw in 1962 the then-future possibilities of "microanalysis" [16,17].

See Section 1.H, Table 1.3, a, b, d.

RICHARD L. M. SYNGE
(1914–1994)
Rowett Research Institute, Aberdeen, Scotland, UK
Nobel Prize in Chemistry, 1952

Richard L. M. Synge (Fig. 3.3) was born in Liverpool in a prominent English family, attended Winchester College for classics and then Trinity College, University of Cambridge (B.A. in 1936 and Ph.D. in Biochemistry in 1941). At that time, the University Biochemical Laboratory was headed by Sir Frederick G. Hopkins (an earlier Nobel Laureate—1929). Synge served as a research student under N. W. Pirie. After taking an advanced course by F. G. Hopkins, he recalls, "Here, immediately one came in contact with exciting facts and exciting ideas. The latter outnumbered the former, usually. However at an early stage in the course, the student engaged in some quite rigorous isolative work under the guidance of N. W. Pirie."

Figure 3.3. Richard L. M. Synge.

After serving on the staff of the Wool Industries Research Association (1939–1943) and the Lister Institute of Preventive Medicine (1943–1948), he became Head of the Department of Protein Chemistry, Rowett Research Institute, Aberdeen, Scotland (1948–1967) [18–20].

Up to the early 1940s, the methods of amino acid analysis included primitive fractional crystallization, the tedious fractional distillation of amino acid esters and bacterial growth turbidometric assay. Synge and A. J. P. Martin worked together at the Wool Industries Research Association (1938–1944), initially on countercurrent distribution (CCD) for amino acid analysis. Simplification from a long series of CCD tubes to one column led to the development of partition chromatography of acetylated amino acids on a column of silica-gel support with water as the stationary phase and a mobile phase of chloroform/ethanol. While amino acids, dipeptides and one tripeptide (glutathione) had been separated by a two-dimensional paper chromatogram and visualized with a ninhydrin spray, gramicidin S, a then-new antibiotic, was selected as a test case for a larger peptide. Amino acid analysis showed a molar ratio of 1 : 1 : 1 : 1 : 1 and their identity; partial hydrolysis of gramicidin S yielded four dipeptides and three tripeptides, which, on analysis and comparison, provided the structure of gramicidin S, a cyclic decapeptide. The structure of another peptide antibiotic, tyrocidin, was resolved in a similar manner. This intellectual strategy was later followed by Fred Sanger to decipher the sequence of amino acids in the larger peptide hormone, insulin (51 amino acids), and then later by Moore and Stein for the protein, ribonuclease (124 amino acids) (see later in this chapter).

A. J. P. Martin and R. L. M. Synge shared the 1952 Nobel Prize in Chemistry "for their invention of partition chromatography" [18].

Professor Synge was known for his extensive scientific reading and detail of scientific information. To illustrate, he wrote with A. J. P. Martin an 83-page review that contained 771 references [21]. His Nobel lecture [19] also covered the early advances in using paper chromatography with the ^{131}I isotope in thyroid biosynthesis by J. Gross and R. V. Pitt Rivers. Another example for paper chromatography was the [14]CO_2 fixation in photosynthesis by M. Calvin. Chromatography plus a microbiological assay led to the practice of bioautographs for antibiotics and growth factors. Synge has been designated as one of the founding fathers of chromatography as a branch of science [22].

While the comments so far indicate a precise, focused research, Synge was keenly aware of the then theories of protein structure [19] and related his research findings to the yet-to-be described primary, secondary, and tertiary structures of proteins [23,24].

See Section 1.H, Table 1.3, a, b.

S. M. PARTRIDGE

Soon after Tiselius proposed displacement analysis in chromatography (see earlier in this chapter), S. M. Partridge and coworkers [25] separated several amino acids and bases on ion-exchange resins by the displacement approach, and led to the isolation of amino acids on a larger scale, but not on an accurate quantitative scale. Partridge's lifetime work [26] on elastin and connective tissue included the recognition and isolation by ion-exchange chromatography of the then new amino acids, desmosine and isodesmosine, cyclic amino acids derived from lysine [27]. These cyclic amino acids provide the covalent crosslinks between the long, flexible peptide backbone of elastin [28].

With the background of Martin, Synge, and Sanger in paper chromatography (see earlier in this chapter), plus the abovementioned Partridge research, the well-known team of Drs. Stanford Moore and William Stein, then at the Rockefeller Institute of Medical Research and later Rockefeller University, began their research on amino acid analysis by ion-exchange chromatography.

STANFORD MOORE (1913–1982) AND WILLIAM H. STEIN (1911–1980)
Rockefeller University, New York
Nobel Prize in Chemistry, 1972

Figure 3.4. Stanford Moore.   Figure 3.5. William Stein.

Professors Stanford Moore (Fig. 3.4) and William Stein (Fig. 3.5) shared research together for over 40 years at the Rockefeller Institute (now University). Their unique teamwork led to their recognition as the recipients of the 1972 Nobel Prize "for their contribution to the understanding of the connection between chemical structure and catalytic activity of the active center of the ribonuclease molecule" [29–33]. Christian B. Anfinsen also shared in this Nobel Prize in Chemistry, 1972; however, for clarity in presentation, his contributions are presented in Section 3.C (Table 3.1) and Section 4.E (on affinity chromatography).

Stanford Moore was born in Chicago, and followed his father, a law professor, on several moves prior to their arrival at Nashville, TN, and then was an outstanding chemistry student at Vanderbilt University (B.A., 1935). His graduate thesis at the University of Wisconsin focused on carbohydrate/benzimidazole derivatives (Ph.D., 1938). He then joined the strong protein chemistry research group of Max Bergmann at the Rockefeller Institute for Medical Research to develop quantitative gravimetric methods for precipitated amino acid salts. Bergmann introduced Stanford Moore to William Stein, and subsequently Stein and Moore devised the gravimetric measurement of several amino acids as their insoluble aromatic sulfonic acids. At this time, the main method to measure individual amino acids was by microbiological turbidometric assay; this method was tedious and had other known limitations.

William H. Stein was born in New York City, educated at Lincoln School within the Teachers College of Columbia University and at Phillips Exeter Academy, and then college work in science at Harvard University. His graduate work in the Department of Biochemistry, College of Physicians and Surgeons, Columbia University was a study of the composition of elastin (Ph.D., 1937). He then joined the Max Bergmann research group at Rockefeller Institute and began his lifelong scientific collaboration and genuine friendship with Stanford Moore.

Moore and Stein (or also Stein and Moore) undertook different research directions under the Office of Scientific Research Development headed by Vannevar Bush during the World War II years. Soon after Bergmann's death in 1944, the Rockefeller Institute appointed the two as members, and then, after changing to Rockefeller University in 1955, they became Professors of Biochemistry. Knowing the less than quantitative nature of the earlier gravimetric method, they shifted to chromatography of amino acids on potato starch columns using a drop-counting fraction collector (1948–1949). With the then known and available ion-exchange resins, this team developed in the mid-1950s a sensitive, quantitative method of amino acids by an automated amino acid analyzer; it was based on the detection of amino acids in the eluate by the ninhydrin colorimetric reaction (1948, 1954) [29–33]. In the late 1950s, D. H. Spackman, W. H. Stein, and S. Moore automated the instrument for amino acid analysis with ion-exchange resins and modifications of pumps, stepwise buffer changes, color detection, and chart recording [34,35]. A series of journal articles in 1951–1963 led to the measurement of all 20 amino acids in a 6.0-h run on a single column of sulfonated polystyrene— an amazing analysis at that time. Subsequently many instrument companies followed Moore and Stein to build and distribute improved models of automated amino acid analyzers. Though many applications of this ion-exchange chromatography (IEC) method followed (their published research on the amino acids in urine, plasma, tissues, etc. in the mid-1950s), Stein and Moore focused on their main goal to determine the structure of a protein [29–33].

They selected ribonuclease, a small stable protein containing 124 amino acids or 1876 atoms of C, H, N, O and S. With W. Hirs, they separated two ribonucleases, RNase A and B, with their IEC columns (polymethacrylate resins), and then determined the amino acid composition of RNase A. Following part of the earlier strategy of Sanger with insulin, they prepared peptides from RNase by performic acid oxidation of covalent disulfide bonds, followed by cleavage with specific proteolytic enzymes to produce peptides, which were isolated by IEC. The amino acid composition of these peptides was ascertained by IEC. By comparing the overlapping amino acid sequences, they were

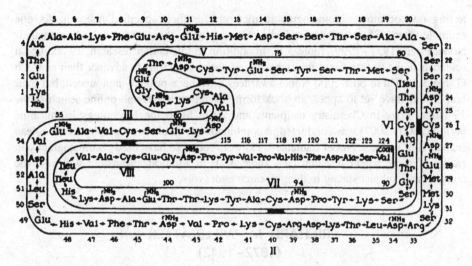

Figure 3.6. The structure of pancreatic ribonuclease A with the sequence of its 124 amino acid residues [37].

able to decipher the overall sequence of the 124 amino acids of pancreatic RNase A (Fig. 3.6). Stepwise sequencing of the amino terminal end of the peptides with phenyl-isothiocyanate (the P. Edman degradation reaction), a then-recent method from Sweden, also facilitated this first elucidation of the primary structure of proteins (i.e., RNase) (completed in 1963) [36,37].

Their subsequent investigations relied on specific chemical reagents to destroy the catalytic activity of the pancreatic RNase. Since RNase has no free-SH groups, the inhibitor, iodoacetate, was shown to act by carboxymethylation of methionine, histidine, and/or lysine residues, of which the latter two were found in the active site. This structure was confirmed later by X-ray crystallographic determination of the RNase by others in 1967. Later, R. B. Merrifield synthesized RNase and found activity for this synthetic (human-made) structure of RNase; see Merrifield presentation in Section 3.C.

This team centered their next research on the structural features of bovine pancreatic deoxyribonuclease, streptococcal proteinase, and pepsin [30–33; cf. 38]. To summarize, the distinctive research of these five Nobel Laureates led to their recognition as *builders of chromatography.*

See Section 1.H, Table 1.3, a, b, c.

3.C. NOBEL AWARDEES WHO USED CHROMATOGRAPHY

The research of the *pioneers and builders of chromatography* (see previous section) led to the prompt dissemination of their methods and results in the scientific press, as well as in their published Nobel lectures. Thus it is not surprising to find many other scientists testing one or more chromatographic methods for their specified research. As one example, some 25 scientists used chromatography in their research leading to their Nobel Prize (1937–1999, Table 3.1). In addition to this earlier research, 19 additional scientists in the twenty-first century have used chromatography to advance their research (Table 3.2). These post-2000 Nobel awardees have been publicly announced, but their oral reports have yet to appear in book form. Therefore, an overall online search of the 19 Nobel Prize in Chemistry recipients and 20 in Medicine/Physiology in this time period (2000–2007) was conducted in combination with chromatography as a Boolean search. Half of these Nobel Laureates (19) were found to have used chromatography in earlier or more recent research (see Table 3.2). In addition, the reader will note the considerable subject spread in their research endeavors.

RICHARD M. WILLSTÄTTER
(1872–1942)
Munich University, Germany
Nobel Prize in Chemistry, 1915

Richard M. Willstätter, Munich University, was a prominent organic chemist for many decades. He received the Nobel Prize in 1915 for "his researches on plant pigments, especially chlorophyll." His advanced chemistry training was with Adolph von Baeyer (Nobel Prize in Chemistry, 1905), receiving his doctorate in 1894 from the University of Munich. Later, Willstätter and his students showed that magnesium is an integral part of native chlorophyll; that chlorophyll was the same entity in more than 200 tested, different plants; prepared the pure forms of α- and β-chlorophyll; and demonstrated that chlorophyll is an ester containing the then new long-chain alcohol (phytol) and showed some similarities of chlorophyll to hemoglobin with the hemin ring containing iron. He also examined many plant pigments—anthocyanins and carotenes found in flowers and berries. Most of this pre-1915 work was based on selective solvent extraction [39,40].

    R. Willstätter and coworkers tried without success to purify chlorophyll on an alumina column and destroyed the pigment; he ignored Tswett's comment that the chlorophylls are unstable and should be chromatographed on a "gentle adsorbent", i.e. powdered sucrose or inulin [41]. Thus, he became a critic of chromatography and M. S. Tswett's work in the 1910s–1920s. Although Willstätter received the 1915 Nobel Prize in Chemistry at age 43, he continued investigations into photosynthesis and the nature of enzymes. In the 1920s–1930s, when enzyme purification was difficult, he conducted key studies on adsorbents, metal hydroxides, hydrogels, and silica acid. While having relatively crude results, these studies were precursory to the more definitive steps of the 1940s–1960s. He and others held the view that an enzyme was some active group or molecule attached to a colloid carrier (proteins at that time were viewed as colloids). To the contrary, James B. Sumner (Cornell University) crystallized urease from jack bean meal in 1926 and showed that it was a pure protein (Nobel Prize in Chemistry, 1946). Similarly, John Northrup (Rockefeller University) worked from 1920 to 1930 to purify and crystallize pepsin, the gastric proteolytic enzyme (Nobel Prize in

TABLE 3.1. Nobel Awardees Who Used Chromatography in Part of Their Research (1937–1999)

Year/Discipline	Awardee, Key Dates, Affiliation, Country, Basis for the Award, and Abbreviated Key References	Subject Area
1937/chem.	**Paul Karrer** (1889–1971), Univ. Zurich, Switzerland: chemistry of carotenoids, flavins, and vitamins A and B_2—relied on adsorption chromatography and spectroscopy *Nobel Lectures—Chemistry* **2** (1937), 410–413, 433–450 *Nobel Laureates in Chemistry* (1993), 242–247 *Biograph. Mem. Fellows R. Soc.* **24**, 245–321 (1978) *Comprehensive Biochemistry*, Vol. 33A, 1979, Chap. 61, The isoprenoid pathway, pp. 277–282; Vol. 36, 1986, Chap. 9 (part), pp. 444–448	b
1938/chem.	**Richard Kuhn** (1900–1967), Heidelberg Univ. and Max Planck Institut fur Medizinische Forschung, Heidelberg, Germany: chemistry of carotenoids and vitamins—see Section 3.C for a longer description	b
1939/chem.	**Leopold Ruzicka** (1887–1976), Federal Inst. Tech., Zurich, Switzerland: chemistry of polymethylenes, higher terpenes, and isoprene rule—based partly on adsorption chromatography *Nobel Lectures—Chemistry* **2** (1939), 459–496 *Nobel Laureates in Chemistry* (1993), 259–265 *Biograph. Mem. Fellows R. Soc.* **26**, 410–501 (1980) *Comprehensive Biochemistry*, Vol. 33A, 1979, Chap. 61, The isoprenoid pathway, pp. 239–324	b
1939/chem.	**Adolph F. J. Butenandt** (1903–1995), Univ. Berlin: chemistry and isolation of sex hormones—based on isolation, purification, and structure of estrone, progesterone, and androsterone *Nobel Lectures—Chemistry* **2** (1939), 462–465 *Nobel Laureates in Chemistry* (1993), 253–258 *Biograph. Mem. Fellows R. Soc.* **44**, 77–92 (1998)	b
1950/physiol., med.	**Tadeus Reichstein** (1897–1996), Univ. Basel, Switzerland (Nobel Prize shared with Edward C. Kendall and Philip S. Hench): for their discoveries relating to the hormones of the adrenal cortex, their structure and biological effects—based on isolation of the hormones (~29) of the adrenal cortex by adsorption chromatography, determined their structure plus synthesis of cortisone and desoxycortisone *Nobel Lectures—Physiol./Med.* **3** (1942–1962), 261–269, 291–310 *Nobel Laureates in Med./Physiol.* (1990), 457–462 T. Reichstein and C. W. Shoppee, Chromatography of steroids and other colourless substances by the method of fractional elution, *Discuss. Faraday Soc.* **7**, 305–311 (1949)	

(Continued)

TABLE 3.1. Nobel Awardees Who Used Chromatography in Part of Their Research (1937–1999) (*Continued*)

Year/Discipline	Awardee, Key Dates, Affiliation, Country, Basis for the Award, and Abbreviated Key References	Subject Area
	D. H. Nelson, The passing of an era—T. Reichstein, *Endocr. News* **21**(5) (1966)—http://www.endo-society.orgnews/december/dec.him W. J. Haines and J. N. Karamaat, Chromatographic separation of the steroids of the adrenal gland, *Meth. Biochem. Anal.* **I**, 171–204 (1954) I. E. Bush, *The Chromatography of Steroids*, Pergamon Press, New York, 1961 *Biograph. Mem. Fellows R. Soc.* **45**, 448–469 (1999)	b
1951/chem.	**Edwin M. McMillan** (1907–1991), Univ. California, Berkeley: chemistry of the transuranium elements—separations based on ion-exchange chromatography *Nobel Lectures—Chemistry* **3** (1951), 309–324 *Nobel Laureates in Chemistry* (1993), 338–343 *Biograph. Mem. Natl. Acad. Sci.* **69**, 214–241 (1996)	c
1951/chem.	**Glenn T. Seaborg** (1912–1999), Univ. California, Berkeley: chemistry of the transuranium elements—separation based on ion-exchange chromatography *Nobel Lectures—Chemistry* **3** (1951), 309–313, 325–352 *Nobel Laureates in Chemistry* (1993), 344–351 L. S. Ettre and A. Zlatkis (Eds.), *75 Years of Chromatography*, 1979, pp. 405–406 *Nature* **398**, 472–473 (1999)	c
1955/chem.	**Vincent du Vigneaud** (1901–1978), Cornell Medical College, New York City, NY: posterior pituitary hormones and the first synthesis of a peptide hormone—based on chromatographic purification of oxytocin and vasopressin, identified component amino acids by starch gel chromatography *Nobel Lectures—Chemistry* **3** (1955), 441–467 *Nobel Laureates in Chemistry* (1993), 380–385 *Biograph. Mem. Natl. Acad. Sci.* **56**, 542–595 (1987) V. Du Vigneaud, *A Trail of Research in Sulphur Chemistry and Metabolism*, Cornell Univ. Press, Ithaca, NY, 1952 See more detailed longer description in Section 3.E	b
1958/chem.	**Frederick Sanger** (1981–), Cambridge Univ., Cambridge, UK: structure of proteins, especially that of insulin—see Section 3.C for a more detailed account of his research	b

		b

1961/chem. **Melvin Calvin** (1911–1997), Univ. California, Berkeley: the chemical reactions during photosynthesis in plants—isolation of ^{14}C metabolites by paper chromatography   b

Nobel Lectures—Chemistry (1961), 613–646

Nobel Laureates in Chemistry (1993), 422–427

Biograph. Mem. Natl. Acad. Sci. **75**, 96–115 (1998)

Comprehensive Biochemistry, Vol. 33A, 1979, Chap. 56, The photosynthetic cycle of carbon reduction, pp. 81–109

M. Calvin, *Following the Trail of Light: A Scientific Odyssey*, ACS, Washington, DC, 1992

1964/med., physiol. **Konrad E. Bloch** (1912–2000), Harvard Univ., Cambridge, MA: biosynthesis of squalene and cholesterol—based in part on administration of ^{14}C acetate to rats and subsequent isolation of squalene, other intermediates, and cholesterol by Al_2O_3 adsorption chromatography   b

Nobel Lectures—Physiol./Med. **4** (1963–1970), 75–102

Nobel Laureates in Med./Physiol. (1990), 51–56

K. Bloch, Summing up, *Ann. Rev. Biochem.* **56**, 1–19 (1987)

Comprehensive Biochemistry, Vol. 20, 1968, Chap. 1, pp. 1–61; Vol. 33A, 1979, Chap. 59, Biosynthesis of fatty acids and glycerides, pp. 171–192; Vol. 33A, 1979, Chap. 61, The isoprenoid pathway, pp. 239–324

D. E. Vance and H. Vanden Bosch (Eds.), Cholesterol in the year 2000: Dedicated to the Memory of K. Bloch, *Biochim. Biophys. Acta (Mol. Cell Biol Lipids)* **1529**(1–3), 1–8 (Dec. 15, 2000)

Chem. Eng. News **78**(45), 51–52 (Nov. 6, 2000)

Nature **409**, 779 (2001)

Biograph. Mem. F. Roy Soc. **48**, 44–49 (2002)

1964/med., physiol. **Feodor Lynen** (1911–1979), Max Planck Inst. fur Zellchemie, Germany: mechanism and regulation of cholesterol and fatty acid metabolism—[14]C intermediates and products by paper chromatography   b

Nobel Lectures—Physiol./Med. **4** (1963–1970), 75–77, 103–140

Nobel Laureates in Med./Physiol. (1990), 372–375

Biograph. Mem. Fellows. R. Soc. **28**, 260–317 (1982)

Nature **285**, 177 (1980)

Comprehensive Biochemistry, Vol. 33A, 1979, Chap. 61, The isoprenoid pathway, pp. 239–324; Vol. 38, 1995, Chap. 1, Life, luck and logic in biochemical research (by F. Lynen), pp. 1–19

1965/chem. **Robert B. Woodward** (1917–1979), Harvard Univ., Cambridge, MA: outstanding achievements in the art of organic synthesis—see a more detailed account of his research in Section 3.C   b, h

(Continued)

37

TABLE 3.1. Nobel Awardees Who Used Chromatography in Part of Their Research (1937–1999) (Continued)

Year/Discipline	Awardee, Key Dates, Affiliation, Country, Basis for the Award, and Abbreviated Key References	Subject Area
1970/chem.	**Luis F. Leloir** (1906–1987), Inst. Biochemical Research, Buenos Aires, Argentina: discovery of sugar nucleotides and their role in the biosynthesis of carbohydrates, including the isolation of several such nucleotides by IEC *Nobel Lectures—Chemistry* **4** (1963–1970), 334–352 *Nobel Laureates in Chemistry* (1901–1992), 520–524 *Biograph. Mem. Fellows R. Soc.* **35**, 201–208 (1988) E. Cabib, Two decades on the biosynthesis of polysaccharides, *Science* **172**, 1299–1303 (1971) *Comprehensive Biochemistry*, Vol. 33A, 1979, Chap. 58, Biosynthesis of complex saccharides from monosaccharides, pp. 152–169; Vol. 35, 1983, Chap. 2, The discovery of sugar nucleotides, pp. 25–42 *The Excitement and Fascination of Science*, compiled by Joshua Lederberg; Annual Reviews, Palo Alto, California, 1990; Vol. 3, pp. 367–381	c
1970/med., physiol.	**Julius Axelrod** (1912–2004), Natl. Inst. Mental Health, NIH, Bethesda, MD: transmitter substances at nerve synapses—followed the biosynthesis of noradrenaline, and used methods based in part on paper chromatography of compounds related to noradrenaline *Nobel Lectures—Physiol./Med.* **4** (1963–1970), 439–469 *Nobel Laureates in Med./Physiol.* (1990), 7–12 Noradrenaline: Fate and control of its biosynthesis, *Science* **173**, 598–606 (1971) *Comprehensive Biochemistry*, Vol. 33B, 1979, Chap. 70, The biosynthesis of epinephrine compounds, pp. 207–213 *Biograph. Mem. Natl. Acad. Sci.* **87**, 1–19 (2005)	b
1970/med., physiol.	**Ulf S. Von Euler** (1905–1983), Karolinska Inst., Stockholm, Sweden: transmitter substances at nerve synapses—followed adrenergic neurotransmitters and their function by paper chromatography and other methods; see the later continuation of his prostaglandin research by S. K. Bergstrom, B. I. Samuelsson, and J. R. Vane, 1982 *Nobel Laureates* *Nobel Lectures—Physiol./Med.* **4** (1963–1970), 441–443, 485–494 *Nobel Laureates in Med./Physiol.* (1990), 174–177 *Biograph. Med. Fellows R. Soc.* **31**, 143–170 (1985)	b

Adrenergic neurotransmitter functions, *Science* **173**, 202–204 (1971).

The Excitement and Fascination of Science, compiled by Joshua Lederberg; Annual Reviews, Palo Alto, California, 1978, Vol. 2, pp. 675–686

Year/field	Content	
1970/med., physiol.	**Bernard Katz** (1911–2003), Univ. College, London (shared with J. Axelrod and U. S. Von Euler): transmitter substances at nerve synapses—based on release of acetylcholine at the synapse and its diffusion and interaction with receptors (probably did not use chromatography) *Nobel Lectures—Physiol./Med.* **4** (1963–1970), 441–443, 470–484 *Nobel Laureates in Med./Physiol.* (1990), 398–303 *Comprehensive Biochemistry*, Vol. 39, 1995, The role of membranes in excitability (by D. E. Goldman), pp. 307–339 *J. Neurocytol.* **32**, 431–436 (2003)	—
1972/chem.	**Christian B. Anfinsen** (1916–1995), Natl. Inst. of Health, Bethesda, MD (shared with S. Moore and W. H. Stein; see Section 3.B): studies on ribonuclease, particularly the relation of the amino sequence and its biologically active conformation—including the role of four disulfide bonds in RNAase, of which only 1 of the 105 possible pairings of the 8-SH groups occurs *Nobel Lecture—Chemistry* **5** (1971–1980), 49–71 *Nobel Laureates in Chemistry* (1993), 532–537. [Other research led to the first publication on affinity chromatography, viz., P. Cuatrecasas, M. Wilchek and C. B. Anfinsen, *Proc. Natl. Acad. Sci. USA* **61**, 636–643 (1968); see Section 4.E for a review of this key research] *Nature* **376**, 19–20 (1995)	i
1972/med., physiol.	**Gerald M. Edelman** (1929–), Rockefeller Univ., New York: chemical structure of antibodies and molecular immunology—followed in part by cleavage of disulfide bonds, separation of H and L chains, and amino acid sequencing, based in part on SEC, IEC, and affinity chromatography *Nobel Lectures—Physiol./Med.* **5** (1971–1980), 23–54 *Nobel Laureates in Med./Physiol.* (1990), 148–151	c, g, i
1972/med., physiol.	**Rodney R. Porter** (1917–1985), Univ. Oxford, Oxford, UK: structure of immunoglobulins (antibodies)—based in part on papain cleavage, DNP derivatives, paper chromatography, SEC, and IEC *Nobel Lectures—Physiol./Med.* **5** (1971–1980), 25–26, 56–87 *Nobel Laureates in Med./Physiol.* (1990), 451–454 *Biograph. Med. Fellows R. Soc.* **33**, 443–489 (1987)	b, c, q
1982/med., physiol.	**Sune K. Bergstrom** (1916–2004), Karolinska Institute, Stockholm, Sweden (shared with B. I. Samuelsson and J. R. Vane, built on the earlier foundations of U. S. von Euler): discoveries concerning prostaglandins and related biologically active substances—relied on CCD and partition chromatography to purify the prostaglandins—	b

(Continued)

39

TABLE 3.1. Nobel Awardees Who Used Chromatography in Part of Their Research (1937–1999) (Continued)

Year/Discipline	Awardee, Key Dates, Affiliation, Country, Basis for the Award, and Abbreviated Key References	Subject Area
	PGE, PGF, and others; showed with B. J. Samuelsson their common origin in arachidonic acid (a C_{20} unsaturated fatty acid) *Nobel Lecture—Physiol./Med.* **6** (1983), 87–112 *Nobel Laureates in Med./Physiol.* (1990), 42–45 J. A. Oates, The 1982 Nobel Prize in Medicine or Physiology, *Science* **218**, 765–768 (1982) *Comprehensive Biochemistry*, Vol. 33A, 1979, Chap. 61, The isoprenoid pathway, pp. 301–311 *Science* **305**, 1237 (2004)	d, f
1982/med., physiol.	**Bengt I. Samuelsson** (1934–), Karolinska Institute, Stockholm, Sweden (see **Bergstrom** citation above): utilized GC–MS on tissue extracts (100 kg of sheep seminal glands used per extract for GC) to isolate and determine the structure of other members of the arachidonic acid/cyclooxygenase pathway, leading to the prostaglandin group PGD_2, PGE_2, and PGF_{2e}; the intermediate endoperoxides of PGG_2, and PGH_2, and other final products of the leukotrienes and thromboxanes *Nobel Lecture—Physiol./Med.* **6** (1982), 87–89, 114–138 *Nobel Laureates in Med./Physiol.* (1990), 482–485 (See also **Bergstrom** references cited above)	
1982/med., physiol.	**John R. Vane** (1927–2004), The Welcome Research Laboratories, Brechenham, UK (See **Bergstrom** citation above): used bioassays/physiological methods to discover prostacyclin, the mechanism of action of acetylsalicylate (aspirin), which inhibits prostaglandin formation (with B. I. Samuelsson) (probably did not use chromatography) *Nobel Lecture—Physiol./Med.* **6** (1982), 87–89, 140–170 *Nobel Laureates in Med./Physiol.* (1990), 538–541	—
1984/chem.	**Robert B. Merrifield** (1921–2006), Rockefeller Univ., New York: development of methodology for chemical synthesis on a solid matrix, synthesized peptides on a polystyrene–divinylbenzone resin, leading to the first synthesis of the protein, ribonuclease (124 amino acids); showed proof of purity by gel filtration, ion-exchange chromatography, and 5 other criteria *Nobel Laureates—Chemistry* (1981–1990), 143–175 *Nature* **441**, 824–825 (2006) (See a lengthier description in Section 3.C)	b, c, h
1995/chem.	**Paul J. Crutzen** (1933–), Max Planck Inst. Chemistry, Mainz, Germany (Nobel award was shared with Mario Molina and F. Sherwood Rowland): Atmospheric chemistry, particularly concerning the formation and	d

Year/field			

decomposition of ozone—based in part on measurement of nitrogen oxides by GC

Nobel Lectures—Chemistry 7 (1991–1995), 179–242

J. Hecht, Ozone prophets reach rarefied heights, *New Scientist* (*This Week* section), p. 10 (Oct. 21, 1995)

1995/chem.   **Mario J. Molina** (1943–), MIT, Cambridge, MA (his Nobel citation is the same as that for P. J. Crutzen)   d

Nobel Lectures—Chemistry 7 (1991–1995), 179–189, 244–264

Ann. Rev. Phys. Chem. **42**, 731–768 (1991), Stratospheric ozone depletion

Angew. Chem. **35**, 1786–1790 (1996), Stratospheric ozone depletion by chlorofluorocarbons (Nobel lecture)

J. Hecht—see the *New Scientist* reference above

1995/chem.   **F. Sherwood Rowland** (1927–), Univ. California; Irvine (his Nobel citation is the same as that for P. J. Crutzen):   d, f

Molina and Rowland worked together to measure the chlorofluorocarbons that deplete the ozone layer of the atmosphere, by chromatography with electron capture detection

Nobel Lecture—Chemistry 7 (1991–1995), 179–180, 266–296

Ann. Rev. Phys. Chem. **42**, 731–768 (1991), Stratospheric ozone depletion

Angew. Chem. **35**, 1786–1789 (1996), Stratospheric ozone depletion by chlorofluorocarbons (Nobel lecture)

J. Hecht—see the above *New Scientist* reference cited above

Science **270**, 381–382 (1995)

1999/med., physiol.   **Günter Blobel** (1936–), Rockefeller Univ., New York: discovery that proteins have intrinsic signals that govern their transport and localization in the cell—see the account of his research utilizing hydrophobic chromatography, adsorption chromatography, affinity chromatography, IEC, SEC, and HPLC, along with other cell biology methods; see the longer account of his research in Ref. 1, pp. 142–144.   b, c, g, h, i, t

---

Notes:

1. Abbreviations that have been used are self-explanatory—Univ., Inst., Corp., Co., Inc., UK, and US postal zipcode abbreviations for the states in USA.

2. The abbreviations used for journals or serial books are given in the reference list in Section 3.A. To conserve space in this table, references are listed in a more concise style.

3. Part of this table (from 1937 to 1972) was modified and extended from earlier tables in the reviews by I. M. Hais, *J. Chromatogr.* **86**, 283–288 (1973) and L. S. Ettre, pp. 1–73; in C. Horvath (Ed.), *HPLC—Advances and Perspectives*, Vol. 11, Academic Press, New York, 1980.

4. Many other similar references (both longer and shorter) are available in scientific journals or reference books.

5. The significance of these Nobel Laureates is described at the end of Section 3.F. Several are described in the text of this chapter in somewhat greater detail to illustrate the close association between their main avenue of research and the use of chromatography. Please note that each "climbed out of the usual box" and/or developed new experimental approaches and their related concepts.

TABLE 3.2. Nobel Laureates (Post-2000)[a] Who Used Chromatography in Part of Their Research

Year/Discipline	Awardees, Key Dates, Affiliations, Country, and Abbreviated References
2000/physiol./med.	**Arvid Carlsson**, Dept. Pharmacology, Göteborg Univ., Germany; **Paul Greengard**, Lab. Molecular and Cellular Science, Rockefeller Univ., New York; **Eric R. Kandel**, Center for Neurobiology and Behavior, Columbia Univ., New York: discoveries on signal transduction in the nervous system—the synapse *Science* **290**, 424 (2000) *Nature* **407**(6805), 661 (2000) B. Fornstedt, A. S. Carlsson, et al., *Neurochem.* **54**, 578–588 (1990) F. Onofrio, P. Greengard, et al., *Proc. Natl. Acad. USA* **94**, 12168–12173 (1997) L. S. Eliot, E. Kandel, et al., *Proc. Natl. Acad. Sci. USA* **86**, 9564–9568 (1989)
2001/chem.	**K. Barry Sharpless**, Scripps Research Inst., La Jolla, CA, and two co-awardees: for work on chirally catalyzed oxidation reactions *Science* **294**, 503–505 (2001) *Nature* **413**(6857), 661 (2001) R. Manetsch, K. B. Sharpless, et al., *J. Am. Chem. Soc.* **126**, 12809–12813 (2004)
2001/physiol./med.	**Leland H. Hartwell**, Fred Hutchinson Cancer Research Ctr., Seattle, WA, and two co-awardees: discoveries on key regulation of the cell cycle *Science* **298**, 527–528 (2001) *Nature* **413**(6856), 553 (2001) R. Dulbecco, L. H. Hartwell, et al., *Proc. Natl. Acad. Sci. USA* **53**, 403–410 (1965)
2002/chem.	**Koichi Tanaka**, Shimadzu Corp., Kyoto, Japan; **Kurt Wüthrich**, Swiss Federal Inst. Technology, Zurich, Switzerland, and one co-awardee: for development of MS and NMR for the 3D structure of biological macromolecules *Science* **298**, 527–528 (2002) *Nature* **419**(6908), 659 (2002) K. Tanaka, Nakanishi, *Biol. Mass Spectrom.* **33**, 230–233 (1994) K. Wüthrich, *Nature Struct. Biol.* **8**, 923–925 (2001)
2002/physiol., med.	**Sydney Brenner**, Univ. California, Berkeley; **H. Robert Horvitz**, Cambridge, MA; and one co-awardee: discoveries on the genetic regulation of organ development and programmed cell death *Science* **298**, 526 (2002) *Nature* **419**(6907), 548–549 (2002) I. N. Maruyama, S. Brennen, et al., *Proc. Natl. Acad. Sci. USA* **91**, 8273–8277 (1994) M. J. Akena, H. R. Horvitz, et al., *Neuron* **46**, 247–260 (2005)

(Continued)

TABLE 3.2. (Continued)

Year/Discipline	Awardees, Key Dates, Affiliations, Country, and Abbreviated References
2003/chem.	**Peter Agre**, John Hopkins Univ., Baltimore, MD: discovery of water channels
	Roderick MacKinnon, Howard Hughes Memorial Inst., Rockefeller Univ., New York: structural and mechanistics studies of ion channels
	Science **302**, 383–384 (2003)
	Nature **425**(6959), 651 (2003)
	Nature Struct. Biol. **10**, 973 (2003)
	Z. Valda, P. Agre, et al., Proc. Natl. Acad. Sci. USA **99**, 13131–13136 (2002)
	L. Seok-Yong and R. Mackinnon, Nature **430**(6996), 232–235 (2004)
2004/chem.	**Aaron Ciechanover**, Techicon-Israel, Inst. Technology, Haifa, Israel **Avram Hershko**, Israel Inst. of Technology, Haifa, Israel; **Irwin A. Rose**, Inst. Cancer Research, Fox Chase, PA: discovery of ubiquitin-medicated protein degradation
	Science **306**, 400–401 (2004)
	Nature **431**(7016), 729 (2004)
	A. Herihko and A. Aechanorer, Ann. Rev. Biochem. **67**, 425–479 (1998)
	I. A. Rose, Ubiquitin at Fox Chase, Proc. Natl. Acad. Sci. USA **402**(33), 11575–11577 (2005)
	A. Hershko, A. Ciechanover, and I. A. Rose, J. Biol. Chem. **256**, 1525–1528 (1981)
	For a lengthier description of their research that included affinity chromatography, see Section 4.E.
2004/physiol., med.	**Richard Axel**, Columbia Univ., Cancer Research Center, New York; **Linda B. Buck**, Fred Hutchinson Cancer Research Center, Seattle, WA: discoveries of the odorant receptors and organization of the olfactory
	Science **306**, 207 (2004)
	Nature **430**(7009), 616 (2004)
	G. S. B. Suh, R. Axel, et al., Nature **431**(7010), 854–859 (2004)
	K. R. Weiss, L. Buck, et al., Proc. Natl. Acad. Sci. USA **86**, 2913–2917 (1989)
2006/chem.	**Roger D. Kornberg**, Stanford Univ., Palo Alto, CA: studies of molecular basis of eukaryotic transcription
	Science **314**, 236 (2006)
	Nature **443**(7110), 615 (2006)
	J. Griesenbeck, R. D. Kornberg, et al., Mol. Cell. Biol. **23**, 9275–9282 (2003)
2006/physiol., med.	**Craig C. Mello**, Program in Molecular Medicine, Univ. Massachusetts Medical School, Worcester, MA, and one co-awardee: discovery of RNA interference—gene silencing by double-stranded RNA
	Science **314**, 314 (2006)
	Nature **443**(7111), 488 (2006)

(Continued)

TABLE 3.2. (*Continued*)

Year/Discipline	Awardees, Key Dates, Affiliations, Country, and Abbreviated References
2007/physiol., med.	**Mario R. Capecchi**, Univ. Utah, Salt Lake City: **Oliver Smithies**, Univ. North Carolina, Chapel Hill; and one co-awardee: discovery of principles for introducing specific gene modifications in mice by the use of embryonic stem cells *Science* **318**, 178 (2007) *Nature* **448**(7163), 642 (2007) R. Westerberg, R. J. Capeechi, et al., *Biol. Chem.* **286**, 4958–4968 (2006) O. Smithies, *Proc. Natl. Acad. Sci. USA* **100**, 4108–4113 (2003)

Source: The usual lecture citation for these Nobel awardees is not yet in book form. Usually the Nobel Foundation publishes the accumulated Nobel lectures about every 5 years. The information presented in this table was derived from an online Boolean search in PubMed by Reference Librarians, Kate Anderson and Rachel Schaff, Univ. Missouri Health Sciences Library.

[a]These post-2000 Nobel Awardees were identified by direct knowledge, their Nobel lecture, a news report in *Science* and *Nature* (cited above, in the table), or a Boolean online search for their name and the word chromatography. The Boolean search showed an occasional single citation and up to 21 citations for the scientists cited in this table. Those with two or more Boolean citations are listed above.

Chemistry, 1946). The latter two findings are part of the experimental basis of the present-day view that enzymes are usually proteins with or without a small cofactor (coenzyme or metal ion) [42].

Personal histories are sometimes, (or frequently in some individuals) affected by larger national patterns. When the Munich faculty opposed in 1924 the appointment of a Jewish academic, Willstätter resigned his university post to protest the anti-Semitism at his age of 42. As a member of the Bavarian Academy of Science, he had a lab and assistants at his disposal. He kept his research in progress by assistants reporting on their work to him in his large home (which included a library with some 10,000 books and journals). When the Gestapo came to his home to arrest him in 1938, he left soon for Switzerland; almost all of his belongings were confiscated. These adverse events are mentioned here because they also affected other scientists, including Richard Kuhn (see next presentation), Edgar Lederer, Hans A. Krebs, and others.

See Section 1.G, Table 1.3, b.

RICHARD KUHN
(1900–1967)
Heidelberg University and Max Planck Institüt für medizinische Forschung, Heidelberg, Germany
Nobel Prize in Chemistry, 1938

Richard Kuhn was born near Vienna, Austria; as the son of an engineer, he was home-schooled by his mother to age 9 and then entered the Gymnasium for 8 years. After service in World War I signal corps, he studied for 2 years at the University of Vienna and then moved to the University of Munich to work in Richard Willstätter's laboratory; his thesis concerned the specificity of enzymes in carbohydrate metabolism (Ph.D., 1922). In 1926, he became Professor of General and Analytical Chemistry at the Federal Institute of Technology, Zurich, and in 1929 became Head of the Department of Chemistry of the then new Kaiser Wilhelm Institute, which after World War II became the Max Planck Institute for Medical Research at the University of Heidelberg. By 1937, he became the chief administrator of the Institute, and in 1950, a Professor of Biochemistry at the University of Heidelberg [43–45].

Kuhn's early research concerned polyenes and their optical, dielectric, and magnetic properties and organic synthesis (one report stated his synthesis of over 300 new materials). His mentor, R. M. Willstätter, had determined the elementary composition of carotene, $C_{40}H_{56}$, and P. Karrer had ascertained the constitution of carotene.

Although R. Willstätter was a critic of M. S. Tswett and his chromatography, he gave his manuscript of a German translation of Tswett's 1910 book to R. Kuhn, who in turn passed it on to Edgar Lederer, one of his research assistants [45]. Lederer read the book and then applied Tswett's chromatographic methods to separate egg lutein into two pure pigments (leaf xanthophyll and zeaxanthin) on a calcium carbonate column [45]. Further details of Kuhn and Lederer's research, their subsequent interaction with Paul Karrer (Nobel Laureate in Chemistry for 1938), László Zechmeister (then Pécs, Hungary), and Alfred Winterstein and Hans Brockmann (in R. Kuhn's lab) may be found in E. Lederer's review [46]. Thus R. Kuhn with E. Lederer and P. Karrer rediscovered adsorption chromatography and advanced this method to a higher level for purification and analysis.

Kuhn and Lederer showed in 1931 that carotene in carrots had both the α and β components, and then found in 1933 the third isomer, γ-carotene. Subsequently, his research group isolated carotenoids by adsorption chromatography from many natural sources: rose hips, saffron, crocus, palm oil, lobster shells, and human placenta, along with several other plant pigments. Besides their isolation, structural determination and organic synthesis, Kuhn was very alert to the then relatively new consideration of their steroisomerism [43–45].

Kuhn's next major area of research concerned the vitamin B complex, particularly riboflavin (then known as vitamin B_2 or lactoflavin). With today's present emphasis on microliters or micromoles (μL or μmol), one stands in awe of Kuhn's achievement to isolate 1.0 g of pure lactoflavin from 5300 liters of skim milk! The composition was demonstrated to be $C_{17}H_{20}O_6N_4$, and led next to the chemical structure of present day riboflavin. Kuhn and colleagues isolated and deciphered the structure of vitamin B_6 (the antidermatitis vitamin) in the late 1930s, and then identified panthothenic acid and p-aminobenzoic acid, leading to the antivitamin role of sulfanilamide and other similar inhibitors. During his later research years, he examined plant glycosides, the nitrogenous oligosaccharides in human milk (with Paul György), the phenomena of resistance to influenza virus (and the role of the receptor), and brain gangliosides [45].

The Nobel Prize in Chemistry was designated for Richard Kuhn in 1938 "for his work on carotenoids and vitamins." However, Hitler forbade all Germans to accept their Nobel Prizes. After World War II, Kuhn received his gold medal and diploma in 1949.

The emphasis here on chromatography in the 1930s–1940s is an integral part of the overall research areas of the early 1900 decades. Before and after Richard Kuhn's and Paul Karrer's research, considerable research prevailed to ascertain the nutritional requirements of laboratory animals, humans, and certain microorganisms for vitamins and essential amino acids: their deficiency signs, isolation, purification, structure determination, organic synthesis, and metabolic role(s). This intense research activity is embodied in the research of several Nobel awardees (giving their name, year of Nobel award, and brief description): Christiaan Eijkman (1929, thiamin), George H. Whipple et al. (1934, vitamin B_{12} and pernicious anemia), Albert von Szent-Györgyi (1937, vitamin C), Paul Karrer (1937, carotenoids, flavins, etc.), Hendrik C. P. Dam and Edward A. Doisy (1944, vitamin K_1), Fritz A. Lipmann (1955, role of pantothenic acid and coenzyme A), and Hans A. Krebs (1953, role of CoA in the citric acid cycle). These highlights from a vast sea of related research reports and books reminds us of the milieu that surrounded the early chromatographers in the 1900–1950s period.

See Section 1.H, Table 1.3, b.

[*Note*: R. Kuhn's research overlaps that of E. Lederer, his student [43]. The reader is also referred to the contribution in Chap. 5 of Ref. 1 (Gehrke 2001 book) by Ernst Bayer, who, as a student of Kuhn, knew R. Kuhn personally. Other chapters on the Heidelberg research investigators are also available in Refs. 46 and 47.]

FREDERICK SANGER
(1918–)
Cambridge University, Cambridge, UK
Nobel Prize in Chemistry, 1958 and 1980

Frederick Sanger was born in Gloucestershire, UK, educated at Bryanston School and St. John's College, Cambridge University, UK (B.A. in 1939 and Ph.D. in 1943 under the guidance of A. Neuberger). After a fellowship period, he was a member of Medical Research Council (1951–1983) and, then building on the then-known findings, he proceeded stepwise to ascertain the structure of bovine insulin (a peptide of 51 amino acids with two chains linked with two disulfide bonds); he determined the N-terminal amino acids with 1,2,4-fluorodinitrobenzene (FDNB), identified the yellow-colored N-terminal DNB amino acids by two-dimensional (2D) paper chromatography ("end-group analysis"), performed a performic acid/oxidation of the sulfide residues, used acid or enzymes to partial hydrolyze the DNB peptides, and next sequenced these DNB peptides (again using selected paper chromatography). Finally, the overlapping sequences of amino acids in the resultant small peptides led to the overall structure of insulin [48,49].

In addition to these straightforward remarks of 15 years' prior research, Sanger's precise methods for the structure of insulin, then the largest polypeptide studied, was developed out of the background of nebulous ideas of proteins as colloids. Sanger was credited with "intelligence, knowledge, skill and wholehearted devotion to the task" [48] for this first identification of a large peptide (or small protein), which has a crucial role in normal carbohydrate metabolism and a devastating effect in its loss in diabetes. While the specifics of protein sequencing have changed over the years, the general intellectual model proposed by Sanger has been followed many times over with other proteins since the mid-1950s years.

In 1962, Sanger moved to the new Laboratory for Molecular Biology at Cambridge University that was organized by their Medical Research Council and began studies on the structures of nucleic acids. Again he applied methods based on radioactive ^{32}P, two-dimensional ionophoresis, use of specific ribonucleases for cleavage of the nucleotide chain, use of $2',3'$-dideoxynucleoside triphosphates for chain termination methods. In 1980 Sanger wrote "a knowledge of sequences could contribute much to our understanding of living matter" [50]—a comment that has subsequently been confirmed many times.

Frederick Sanger shared the 1980 Nobel Prize in Chemistry with Paul Berg (Stanford University) and Walter Gilbert (Harvard University). Sanger's second Nobel citation recognized his "contributions concerning the determination of base sequences in nucleic acids" [50–52]. These discoveries helped initiate the current research emphasis on genomics and proteomics (see Section 11.F).

See Section 1.H, Table 1.3, b.

ROBERT BURNS WOODWARD
(1917–1979)
Harvard University, Cambridge, MA
Nobel Prize in Chemistry, 1965

Robert B. Woodward was born in Boston, and had a quite unique undergraduate education and then graduate education at Massachusetts Institute of Technology (Ph.D., 1937 at the youthful age of 20 years). His subsequent research years at Harvard University were characterized by research in four areas:

- Analysis of structure (alkaloids, 11 antibiotics)
- Analysis of organic reaction mechanisms
- Organic synthesis of a long list of natural products, such as alkaloids, sterols, antibiotics and porphyrins (including chlorophyll and vitamin B_{12})
- Elucidation of theoretical organic principles—the "octant rule" and "orbital symmetry" [53,54].

Others [54–56] have emphasized Woodward's brilliance, persevering work at unusual hours, unique lecture style, and intellectual assets for multistep organic synthesis of complicated molecules. This versatility included reliance on chromatography on silica gel to separate late intermediates and final product in the organic synthesis of cephalosporin [53] and its identification by paper chromatography in several solvent systems, plus its antibacterial activity. High-pressure liquid–liquid chromatography (now HPLC) was utilized for purification of the substituted porphyrin rings; the latter was a few steps prior to the final synthesis of vitamin B_{12}—a large and complex porphyrin structure [54]. He received the Nobel Prize in Chemistry (1965) "for his outstanding achievements in the art of organic synthesis" [53] and many other distinguished awards [55–57].

See Section 1.H, Table 1.3, b, h.

ROBERT B. MERRIFIELD
(1921–2006)
Rockefeller University, New York
Nobel Prize in Chemistry, 1984

Robert B. Merrifield developed the sequence of organic reactions—"solid-phase synthesis"—and thereby was able to reproduce the sequence of amino acids in proteins. His 1984 Nobel award cited "his development of methodology for chemical synthesis on a solid matrix" [58]. After the organic synthesis of small peptides, he synthesized the protein/enzyme, ribonuclease—a 124-amino-acid chain [59]. He has written that he continually used paper chromatography, TLC, IEC, and reversed-phase HPLC, and that they "were indispensable to my work" [59]. His research paved the way for preparing synthetic nucleic acids—a crucial step for molecular biology, biotechnology, medicinal chemistry, and drug discovery [60].

Attention will now focus on the Nobel Laureates in the 2000–2007 period.

So far, one of the major roles of the stated Nobel Prize awardees has been coupled with their utilization of chromatography (Table 3.1) and then similarly for the post-2000 awardees (Table 3.2). This application of chromatography in their respective quite different research areas showed the general versatility of chromatography. In addition to the earlier brief scientific biographies, several essays and books provide the additional history of the Nobel award, including its criteria, problems, rules, and significant achievements (see references cited at the beginning of this chapter, Section 3.A, pp. 24–25).

[*Note*: The research contributions of two living Nobel awardees, Günter Blobel (1999) and Avram Hershko (2004), who used chromatography extensively in their investigations, are described in Chapter 4. The significance of the research of the just-mentioned Nobel Laureates is summarized at the end of the next section.]

3.D. NATURE OF PARADIGM SHIFTS

The reader may have noted the major shifts in subject matter and chromatography methods described earlier (see also Tables 3.1 and 3.2). Our next approach is to examine the cause and/or nature of these shifts.

Science involves the organized pursuit of discovery, the clarification of the unknown, and the achievement of understanding. Science is far more than the organized link between question(s) and answer(s); it involves both intuitional and rational effort by scientists. The intuitional aspect includes creativity—inspiration—generation of ideas—a sense of skepticism—a gift to see the whole, as well as the minute features. The distinction between the institutional and the rational is fuzzy, as the former frequently flows into the rational aspect, which includes perceptive questioning; searching for hidden relationships; puzzling for cause and effect; rigorous testing by experiments; intentional planning in the use of resources (laboratories, funds, instruments, etc.); check, recheck, and confirmation; and finally, sharing the results in the form of an original research paper or thesis. All of the above precedes the public aspects of science. Many historians of science have probed these areas that, with one exception, are outside the focus of this chapter and this book. The exception is the book by T. S. Kuhn, *The Structure of Scientific Revolutions* [61].

THOMAS S. KUHN
(1922–1998)

Thomas S. Kuhn was educated at Harvard University (B.Sc., 1943; M.A., 1946; Ph.D., 1949 in Physics). On the basis of his knowledge of the history of physics, he wrote the first edition (1962) of *The Structure of Scientific Revolutions* [61]. From physics, he gradually shifted to the history and philosophy of science with faculty appointments at Harvard University, University of California at Berkeley, Princeton University, and then Massachusetts Institute of Technology, 1968–1991 as Professor of Philosophy. He retired in 1991 and died in 1998. While he had other publications, our focus here is on his seminal book, *The Structure of Scientific Revolutions*, and its three editions (1962, 1970, 1996) [61].

Kuhn writes, "The aim of textbooks, from which each new scientific generation learns to practice its trade (profession), is persuasive and pedagogic. . . . If science is the constellation of facts, theories and methods collected in current textbooks, then scientists are the men/women who, successfully or not, have striven to contribute one or another elements to that constellation. Scientific development becomes the piecemeal process by which these items have been added, singly or in combination, to the ever-growing stockpile that constitutes scientific technique and knowledge" [61].

Kuhn suggests that there are two patterns in science, "*Normal science* means research firmly based upon one or more past scientific achievements, achievements that some particular scientific community acknowledge for a time as supplying the foundation for its further practice . . . . Achievement or *paradigm* means "actual scientific practice, including law, theory, application and instrumentation together, and provides models from which spring particular coherent traditions of scientific research" [61]. Normal science is the usual, prevailing scientific enterprise and includes vigorous attempts to place natural events into a conceptual box or boxes supplied by current scientists; it has relatively long periods of mainly questions and answers, or puzzle-solving activity. Most of his examples come from the eighteenth–nineteenth century of physics. However "Major turning points in scientific development are associated with Copernicus, Newton, Lavoisier, Einstein and others. Each investigator necessitated the community's rejection of one time-honored scientific theory (or hypothesis) in favor of another incompatible with it. Each produced a consequent shift in the problems available for subsequent scientific scrutiny . . . and a transformation of the world within which scientific work was done"—i.e., a scientific revolution [61], or a *paradigm shift*. Following this shift, a new sequence of normal (or mainstream) science prevails until a new anomaly occurs.

An analogy of *paradigm shifts* might be introduced. Henry Ford developed the Model A Ford car in 1903. Since other cars were being produced at this time, the Model A was not a paradigm shift. However, when Henry Ford introduced the assembly line in 1913, the mass production enabled him to sell his Model T Ford for $500! Then America changed from dirt roads to paved highways and now multilane "interstate highways," from ferries to major suspension bridges, and so on. Furthermore, the interstates have been a factor of change in the character of cities to megalopolis with extensive suburbs. Some knowledge of history is needed to detect such slow, but major changes, or *paradigm shifts* are usually noticed. Hence the following graphic (Fig. 3.7) of the bridge of chromatography condenses time, whether months or decades, and highlights the turbulence at the period where experimental results cannot be interpreted readily by the then existing hypotheses [61].

The "bridge" as a connection, appears in the subtitle in the frontispiece at the beginning of this book. The graphic depicting the evolution of chromatography shows some of the early scientists who invented, rediscovered, and/or advanced chromatography; they include M. S. Tswett, L. S. Palmer, R. Kuhn, A. W. K. Tiselius, A. J. P. Martin, R. L. M. Synge, F. Sanger, S. Moore, and W. H. Stein, and the awardees who are described in this chapter.

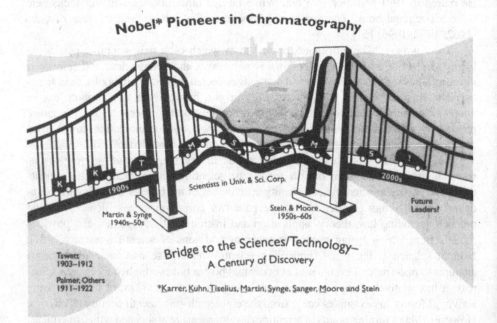

Figure 3.7. The bridge of chromatography under stress. (Chromatographers' initials were defined in the legend to the frontispiece on p. ii.)

The preceding thoughts of Kuhn and others are expressed in Figure 3.8, where "normal science" or "mainstream science" that the authors prefer, is depicted on the left side; this flow of ideas → questions → experiments → papers → yes, hundreds of theses → critical reviews → scholarly books → treatises is a widely accepted phenomena.

Who is the next farsighted scientist?

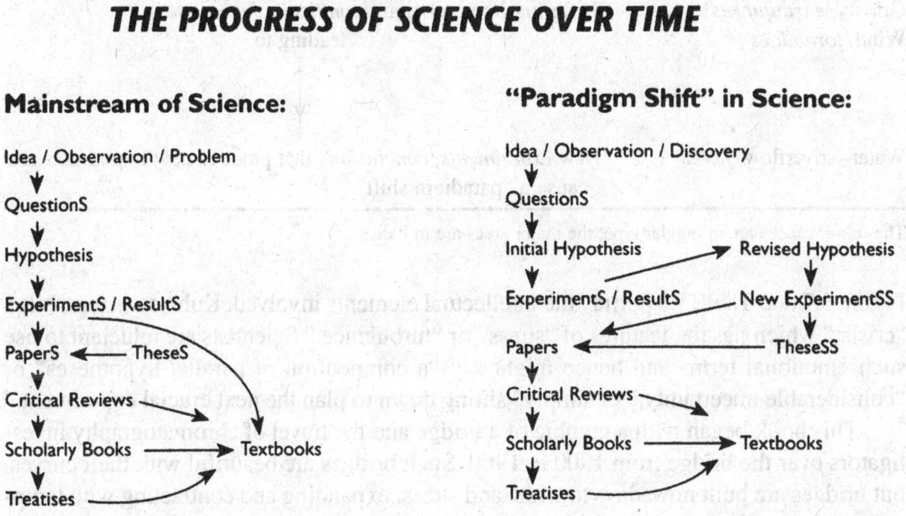

THE PROGRESS OF SCIENCE OVER TIME

Mainstream of Science: **"Paradigm Shift" in Science:**

Idea / Observation / Problem   Idea / Observation / Discovery
        ↓   ↓
QuestionS   QuestionS
        ↓   ↓
Hypothesis   Initial Hypothesis ——→ Revised Hypothesis
        ↓   ↓   ↓
ExperimentS / ResultS   ExperimentS / ResultS ←—— New ExperimentSS
        ↓   ↓   ↓
PaperS ←—— TheseS   Papers ←——————————— ThesesSS
        ↓   ↓
Critical Reviews ——————┐   Critical Reviews ———┐
        ↓   │   ↓   │
Scholarly Books ——→ Textbooks   Scholarly Books ——→ Textbooks
        ↓   ↓
Treatises ——————————┘   Treatises ——————————┘

Figure 3.8. The progress of science over time. (Where a large number of experiments or theses prevail, a capital S or SS follows.)

[*Note*: An online search in Chemical Abstracts identified 166,652 papers on chromatography (in the title or abstract) for the 1967 through 1979 period; this calculates to a rate of 12,887 papers per year [62]. Our next online search found 263,895 similar papers on chromatography from 1980 to 1994 (i.e., 18,850 per year) [62]. The third online search retrieved 181,038 chromatography papers for 1995 to Jan. 1, 2000 (20,208 per year). The publishing rate is accelerating! The *Journal of Chromatography* Library series (Elsevier Scientific Publishers) had 45 volumes from 1973 to 1990 (2.5/year), 18 volumes from 1990 to 2000 (1.8/year), and 8 volumes from 2001 to 2006 (1.1/year). In about the same period, Chromatographic Science Series (Marcel Dekker, Inc.) has grown from 46 volumes from 1965 to 1990 (1.8/year), 35 volumes from 1990 to 1999 (3.5/year), and 18 volumes from 2000 to 2007 (2.5/year). As examples of treatises, one of the editors (CWG) has a 3 volume treatise on the GLC chromatography of amino acids [63] and another 3 volumes on HPLC of nucleosides [64]. C. Horváth has a five-volume treatise on HPLC [65]. *Yes, science is additive; science is cumulative!* But science is also continuously evaluating, selecting, and pruning. That is the usual function of editors, reviewers, and treatise writers.]

Does the left column of Figure 3.8 show all of the facets of science? No, according to T. S. Kuhn. When the experiments do not confirm the initial hypothesis, scientists find a discontinuity; they have "to stop short" and "have to rethink." The right side of

TABLE 3.3. Causes of Stress in Bridges . . . or Causes of Stress in Science

Minor/*Major* Causes for Bridges	Minor/*Major* Causes for Science
Changing total load	Unresolved questions/*nonrepeatable experiments*
Varying load distribution	Poor data/*incomplete evidence*
Air content—oxygen/*rust*	Instrumental errors/*human mistakes*
Gravity/*earthquakes*	Differing interpretations/*conflicting hypotheses*
Wind/*tornadoes*	leading to
	↓
Water—riverflow/*floods*	*New experiments/conclusions* that produce new hypotheses and cause a "paradigm shift"

The minor causes are in regular type; the major areas are in italics.

Figure 3.7 and Table 3.3 portray the intellectual elements involved. Kuhn calls this shift a "crisis," which has the features of "stress" or "turbulence." Scientists are reluctant to use such emotional terms and hence might say "a competition of parallel hypotheses" or "considerable uncertainty," or simply "sitting down to plan the next crucial experiment."

This book began with a graphic of a bridge and the travel of chromatography investigators over the bridge from 1900 to 1960. Such bridges are beautiful with their curves, but bridges are built nowadays to withstand stress: expanding and contracting with temperature, sideways oscillation with tornadoes, or as here, vertical oscillation with a major earthquake (Fig. 3.7). Our Nobel Laureates have ventured beyond "normal" or "mainstream science," and may have encountered some "turbulence" or "stress"; more importantly, they have driven over the bridge to safety and enlarged our understanding for the benefit of all scientists and to society at large.

Science has had lively disputes in the past and present [66,67], although these disputes rarely reappear in the textbooks of science. Science and society at large has benefitted by the persistence of scientists, who in the face of stress, turbulence, and disputes, have demonstrated creativity and initiative to do the experimental work that led to the "paradigm shifts" in their respective areas.

The stated thesis of T. S. Kuhn has provoked discussion and debate, both favorable and critical. The subject has been reviewed by writers in the philosophy of science and history of science [68–88], the many social sciences [77,78], and the library sciences, but seems to be ignored or is relatively unknown by the working, laboratory-oriented research scientists. One scholar, H. Margolis, adds to Kuhn's thesis by emphasizing that "habits of mind" reflect entrenched responses that prevail unconsciously, are difficult to change and thus become barriers to new interpretations; anomalies may lead to cognitive stress, that are dependent on norms, institutions, technology and other social factors [78].

I. B. Cohen writes on four kinds of revolution in science [81]:

1. The marked change in the analysis of a subject (T. S. Kuhn's areas)
2. The rise of the scientific community, whether professional societies or specialized institutions

3. The time of flowering of specialized societies in contrast to overall disciplinary organizations
4. The arrival of new-found scientific communication—short notes, preprints, email, the invisible college, etc. [81]

With such a widened meaning of the term "revolution in science," the writer (and the reader) must be precise in his/her description.

Cohen has also provided the evidence for the stages in a scientific revolution: (1) the *intellectual revolution*, (2) the *paper revolution*—journals and books, and (3) the *revolution in science* [81]. Some of the above descriptions are present in these and subsequent pages, but the first one was predominate in the present texts. The above comments are included to recognize the evaluation by others of the *paradigm shift* approach. They contain applications to other areas of science, whether concurrence, partial acceptance and modification, and/or significant criticism [68–88]. However, the purpose of this chapter is, not to write a critique of the above philosophical considerations, but to explore the history of chromatography with respect to *paradigm shifts*.

To summarize, some common denominators of the five Nobel scientists described above are

- Distinctive advances in chromatography, or one of its branches as a method
- Reliance on the specific chromatography to solve major research problem(s)
- Guiding young scientists in the abovementioned areas, so that they may later make their scientific contributions
- Last but not least, "turned a corner," "broke new ground," or made a major inflection in research direction from the previous prevailing pattern.

Thus, these five Nobelists induced a *paradigm shift* in the research directions as described earlier.

3.E. PARADIGM SHIFT FOR ONE NOBEL AWARDEE

VINCENT DU VIGNEAUD
(1901–1978)
Cornell University, Medical College, New York, NY
Nobel Prize in Chemistry, 1955

Dr. **Vincent du Vigneaud** received his B.Sc. degree in 1923 and his M.Sc. in 1924 at the University of Illinois, which were followed by his Ph.D. in 1927 at Rochester University. After several fellowships, he joined the Department of Physiological Chemistry at the University of Illinois (1927–1930), then George Washington University (1932–1938), and followed as Chair of the Biochemistry Department of Cornell University (1938–1964) [89,90].

During the World War II years, du Vigneaud and others contributed extensively to the chemistry of the then new antibiotic, penicillin [91]. The completed research of du Vigneaud involved the metabolism of methionme, cystine, cystathionine, choline, betaine, the "active methyl group," the tripeptide—glutathione and the hormone—insulin. However, both before and after World War II, du Vigneaud's focus was on the hormones of the posterior pituitary gland, oxytocin, and vasopressin. They were purified by countercurrent distribution and starch gel column chromatography. Both hormones have a peptide ring with eight amino acids and a side chain; in place of leucine and isoleucine in oxytocin, vasopressin posses phenylalanine and arginine. Their analysis and degradative results were confirmed by the peptide synthesis of oxytocin and the similar results with isolated oxytocin in a variety of tests.

Turning to their biological properties, the "oxytocic effect" (i.e., uterine contraction) was recognized in 1906, and the "milk-ejecting effect" in 1910 and the "blood pressure lowering effect" in 1912 [89]. Without slighting the brilliance of the above investigations, let us examine the subsequent research knowledge of oxytocin. The 73 references in du Vigneaud's 1955 Nobel lecture may be divided into 24 earlier references by du Vigneaud on oxytocin, eight references on oxytocin by other scientists, and 41 other nonoxytocin references. du Vigneaud and his coworkers conducted the research for 58 more research papers that were published in the 1955 to 1978 years. What does the future hold for these oxytocin references? A search was made of the number of citations in the MEDLINE database for oxytocin over the later decades (see Fig. 3.9).

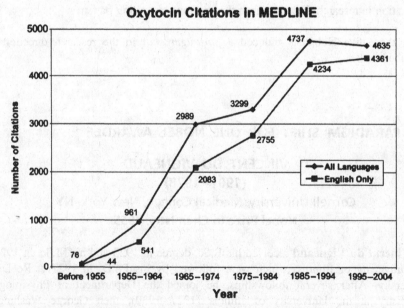

Figure 3.9. Growth of oxytocin citations in MEDLINE.

The growth from du Vigneaud's 24 oxytocin references (1955) to the 17,881 total is explosive—a result in large part of du Vigneaud's thoroughness and penetrating insights!! This analysis might be performed also with other scientists and their key investigations and/or concepts; it would probably lead to similar results. Furthermore this growth pattern is consistent with a *paradigm shift* according to T. S. Kuhn's description.

3.F. SUMMARY

The text for this chapter describes the research achievements of five Nobel Prize Laureates, who are designated as the "builders of chromatography"—A. W. K. Tiselius, A. T. P. Martin, R. L. M. Synge, S. Moore, and W. H. Stein, each of whom made major shifts in chromatography and that were then followed by others. The next section described 37 additional Nobel recipients who used chromatography in part of their research, and were leaders in their respective areas of chemistry, biochemistry, biology, and medicine. To comprehend this extensive flow of results, alterations, and anomalies in research knowledge over the decades, a description of Thomas Kuhn's analysis of scientific revolutions leading to *paradigm shifts* was presented. Consistent with the *paradigm shift* analysis was the demonstration of a 774-fold increase in numbers of research papers on oxytocin in the 52 years after V. du Vigneaud's 1955 Nobel lecture.

For a humorous depiction of the progress in chromatography, see Figure 3.10.

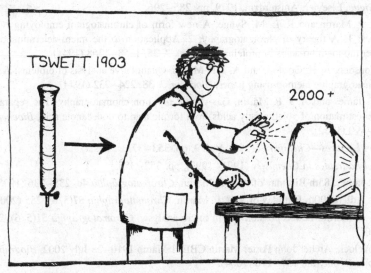

Figure 3.10. Chromatography from Tswett to the twenty-first century.

REFERENCES

(See reference list at end of Section 3.A for book title abbreviations.)

References on A. W. K. Tiselius

1. C. W. Gehrke, R. L. Wixom, and E. Bayer (Eds.), *Chromatography: A Century of Discovery (1900–2000)*, Vol. 64, Elsevier, Amsterdam, 2001.
2. *Nobel Lectures—Chemistry* (1948), Vol. 3, pp. 191–217.
3. *Nobel Laureates in Chemistry* (1901–1992), pp. 315–320.
4. A. Tiselius, Reflections from both sides of the counter, in W. C. Gibson, (Ed.), *The Excitement and Fascination of Sciences*, Vol. 2, Annual Reviews, Palo Alto, CA, 1978, pp. 549–572.
5. M. Heidelberger, Perspectives in the biochemistry of large molecules—dedicated to Arne Tiselius, *Arch. Biochem. Biophys. Suppl.* **1**, pp. 1–334, (1961); see Preface, Bibliography of A. W. K. Tiselius (1926–1961), pp. v–xii and article by R. L. M. Synge, pp. 1–11, and J. H. Northrop, pp. 7–11.
6. K. O. Pedersen, The Svedberg and Arne Tiselius—The early development of modern protein chemistry at Uppsala, A. Neuberger, L. L. M. Van Deenen, and G. Semenza (Eds.), *Comprehensive Biochemistry*, Vol. 35, 1983, Chap. 8, pp. 233–281.

References on A. J. P. Martin

7. A. J. P. Martin and R. L. M. Synge, Separation of the higher monoamino-acids by counter current, liquid–liquid extraction: The amino acid composition of wool, *Biochem. J.* **35**, 91–21, (1941).
8. A. J. P. Martin, in L. S. Ettre and A. Zlatkis (Eds.), *75 Years of Chromatography—a Historical Dialogue*, Elsevier, Amsterdam, 1979, pp. 285–296.
9. A. J. P. Martin and R. L. M. Synge, A new form of chromatogram employing two liquid phases; 1. A theory of chromatography; 2. Application to the micro-determination of the higher monoamino-acids in proteins, *Biochem. J.* **35**, 1358–1368 (1941).
10. R. Consden, A. H. Gordon, and A. J. P. Martin, Quantitative analysis of proteins: A partition chromatographic method using paper, *Biochem. J.* **38**, 224–232 (1944).
11. A. T. James and A. J .P. Martin, Gas–liquid partition chomatography: The separation and micro-estimation of volatile fatty acids from formic acid to dodecanoic acid, *Biochem. J.* **50**, 679–690 (1952).
12. *Nobel Lectures—Chemistry* (1952), Vol. 3, pp. 353–373.
13. *Nobel Laureates in Chemistry* (1901–1992), pp. 352–355.
14. A. T. James, 85th Birthday of A. J. P. Martin, *Chromatographia* **40**, 235–236 (1995).
15. E. R. Adlard, 90th Birthday of A. J. P. Martin, *Chromatographia* **57**(5/6), 255 (2000).
16. A. J. P. Martin, Future possibilities in micro-analysis, *Chromatographia* **51**(5/6), 256–259 (2000).
17. J. Lovelock, Archer John Porter Martin CBE. 1 March 1910–28 July 2002, *Biograph. Mem. Fellows R. Soc.* **50**, 157–170 (2004).

References on R. L. M. Synge

18. *Nobel Lectures—Chemistry* (1952), Vol. 3, pp. 353–358, 374–389.

19. *Nobel Laureates in Chemistry* (1901–1992), pp. 356–358.

20. H. Gordon, Richard Lawrence Millington Synge, *Biograph. Mem. Fellows R. Soc.* **42**, 455–479 (1996).

21. A. J. P. Martin and R. L. M. Synge, Analytical chemistry of proteins, *Advances in Protein Chemistry*, 1945, Vol. 2, pp. 1–83.

22. L. S. Ettre, *Chromatographia* **39**, 517–518 (1994).

23. L. S. Ettre and A. Zlatkis (Eds.), *75 Years of Chromatography* (1979), pp. 447–451.

24. See also shared publications with A. J. P. Martin and Ref. 23 above.

References on S. M. Partridge

25. J. R. Bendall, S. M. Partridge, and R. G. Westall, Displacement chromatography on cation-exchange materials, *Nature (Lond.)* **160**, 374–375 (1947).

26. A. J. Bailey, Stanley Miles Partridge, *Biograph. Mem. F. Roy. Soc.* **40**, 129–346 (1994).

27. S. M. Partridge, D. F. Elsden, and J. Thomas, Constitution of the cross-linkages in elastin, *Nature* **197**, 1297–1298 (1963).

28. J. Thomas, D. F. Elsden, and S. M. Partridge, Partial structure of two major degradation products from the cross-linkages in elastin, *Nature (Lond.)* **200**, 651–652 (1963).

References on S. Moore and W. H. Stein

29. S. Moore, *Nobel Lectures—Chemistry* (1972), Vol. 5, pp. 49–50, 72–74, 80–95.

30. W. H. Stein, *Nobel Lectures—Chemistry* (1972), Vol. 5, pp. 49–50, 76–79, 80–95.

31. S. Moore and W. H. Stein, *Nobel Laureates in Chemistry* (1993), pp. 538–545, 546–550.

32. S. Moore, *Biograph. Mem. Natl. Acad. Sci. USA* **56**, 353–385 (1987).

33. W. H. Stein, *Biograph. Mem. Natl. Acad. Sci. USA* **56**, 415–442 (1987).

34. S. Moore, D. H. Spackman, and W. H. Stein, Chromatography of amino acids on sulfonated polystyrene resins: An improved system, *Anal Chem.* **30**, 1185–1190 (1958).

35. D. H. Spackman, W. H. Stein, and S. Moore, Automatic recording apparatus for use in chromatography of amino acids, *Anal. Chem.* **30**, 1190–1206 (1959).

36. S. Moore and W. H. Stein, Chemical structures of pancreatic RNAase and DNAse, *Science* **180**, 458–464 (1973).

37. D. G. Smyth, S. Moore, and W. H. Stein, The sequence of amino acid residues in bovine pancreatic ribonuclease: Revisions and confirmations, *J. Biol. Chem.* **238**, 227–234 (1963).

38. L. S. Ettre and A. Zlatkis (Eds.), *75 Years of Chromatography* (1979), pp. 297–308.

References on R. M. Willstätter

39. *Nobel Lectures—Chemistry* (1901–1921), Vol. 1, pp. 295–314.

40. *Nobel Laureates in Chemistry* (1992), pp. 108–113.

41. *Comprehensive Biochemistry*, Vol. 36, 1986, Chap. 9, Adventures and research (in chromatography), by E. Lederer, pp. 437–490.

42. *Comprehensive Biochemistry*, Vol. 30, 1972, Chap. 13, Biocatalysis and the enzymatic theory of metabolism, pp. 265–278 and Vol. 34A, 1986, Chap. 34, Crystallization of enzymes, pp. 443–455.

References on R. Kuhn

43. *Nobel Lectures—Chemistry* (1922–1941), Vol. 2, pp. 451–455.
44. *Nobel Laureates in Chemistry* (1993), pp. 248–252.
45. H. H. Baer, Richard Kuhn 1900–1967, *Adv. Carbohydr. Chem. Biochem.* **24**, 1–12 (1969).
46. *Comprehensive Biochemistry*, Vol. 36, 1986, Chap. 9, Adventures and research in chromatography, by E. Lederer, pp. 437–490.
47. *Comprehensive Biochemistry*, Vol. 38, 1995, Chap. 2, Memories of Heidelberg, by T. Wieland, pp. 21–108.

References on F. Sanger

48. *Nobel Lectures—Chemistry* (1942–1962), Vol. 3, pp. 541–557.
49. *Nobel Laureates in Chemistry* (1993), pp. 406–411.
50. *Nobel Lectures—Chemistry* (1971–1980), Vol. 5, pp. 377–380, 428–447.
51. *Nobel Laureates in Chemistry* (1993), pp. 633–638.
52. *Comprehensive Biochemistry*, Vol. 34A, 1986, Chap. 12, The polypeptidic nature of proteins, pp. 185–208, and Chap. 16, Sequence analysis, pp. 289–309; Vol. 35, 1983, A French biochemists life (part), pp. 305–306.

References on R. B. Woodward

53. *Nobel Lectures—Chemistry* (1963–1970), Vol. 4, pp. 95–123.
54. *Nobel Laureates in Chemistry* (1993), pp. 462–470.
55. A. R. Todd and J. W. Cornforth, R. B. Woodward, *Biograph. Mem. Fellows R. Soc.* **27**, 629–695 (1981).
56. J. A. Berson, *Chemical Creativity: Ideas from the Work of Woodward, Hückel, Meerwein and Others*, Wiley-VCH, New York, 1999.
57. M. E. Bowden and O. T. Benfey, *Robert B. Woodward and the Art of Organic Syntheses*, Chemical Heritage Foundation, Philadelphia, 1992.

References on Robert B. Merrifield

58. *Nobel Laureate—Chemistry* (1981–1990), pp. 143–175.
59. *Nature* **441**, 824–825 (2006).
60. R. B. Merrifield, in C. W. Gehrke, R. L. Wixom, and E. Bayer (Eds.), *Chromatography: A Century of Discovery (1900–2000)*, Vol. 64, Elsevier, Amsterdam, 2001, pp. 412–413.

References on T. S. Kuhn—The Paradigm Shift and the Nature of Science

61. T. S. Kuhn, *The Structure of Scientific Revolutions*, Univ. Chicago Press (1st ed., 1962; 2nd ed., 1970; 3rd ed., 1996).
62. R. S. Graves and C. Scoville, *Searches in Chemical Abstracts Online* (CASONLINE); now *SciFinder Scholar*.
63. R. W. Zumwalt, K. C. T. Kuo, and C.W. Gehrke (Eds.), *Amino Acid Analysis by Gas Chromatography*, Vols. 1–3, CRC Press, Boca Raton, FL, 1987.

64. C. W. Gehrke and K. C. T. Kuo (Eds.), *Chromatography and Modification of Nucleosides*, 3 vols., *J. Chromatogr.* Library Series, Vols. 45A–C, Elsevier, Amsterdam, 1990.

65. C. Horváth (Ed.), *High Performance Liquid Chromatography*, Academic Press, New York, 1980–1988, 5 vols.

66. H. Hellman, *Great Feuds in Science* (10 earlier disputes described), Wiley, New York, 1998.

67. *Comprehensive Biochemistry*, Vol. 34A, 1986, Chaps. 7 and 8, Molecular correlates of biological concepts, by P. Laszlo, pp. 88–159.

68. K. R. Popper, *The Logic of Scientific Discovery*, Basic Books, New York, 1959.

69. M. Kochen, *The Growth of Knowledge—Readings on Organization and Retrieval of Information*, Wiley, New York, 1967.

70. D. W. Gotshalk, *The Structure of Awareness—Introduction to a Situational Theory of Truth and Knowledge*, Univ. Illinois Press, Urbana, 1969.

71. V. L. Bullough (Ed.), *Scientific Revolution*, Holt, Rinehart and Winston, New York, 1970.

72. J. J. P. Maquet, *The Sociology of Knowledge—Its Structure and Relation to the Philosophy of Knowledge*, Beacon Press, Boston, 1973.

73. I. Lakatos and A. Musgrave (Eds.), *Criticism and the Growth of Knowledge*, International Colloquium in the Philosophy of Science, Cambridge Univ. Press, Cambridge, UK, 1974.

74. K. R. Popper, *Conjectures and Refutations: The Growth of Scientific Knowledge*, 5th ed., Routledge, London, 1989.

75. G. Gutting (Ed.), *Paradigms and Revolutions—Appraisals and Applications of T. S. Kuhn's Philosophy of Science*, Univ. Notre Dame Press, Notre Dame, IN, 1980.

76. I. Hacking (Ed.), *Scientific Revolutions, Oxford Readings in Philosophy*, Oxford Univ. Press, Oxford, UK, 1981.

77. B. Barnes, *T. S. Kuhn and Social Science*, Columbia Univ. Press, New York, 1982.

78. H. Margolis, *Paradigms and Barriers—How Habits of Mind Govern Scientific Belief*, Univ. Chicago Press, 1993.

79. K. R. Popper, *Realism and the Aim of Science*, Routledge, New York, 1992.

80. S. A. Ward and L. J. Reed (Eds.), *Knowledge Structure and Use: Implications for Synthesis and Interpretation*, Temple Univ. Press, Philadelphia, 1983.

81. I. B. Cohen, *Revolution in Science*, Harvard Univ. Press, Cambridge, MA, 1985.

82. L. Bonjour, *The Structure of Empirical Knowledge*, Harvard Univ. Press, Cambridge, MA, 1985.

83. S. W. Hawking, *A Brief History of Time—from the Big Bang to Black Holes*, Bantam Books, New York, 1988; rev. 1998.

84. K. R. Popper, *In Search of a Better World*, Routledge, London, 1992.

85. P. Horwich (Ed.), *World Changes—T. Kuhn and the Nature of Science*, MIT Press, Cambridge, MA, 1993.

86. P. Hoyningen-Huene, *Reconstructing Scientific Revolutions—T. S. Kuhn's Philosophy of Science* (transl.), Univ. Chicago Press, 1993.

87. P. Kitcher, *The Advancement of Science—Science without Legend, Objectivity without Illusions*, Oxford Univ. Press, Oxford, UK, 1993.

88. J. Conant and J. Haugeland (Eds.), *Thomas S. Kuhn's the Road Since Structure—Philosphical Essays (1970–1993)*, Univ. Chicago Press, 2000.

References on V. du Vigneaud and Oxytocin

89. V. Du Vigneaud, A trail of sulfa research from insulin to oxytocin, *Nobel Lecture—Chemistry* (1955), Vol. 3, pp. 441–465.
90. V. Du Vigneaud, *Biogr. Mem. Nat. Acad. Sci. USA* **56**, 542–545 (1987).
91. National Academy of Sciences (H. T. Clarke, Editorial Board), *The Chemistry of Penicillin*, Princeton Univ. Press, Princeton, NJ, 1949.

4

THE TRAILS OF RESEARCH IN CHROMATOGRAPHY

Robert L. Wixom

Department of Biochemistry, University of Missouri, Columbia

In the world of science, nothing invigorates the mind so much as to watch a concept from some small seed of discovery to a universally applicable technology. In the last four decades, chromatography, the once mysterious and very crude technique, has grown into a very sophisticated and reliable separation technology.

—T. E. Beesley and R. P. W. Scott, *Chiral Chromatography*, 1998.

CHAPTER OUTLINE

4.A. INTRODUCTION

Now that the *pioneers of chromatography* and the *discoverers of chromatography* have been introduced (see Chapter 3), a rapid growth in the numbers of chromatography research papers and number of investigators developed in the 1950s–1960s and led to the flowchart of relationships (see Chapter 1, Fig. 1.1). Hence this magnitude directs the present chapter to follow a subject-based outline, much of which follow *a trail of research* (i.e., fulfilling a pattern, or "normal" science according to T. S. Kuhn), and others start new trails due to a "paradigm shift." These leaders become the continuing *builders of chromatography* in this chapter. Some of the following sections may have characteristics of both of these approaches and thus fall between these two approaches. This chapter presents some additional evidence for the relevance of *paradigm shifts in chromatography*. The sections selected for this chapter were written to cover the period 1950–2000. See Chapters 5–9 for later period information (2000–2008).

Some of the features of chromatography methods were described earlier [1,2]. Several books present a more thorough review of some of the chromatography methods [3–5]. A published glossary of chromatography terms is valuable for communications and writing clarity [6].

4.B. CAROTENOIDS BY CHROMATOGRAPHY

The early carotenoid research of Mikhail Tswett, Leroy S. Palmer, Paul Karrer, and Richard Kuhn, described briefly in Ref. 1, was based on adsorption chromatography, spectral evidence, and selective extraction; the last two investigators demonstrated that the carotenoids are the biological precursors of vitamin A. About 10 years after his Nobel Prize, Karrer described 28 carotenoids of known structure and 37 more with partly known or unknown structure [7]. About the same time, Isler et al., synthesized crystalline vitamin A [8] and Karrer and Eugster synthesized β, β-carotene [9]. These discoveries set the stage for the subsequent marked expansion of knowledge of these long-chain pigments.

These earlier thrusts of carotenoid research have been collated and organized in the books by E. Lederer (1934, 1949, 1952, and 1957), L. Zechmeister (1934, 1938, 1941, and 1948), and E. Heftmann (1961, 1967, and 1975) (cited in Ref. 2); these same books include other areas of chromatography. The next wave of carotenoid research has been described in books by Trevor W. Goodwin et al., Robert Macrae et al., and G. Britton et al. [10–14].

Following the "isoprene rule," most carotenoids have eight isoprenoid groups, being 40 carbons in chain length, but a few are 45 and 50 carbons in length. T. W. Goodwin [10,11] summarizes this growth in carotenoid knowledge that in 1984 included:

- 39+ carotenoids in higher plants
- 67+ carotenoids in fruits
- 46+ pigments in photosynthetic bacteria
- 8+ pigments in chromatia
- 26+ pigments in green photosynthetic bacteria
- 186 partial total [10]

Later, other sources compiled lists of 563 carotenoids [13], or over 600 carotenoids [14], that have been isolated and characterized. More than 100 million tons of carotenoids are produced annually by algae, bacteria, fungi, and higher plants. In addition to such isolations, investigators have been active to search for the intermediate steps for the formation of carotenoids [12,14], where the high separation power of chromatography, along with other techniques, has been amazing. Some of these intermediates are in the biosynthesis of the mature pigments; for instance, the final red pigment, spirilloxanthin, in the photosynthetic bacteria, *Rhodopseudomonas spheroides*, is formed by the following biosynthetic pathway:

Acetate (5 carbons) $\rightarrow \rightarrow \rightarrow$ mevalonate (5 carbons)
↓
Isopentenyl pyrophosphate (5 carbons)
↓
Prephytoene pyrophosphate (40 carbons)
↓
Phytoene (40 carbons)
↓ 3 intermediate carotenoids
Lycopene (40 carbons)
↓ 5 intermediate carotenoids
Spirilloxanthin (40 carbons)

In addition to the other more general chapters on their natural occurrence and structure elucidation, Britton et al. [14] has written five chapters on column chromatography, TLC, HPLC, and SFC of the carotenoids. These approaches are coupled with UV/visible spectroscopy, circular dichroism, infrared spectroscopy, Raman spectroscopy, NMR spectroscopy, mass spectrometry, and X-ray crystallography. From Mikhail Tswett (1900s) to the many later scientists, a pattern of investigation arises: *first*, a series of observations and/or questions arises, *then* many research findings follow, *a new* hypothesis is proposed, and, *again*, new questions emerge to be resolved. In addition to the carotenoids, the "isoprenoid pathway" leads also to the biosynthesis of many terpenes, rubber, steroids, and other end products [15]. These several paragraphs present only part of the highlights of carotenoids, but sufficient to indicate that *a pattern of both "normal science" and "paradigm shift(s)" has prevailed in this subject area since the late 1920s.*

4.C. PARADIGM SHIFTS FOR OTHER NATURAL PRODUCTS AND CHROMATOGRAPHY

In addition to carotenoids, the recognition of many natural products, including their isolation, purification, structure determination, and organic synthesis, and search for their biological properties and roles in life forms is a field of longstanding interest for scientists. When did this area start? The simple amide, urea, was discovered in human urine by Rovelle in 1773, and then synthesized in Wöhler's laboratory (1828); urea became the first organic compound to be synthesized from inorganic substances. As the field has grown, many papers, books, and treatises have been written on this subject, but the older ones do not mention chromatography. The first book on carotenoids was published in 1902 by F. G. Kohl. M. S. Tswett's 1910 book on plant pigments was the second, and L.S. Palmer's 1922 book on carotenoids was the third (see references in cited in Ref. 1 of Chapter 1). The early books by E. Lederer (1934), L. Zechmeister (1934 and others), H. W. Willstaedt (1938), H. H. Strain (1942) and others arrived in the 1940s–1950s [2]. Subsequent broad-spectrum books and multivolume treatises have references to chromatography throughout the text [16–21]. Books mainly on the chromatography of natural products and their biosynthesis may be found in the literature [16–28]. Finally, the thorough, nine-volume treatise, Comprehensive Natural Product Chemistry, edited by D. H. R. Barton, K. Nakanishi, and O. Meth-Cohn, was published in 1999 [27].

In the early 1900s, natural products were isolated and purified by selective extraction with organic solvents, selective fractionation by salts or organic solvents, selective precipitation by added heavy-metal ions (or large anions, or cations), etc. As mentioned earlier with the chromatography of chlorophylls and carotenoids, the post-1930 decades witnessed the application of chromatography (PC, TLC, GC, LC, IEC, HPLC, and hyphenated techniques with MS and NMR, etc.) to studies on many other natural products. The term "natural products" embraces the following diverse groups:

- Polyketides and other secondary metabolites, including fatty acids and their derivatives
- Isoprenoids, including terpenoids, carotenoids, and steroids
- Carbohydrates and their derivatives, including tannins, cellulose, and related lignins
- Amino acids, peptides, proteins, and their many derivatives
- Vitamins and growth factors
- Porphyrins
- Alkaloids

- Antibiotics
- Plant hormones
- Pheromones
- Marine natural products

Such a variety of structures in natural products is a source of continuous amazement for the writer and many others. Even more so is the wonder involved in the frequently long, multistep chain of reactions for their biosynthesis. Biochemical history is replete with examples: the biosynthesis of carotenoids, sterols (see earlier research by Konrad Bloch, Feodor Lynen, and others), porphyrins, vitamins, amino acids, and many alkaloids. Fascinating enzymes, many of which have been isolated and characterized, catalyze these unique biosynthetic steps; however, there are many yet-to-be-identified pathways and enzymes to be explored in the twenty-first century.

Reflecting the steady growth from the late 1800s to the present, there are of course many more books on the natural products than the 13 cited ones [16–28]. However, more significant than the number of books is the number of volumes per citation: 76, 49, 16, 9, 3, 3, 1, 16, 21, 1, a volume with 3432 pages, and 9 [16–27]. The K. Nakanishi et al., 1974 treatise of 3 volumes [20] has expanded to 9 volumes for his 1999 treatise [27]. With this vast volume of treatises, books, journal articles, and their respective authors, it is difficult for the outsider to identify significant "turning points"; maybe most are "normal science." The probable exceptions are the brilliant, stepwise research of Richard Willstätter, Richard Kuhn, Paul Karrer, Vladimir Prelog, and Robert B. Woodward, each of whom received the Nobel Prize for their natural products research.

Chromatography in its various branches has had and will continue to have a key role in all of the above areas; a strong foundation has been built and chromatography will facilitate further research advances. From the 1828 synthesis of urea, to the present nine-volume treatise [27], the natural products area has undergone both steady "normal growth" and *also the sudden, new developments and twists in research directions that is characteristic of "paradigm shifts."*

4.D. PARADIGM SHIFTS IN AMINO ACIDS, PEPTIDES, AND PROTEINS

To provide perspective over time, the first amino acid to be isolated was glycine from an acid hydrolyzate of gelatin (Braconnot, 1820). The last essential amino acid (i.e., that needed for growth of young weanling rats) was threonine (William C. Rose, 1934). Column-partition chromatography of amino acids began with A. J. P. Martin and R. L. M. Synge (1941) and was extended to planar (paper) chromatography by R. Consden, A. H. Gordon, and A. J. P. Martin (1944). The next steps were the extension to two-dimensional paper chromatography to ascertain the structure of peptides (gramicidin and tyrocidine) by R. L. M. Synge, then to insulin by F. Sanger and followed by oxytocin and vasopressin by V. du Vigneaud. This classical research has been described in careful detail in Chapter 3 and by Jesse P. Greenstein and Milton Winitz in 1961 [29]. The next major breakthrough was the quantitative ion-exchange chromatography of amino acid analysis by S. Moore and W. H. Stein in the 1950s; thus, IEC was the foundation for their elucidation of the structure of RNase with its 124 amino acids in the 1960s. Each of these achievements is described in more detail earlier in Chapter 3 with appropriate references with further details and developments in Chapter 7.

The sequence analysis of peptides has been extended to many peptides other than insulin (see Chapter 3 presentation on F. Sanger) and to many other and proteins larger than ribonuclease (see Chapter 3, on S. Moore and W. H. Stein). Key chromatographic methods have been GLC, HPLC, their coupled detectors, and the more recent automated protein sequences. Further discussion of this subject is beyond the scope of this book, except to mention that sequence determination is one of the fundamental methods for proteomics (see Chapter 11, Section 11.F).

Chromatographic methods have contributed to clinical medicine over the years, particularly the inherited diseases. Sir Archibald Garrod (1908) examined alcaptonuria—a defect in tyrosine metabolism, cystinuria, pentosuria, and albinism and introduced the concept of "inborn errors of metabolism" [30]; a half-century later, the defect in alcaptonuria was shown directly to be the loss of activity of homogentisic acid oxidase in the liver of an alcaptonuria patient [31]. After 1950, metabolic defects were found for the metabolism of almost all common amino acids [31,32], and was extended to inherited disorders in the metabolism of carbohydrates, lipids, purines, pyrimidines, steroids, metals (Co^{2+} and Fe^{+}), porphyrins, transport processes, complex carbohydrates, endocrine disorders, connective-tissue defects, etc. [31,32]. In most of these inborn errors of metabolism, end products of biosynthesis and/or intermediates in their catabolic breakdown accumulate in the blood, urine, and probably tissues; chromatography has led a major role in their detection. A gifted chromatographer, Dr. Egil Jellum, has made many excellent contributions in these areas (see Ref. 29, his section in Chap. 5 and the concept of "metabolic profiling").

An early electrophoretic study of hemoglobin showed the distinction between normal adult hemoglobin A and hemoglobin S, the defective protein in sickle cell anemia [33]; Linus C. Pauling, Harvey A. Itano, and others in 1949 showed that HbS was altered in only one amino acid—a change from the normal glutamic acid to an abnormal valine [33]. They coined the descriptive term "molecular disease." L. C. Pauling (1901–1994) was the recipient of the 1954 Nobel Prize in Chemistry for "his research into the nature of the chemical bond and its application to the elucidation of the structure of complex substances." Being a very versatile scientist, the above comments are only a glimpse of his overall research (see description in Refs. 34 and 35).

Hemoglobin—a major long-studied protein, from the perspectives of medicine, hematology, physiology, and biochemistry—has four subunits (2α and 2β chains) with a known amino acid sequence, tertiary structure and mechanism of oxygen transport. Since Pauling's 1949 paper, many abnormal hemoglobins have been identified by 1996: 634 hemoglobin variants with a single amino acid substitution, plus about 51 other variants [36,37]. Electrophoresis and chromatography have contributed to these findings. Thus, the 1949 report of L. C. Pauling was a classical example of a "paradigm shift" that led to the subsequent elucidation by others of the many hemoglobin variants in the usual pattern of "mainstream science," or "normal science."

TABLE 4.1. Milestonses in Quantitative Chromatography of Amino Acids, Peptides, and Proteins

Structure of Peptides/ Proteins	Number of Amino Acids	MW [1]	Nobel Investigators	Experiment Date	Nobel Prize Date
Amino acids and protein hydrolyzates	20+	200±	A. J. P. Martin and R. L. M. Synge	1941	1952
Tyrocidin (A, B, or C)	10	1,308±	R. L. M. Synge	1944	1952
Gramicidin	10	1,141	R. L. M. Synge	1945	1952
Oxytocin	9	1,007	V. du Vigneaud	1953–1954	1955
Vasopressin	9	1,056	V. du Vigneaud	1953–1958	1955
Insulin	51	6,000	F. Sanger	1951	1958
Ribonuclease A (bovine pancreas)	124	14,000	S. Moore, W. H. Stein, and C. B. Anfinsen	1960s	1972
Hemoglobin (human)	574	64,500 (4 subunits)	L. C. Pauling,[a] M. F. Perutz,[a] J. C. Kendrew[a]	1936–1951 1950s 1950s	1954 1962 1962

Source: G. D. Fasman (Ed.), Handbook of Biochemistry and Molecular Biology, 3rd ed., CRC Press, Cleveland, OH, 1976.
[a]The research of these investigators concerned primarily the tertiary structure of hemoglobin and myoglobin.

Furthermore recent details on the chromatographic separation and analysis by MS, NMR and other analytical methods for peptides and proteins is presented in the detailed 1999 review by C. K. Larive and co-authors [38].

The preceding remarks are essentially a sketch, and are concisely portrayed in the Table 4.1.

The reader may have noted a major concentration on peptides and proteins in the chromatography literature (see the titles of many books cited in Apps. 4, 5, and 7 of Refs. 1 and 2, to avoid lengthy repetition, and find references for the purification of proteins and the determination of the sequential structure) [39]. These high-molecular-weight molecules are frequently similar to each other in properties; the secret is to find the property(s) that will lead to chromatographic or other methods of separation. The ease of denaturation of proteins/enzymes (i.e., their lability) was a handicap for the decades of the 1920s–1960s, but now sufficient specific details on how to handle and to purify 100s of proteins to homogeneity are available; the primary chromatography methods are IEC, SEC, HPLC, and affinity chromatography. Another reason for this major emphasis is the central role of proteins in hundreds of biological functions.

4.E. AFFINITY CHROMATOGRAPHY

Affinity chromatography, sometimes called *bioaffinity chromatography* or *bioselective chromatography*, has had a rapid growth since its conception in the late 1960s; it is an isolation/purification chromatography method based on the principle that the molecule under study can interact selectively and reversibly bind with another, structurally related compound, that is immobilized on a chromatographic support column. While several earlier brief reports are known, most authors recognize two initial papers. J. Porath et al. [40] bonded a ligand covalently to an open matrix of cellulose, starch, or crosslinked "Sephadex" for the stationary phase and then added cyanogen bromide or iodide to form an active intermediate. Next, they added a dipeptide as the ligand, but conceivably, it could be an enzyme, a protein, or a nucleic acid. Thus, this synthetic stationary phase has a specific binding affinity for a related, particular analyte. In the case of an enzyme, an insoluble "immobilized enzyme" is formed and is usually capable to function in a column with the substrate becoming the eluent. For a bound protein, it became possible to examine the binding of protein to protein, antibody to antigen, coenzyme to enzyme, biotin to avidin binding, and so on. The next key advance was from Anfinsen's laboratory [41].

Christian B. Anfinsen et al. (see Table 3.1) was one of three 1972 Nobel awardees; the prize was shared with Stanford Moore and William H. Stein for research on the structure of ribonuclease (see Chapter 3). Anfinsen's contribution was slightly different from that of Moore and Stein or, in the words of the Nobel Committee, for "Studies on ribonuclease, particularly the relation of the amino acid sequence and the biologically active conformation." Their original papers reported the use of chromatography, although this evidence is a small part of their overall contributions for RNase. During the 1967–1993 period, Anfinsen and associates published about 36 research papers, many of which included reports on the use of IEC, SEC, TLC, and affinity chromatography (mainly on human lymphoblast interferon and *Staphylococcal nuclease*). By contrast, their key 1968 paper [41] relied on Porath's methods to build selective affinity chromatography columns and to next purify α-chymotrypsin by the bound inhibitor C-aminocaproyl-D-tryptophan methyl ester; then they separated carboxypeptidase A by bound L-tyrosine-L-tryptophan, and finally *Staphylococcal nuclease* by a bound specific inhibitor [3′,4 amino-phenylphosphoryl)-deoxythymidine-5-phosphate]. Modified agarose columns were also employed by this National Institutes of Health group, and have since then been widely used by other investigators. These brilliant specific examples of the general utility of affinity chromatography became the turning point—the "paradigm shift"—that led to many subsequent studies, so great that the following paragraphs constitute a general summary [42–47].

With the subsequent influx of many other investigators, the possible experimental directions have increased. Many solid supports were investigated and include agarose and its crosslinked derivatives, polyacrylamide gels, agarose/polyacrylamide copolymers, hydroxyalkyl methacrylate copolymers, silica-based supports, cellulose, crosslinked dextran, and collagen. Most of these are commercially available from laboratory supply houses along with a proliferation of brand names. Cyanogen bromide continues to be the most frequently used coupling agent, although others are used, such as sulfonyl chloride- and epoxide-containing reagents, carbodiimide-promoted methods, bifunctional reagents (e.g., glutaraldehyde), and reductive alkylation (oxidation of polysaccharides, followed by binding to proteins as Schiff's bases and then $NaBH_4$ reduction). The length of these functional groups vary and will be a factor in the binding and separation of the solutes; they are sometimes called "linkers." *One constant in these variations of affinity chromatography is the specific binding of the analyte to the ligand on the support, leading to the flowthrough of the undesired protein, and followed by selective elution of the desired protein.*

When the column support is large and the immobilized ligand is small, steric hindrance may cause reduced or lack of binding for the substances to be isolated. This problem was resolved by the use of a "spacer arm," usually five to seven carbon atoms or equivalent, that will allow the ligand to be beyond the surface of the support. Spacer arms in use include glutaraldehyde, hexamethylene diamine, ε-aminocaproic acid, *N*-hydroxysuccinimide, 3,3'-diaminodipropylamine, bisoxiran, and hydroxyalkyl derivatives.

Affinity chromatography has been effectively employed for the isolation and purification of enzymes, peptides, plasma proteins, receptors, binding and transfer proteins, nucleotides and nucleic acids, viruses, antibodies, antigens, lectins, and other compounds. The secret is to identify a specific compound to bind to the linker–solid support; as many have been tested, the reader is referred to several books [42–47]. Since affinity chromatography has been most widely employed for the purification of enzymes, the possible affinity ligands may be of low or high molecular weight and include substrates, products, coenzymes (NAD, NADP, AMP, etc.), coenzyme analogs (such as triazine dyes), proteins, peptides, and lectins for glycoprotein enzymes. Many examples and reports on immunoaffinity chromatography are available. The research on ribonuclease by C. Anfinsen led to probing its three-dimensional structure; mild heat caused it to unravel to a long, crooked string. When cooled, the denatured RNase refolded into its normal 3D shape. More recently, other investigators using NMR and supercomputers are very close to deciphering the stepwise organization of a protein's three-dimensional organization [48]. Biotin, a B vitamin, can bind to avidin, an egg-white protein; this pair has been widely used in affinity chromatography [48]. Thus, the early recognition of the flexibility and specificity of affinity chromatography led to its widespread adoption [42–49], which was facilitated by the additional chromatography awardees in this book, who may be identified by the code letter "i" in the Author Index (see also Table 1.3).

With the abovementioned contributions and their major biochemical, biological, and medical significance, research exploded in this area in the 1970s–1990s. Hundreds of research papers and applications followed; then came many reviews of this expanding area. All of which led to symposia and/or books during 1978–1993 period [42–51]. Following the earlier stated literature pattern, we have T. Kline's *Handbook of Affinity Chromatography* (1993) [46]. A series of at least eight International Symposia on Affinity Chromatography and Biological Recognition have facilitated development of the method [51,52] have been published. This groundwork is now being actively investigated in the pharmaceutical field for drug–receptor interactions [52] and related books [53,54].

A Case History of Ubiquitin and Affinity Chromatography.

The preceding general features of affinity chromatography lead to a specific example—namely, the role of ubiquitin in protein degradation. During the 1940s–1950s, investigators showed that proteins undergo "turnover"—the continuous synthesis and breakdown in the body. These rates differ with each protein; hemoglobin has a biological half-life of 120 days, whereas plasma proteins have $t_{1/2}$ of $\sim$7 days. The highly detailed mechanism and specific intermediates in protein synthesis were unraveled in the 1960s–1970s; to over-simplify, the sequence to transmit genetic information is

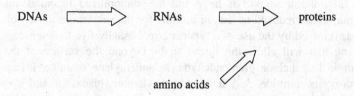

The role of lysosomes in the degradation of some proteins was elucidated in the 1970s by Christian R. de Duve (Nobel Prize in Physiology/Medicine, 1974). However, the action of these hydrolytic enzymes in the lysosomes did not explain the observation that the addition of ATP (adenosine triphosphate) accelerated the rate of in vitro protein breakdown. This anomaly begged for an explanation. At this time, a "paradigm shift" occurred.

To study protein breakdown (subcellular), Avram Hershko et al. [55] selected reticulocytes, the immature red blood cells, that are rapidly degrading subcellular organelles in the erythroblast series—just prior to their release as the circulating erythrocytes. Phenylhydrazine was administered to rabbits to induce anemia and the now circulating reticulocytes (70% of total RBCs) were harvested. Fractionation of the cell-free extracts of reticulocytes showed the necessity for both a heat-stable factor and a heat-labile fraction (the enzymes) to degrade proteins [55]. Further steps by Avram Hershko, Irwin A. Rose, and coworkers with chromatography led to the isolation of the former factor, its purification, and its structure determination to be a polypeptide of 76 amino acids, now called *ubiquitin* (or Ub) [56]. This group proposed that ubiquitin is covalently bound to the substrate protein(s) and that this tag commits the protein(s) for degradation [56]. To resolve the mix of enzymes in the heat-labile fraction, ubiquitin was bound to an affinity chromatography column, which in turn led to the separation of Ub–protein conjugates: E_1—the Ub-activating enzyme (ATP-dependent), E_2—the Ub-conjugating enzymes, or Ub–carrier protein, and E_3—the Ub protein ligase. Polyubiquitinylation of protein substrates also occurs [57,58]. The Ub-dependent proteolysis system responds to glucocorticoids, tumor necrosis factor, metabolic acidosis, viral infection, feeding cycles, and damaged proteins [59], and has been reported in such cellular processes as heat-shock response [58–61], DNA repair [58,60], cell cycle progression [58,60,62], modification of histones and receptors [58], and the pathogenesis of some neurodegenerative diseases [58,60].

The proteins ligated to Ub are then degraded by an "ATP-dependent 26S protease complex" (also called *proteasome*), or ubiquitin-conjugating–degrading enzyme, which has a 20S particle (proteasome), one or two 19S regulatory complexes, and 28 additional polypeptides; it rapidly degrades tagged proteins to short peptides, which in turn are hydrolyzed quickly by cytoplasmic exopeptidases [63–68]. A 20S proteasome has also been recognized in archaebacterium and other bacteria (with 14 α subunits and 14 β subunits) and in eukaryotic cells with a more complicated subunit structure [65, 68]. The Ub system has also been scrutinized from the viewpoint of genetics [69–71], several disease states [71], and potential proteasome inhibitors (lactacystin, epoxomicin, peptide aldehydes, peptide vinyl sulfones, and others to be announced later) that may have therapeutic merit in disease processes ranging from inflammation to cancer [72]. This sketch of key, but not all, reviews [58–72] that each have from 67–414 references indicate the pivotal role of *an early paper that utilized affinity chromatography* [56]. Along with other many research papers and the resultant reviews are more recent articles and books on ubiquitin and its various roles in living cells: 1988, 1994, 1994, and 1998 [57,73–77].

Ubiquitin—an Example of a Paradigm Shift. The unfolding of the ubiquitin roles is a dramatic example of the "paradigm shift" (see Section 3.D), so much so that some figures from an online search in WorldCat are now introduced. A recent online search of WorldCat showed that the number of Ph.D. theses per year rose steadily; 1990—2, 1991—0, 1992—3, 1994—2, 1995—8, 1996—8, 1997—4, 1998—4, 1999—18, 2000–2007—491, for a total of 540 theses. From 1966 to 1999, there were 1535 research articles, journal reviews, or books with "ubiquitin" as the key concept. Of this total, 150 were scholarly reviews (1970s—0, 1980s—23, 1990s—127, and 2000s—0). The titles of selected reviews [58–72] and now books [73–75] also show the magnitude of the recent penetration of the ubiquitin/proteasome function into the life sciences.

(*Note*: The variety in the journal names and the key words in the title of the cited review books reflect the breadth of roles for ubiquitin. To go into further detail would stray beyond affinity chromatography—our starting point for this section.)

Some characteristics of a *paradigm shift* were described in Section 3.D. When a question arises, or an experimental result does not fit the prevailing theory on the subject, the investment in research and doctoral theses frequently undergoes a marked rise. Such was the case for ubiquitin and its several roles, as depicted by the preceding reviews and book titles.

To summarize, ubiquitin should be considered as a posttranslational signal sequence that is attached to a variety of cellular proteins, which are led to one of several fates, depending on the localization of the conjugates and the enzymatic specificities. One of the earliest Hershko papers (in 1980) [56] relied on *affinity chromatography* to separate the enzymes involved in protein degradation (and later other roles). This key turning point unleashed a flood of subsequent investigations, and thus is another example of a *paradigm shift*. Thus these fundamental advances by Avram Hershko, Irwin A. Rose, and Aaron Ciechanover led to their recognition as Nobel Prize awardees (2004); see Chapter 3, Table 3.2.

Molecular Recognition. Parallel to the above development of affinity chromatography has been the growing knowledge over the past several decades of the phenomena called *molecular recognition*. "Molecular recognition depends on a number of elementary interactions from which are built with more complex interactive systems, such as

an enzyme and its substrate, a hormone and its receptor, antibodies and antigens, etc."
[78]. Such recognition has a time scale (i.e., longer than the initial collision time) and
a space scale (i.e., retention of their individual structures along with their association
of other molecules) [78]. This thrust includes *receptors* or ligands for hormones (i.e.,
insulin, estrogen, angiotensin II, thyroxine, and many others), for many drugs (β- and
α-adrenergic receptors, serotonin receptors, etc.), for growth factors (epidermal, enkeph-
alin), for membrane lectins, monoclonal antibodies, etc. Receptors have been identified
for estrogen, progesterone, testosterone, oxytocin, relaxin, prostaglandins, and perhaps
other hormones, and thus have key roles in the physiology of reproduction for both animals and humans [79]. This area also includes intercellular synapses, microbial adher-
ence, embryologic interactions, viral receptors, neurotransmitter receptors, purinergic
receptors, histocompatability antigens, receptor-mediated endocytosis, and other areas
[78,80–84]. Hence appearing on the recent research scene is the term *molecular imprinting*, which refers to the intentional complementary guiding sites having high selectivity
and affinity for the template molecule [85–87]. The biotin–avidin complex has more
recently been used to construct microscopic arrays of active enzyme sites on a carbon
fiber substrate by maskless photolithography; such "biochips" may have potential application in biosensing, cell control and guidance, and nanotechnology [88]. During these
investigations, affinity chromatography has had a minor role, but now with the many
more recent advances in affinity chromatography, this approach will probably be used
more frequently in investigations related to molecular recognition and imprinting. The
earlier turning points or "paradigm shifts" on affinity chromatography, ubiquitin and
receptors will probably continue to grow and expand (see Table 4.2).

TABLE 4.2. From Affinity Chromatography to Receptors

Research Area	1960–1969	1970–1979	1980–1989	1990–1999	2000–2007
From CASLINE on affinity chromatography					
Original papers	5	466	1,314	1,687	3,084
Reviews	3	266	386	440	358
Books	0	2	2	11	2
From MEDLINE on receptors in biological systems					
Original papers	266	11,121	58,309	130,097	245,398
Boolean search of receptors and affinity chromatography	0	260	1,457	1,742	686

Note: The numbers for the research literature on affinity chromatography, being chemically oriented, were
elicited from online searches of *Chemical Abstracts* (American Chemical Society, Washington, DC;
CASLINE). Similar data on research literature for receptors in biological systems were derived from online
searches of *Index Medicus* (National Institutes of Health, Bethesda, MD; MEDLINE). The findings listed in
this table constitute a "paradigm shift" for this area of interactions.

4.F. CHIRAL CHROMATOGRAPHY

The phenomena of asymmetric carbon atoms, optical isomers and rotation of plane polarized light has been known for a long time. Louis Pasteur in 1848 [89] separated physically the isomers of crystalline tartaric acid. J. H. Van't Hoff [90] and J. A. Le Bel [91] suggested that this asymmetry was the basis for optical rotation. Now it is known that all natural amino acids in mammalian proteins, except glycine, are the L-isomer, and all common monosaccharides are the D-isomer. These isomers are optically active, or chiral, and are also called *enantiomers*; they lack centers, planes, or improper axes of symmetry. Emil Fischer ascertained the absolute stereochemistry of (+) glucose, related it to D-(+) glyceraldehydes [92], which was part of the basis of his 1902 Nobel award. During the early years, scientists relied on polarimetry, and later stereospecific enzymes, to study the stereoisomers. At the turn of this millennium, there were 1000s of naturally occurring and synthetic stereoisomers, of which enantiomers are a part; further consideration of this subject is beyond the range of this book [93].

Enantiomers have identical physical properties (melting and boiling points, solubilities, partition coefficients, etc.) and identical chemical properties with respect to nonchiral reagents. However, on reaction of one enantiomer with a chiral reagent, the product may show differences in reaction rates and product stereochemistries, as well as their effect on the rotation of plane polarized light. In addition to one or more asymmetric carbon atoms, molecular asymmetry may be caused by twisted structures (cumulines, spiro compounds), hindered rotation (atropisomers, ansa compounds), or molecular overcrowding (polycyclic aromatics). To return to chromatography, Emanuel Gil-Av was the first to develop a chiral stationary phase for gas chromatography in 1966 [94].

Emanuel Gil-Av (1916–1996), born as E. Zimkin in Pensa, Russia, has been recognized as "the great pioneer of enantiomer separation by chromatography" and shared being the first recipient of the Gold Chirality Medal in 1991. He migrated to Tel-Aviv, Palestine, with his family in 1928 and later studied chemical engineering at the University of Strasbourg, France. With the outbreak of World War II in 1940, he was forced to leave France and worked in London with Dr. Chaim Weizmann, later founder of the Weizmann Institute of Science and later President of the State of Israel. He returned to Israel in 1946 to begin research on the solvent extraction of penicillin and later a variety of petroleum chemistry projects by gas chromatography. He completed his Ph.D. in 1951 under Dr. E. D. Bergmann on syntheses in the oxazolidine series. Since the Weizmann Institute had known strength in peptide chemistry, Gil-Av commenced a 30-year study on chiral discrimination with a group of active younger investigators. His first paper (in 1966) on quantitative enantiomer separation of amino acids on diamide and peptide stationary phases was performed with glass capillary columns [94]. This finding led to the first preparative enantiomer separation by GC, enantioselective GC by complexation, and other significant findings [95–97].

Gil-Av's chromatography had the limitation that the higher operating temperatures of GC caused some racemization of the chiral stationary phase that led to a loss of ability to separate the enantiomers. Later, Frank, Nicholson, and Bayer [98] developed a more stable chiral stationary phase of substituted polysiloxanes that also allowed higher temperatures (175°C) to be used, leading to shorter retention times (30 min) [98]. Then, the advent in the late 1960s of HPLC, along with improved hardware (pumps, injectors, and detectors) and special stationary phases (bonded groups on the silica support), paved the way for the rapid advances of the 1970s. Pirkle et al. [99] provided evidence for his "three-point rule"; namely, for chiral resolution to result, three simultaneous interactions between one solute enantiomer and the chiral stationary phase (CSP) must take place and one or more of which must be stereochemically dependent. About this time, the number of possible CSPs exploded with immobilized chiral agents, such as amino acids, metal complexes (L-proline complexed with copper ions in Davankov's work), synthetic compounds (chiral fluoroalcohols), dinitrobenzoyl phenylglycine, proteins (albumin and α_1-acid glycoproteins), cyclodextrins via inclusion complex formation, and crown ethers (a host-guest interaction within the chiral cavity) [100]. Cellulose, cellulose triacetate and other polysaccharides, synthetic dyes, and substituted methacrylate on silica and over a 100 different chiral HPLC columns are available commercially [101,102].

One or perhaps multiple mechanisms participate in retention in chiral chromatography: hydrogen bonding, electrostatic, lipophilic, charge transfer steric interactions, conformational adjustment, and molecular fit. The reader is also referred to the thorough contributions of D. W. Armstrong, W. H. Pirkle, and C. Welch and their later developments [1]—see Chapter 5.

Chiral chromatography may be indirect, where the enantiomers are derivatized with a chiral derivatizing agent (several dozen are commercially available) to provide a mixture of diastereoisomers, which are then separated on a conventional achiral chromatography column. While many examples are known, the direct approach seems to be preferred; here the enantiomers are separated on a chiral stationary phase with or without prior derivatization.

The most frequently used technique for chiral chromatography is HPLC, although a growing number of reports apply the concepts to GC. The chiral chromatography bed has also been explored with TLC.

Another design feature for the derivatized chiral ligand is to use ultraviolet absorption, and fluorescence reagents to facilitate the detection of the two isomers. One source [101] lists 52 different organic chiral reagents for indirect chiral chromatography: acylating and amine reagents, isocyanates and activated carbonates, isothiocyanates, chloroformates, o-phthaldehyde with chiral thiols, and aryl halides. For further specifics, please refer to the related books [101–110].

Earlier limitations were overcome with the development of a more sensitive chiral detector that is now commercially available [110]; Armstrong et al. [111] have used this laser-based polarimetry detector with cyclodextrin columns to study more than 230 chiral compounds. Other detectors for HPLC have been applied (or developed) for chiral chromatography: UV, diode array, electrical conductivity, fluorescence, light-scattering, and refractive index detectors [110]. In addition to carbon compounds, chiral centers are known for substituted sulfur, phosphorus, nitrogen, and boron compounds [110].

The realm of chirality has been a long preoccupation of organic chemists, biochemists, and many other life scientists. Thus it is not surprising that an International Symposium on Chiral Discrimination has been meeting from 1991 to date, and that a journal, *Enantiomers—a Journal of Stereochemistry*, has been published from Vol. 1

TABLE 4.3. The Chirality Medal Awardees

Year	Awardee, Affiliation, and Country
1991	Jean Jacques, Chimie Interactions Moleculaires, College France, Paris, France
1991	Emanuel Gil-Av, Weizmann Inst. Science, Rehovot, Israel
1992	Vladimir Prelog, Swiss Federal Inst. Technology, Zurich, Switzerland
1993	Kurt M. Mislow, Princeton Univ., Princeton, NJ, USA
1994	William H. Pirkle, Univ. Illinois, Urbana, IL, USA
1995	Koji Nakanishi, Columbia Univ., New York, NY, USA
1996	Ernst L. Eliel, Univ. North Carolina, Chapel Hill, NC, USA
1997	Ryoji Noyori, Nagoya Univ., Nagoya, Japan
1998	Henri Kagan, Univ. Paris-Sud, Paris, France
1999	Vadim A. Davankov, Acad. Sciences, Moscow, Russia
2000	K. Barry Sharpless, Scripps Research Inst., La Jolla, CA, USA
2001	Yoshio Okamoto, Nagoya Univ., Nagoya, Japan
2002	Dieeter Seebach, Swiss Federal Inst. Technology, Zurich, Switzerland
2003	Daniel W. Armstrong, Iowa State Univ., Ames, Iowa, USA
2004	Volker Schurig, Univ. Tubingen, Tubingen, Germany
2005	Kenso Soai, Tokyo Univ. Science, Tokyo, Japan
2006	Meir Lahav, The Weizmann Inst. Science, Rehovot, Israel
2007	Nina Berova, Columbia Univ., New York, NY, USA

in 1996 to Vol. 7 in 2002. This symposium also has a Chirality Gold Medal whose 1991–2007 recipients are described in Table 4.3; many, but not all, scientists studying chirality relied on chromatography. The description of several of these awardees (Emanuel Gil-Av, William H. Pirkle, Vadim A. Davankov, and Daniel W. Armstrong may be found in our earlier book [1].

Emanuel Gil-Av, who shared the first Chirality Gold Medal Award and was described earlier was followed by **Vladimir Prelog** (1906–1998), the 1992 Chirality Gold Medal awardee; Prelog was born in Sarajevo, Yugoslavia, educated at the Institute Technical School of Chemistry, Prague, Czechoslovakia (1928), and earned his Doctorate in Chemical Engineering (1929). For most of his professional life, he was a faculty member at the Swiss Federal Institute of Technology, Zurich, Switzerland (1942–1976). Prelog made major contributions in the determination of structure and stereochemistry of alkaloids (e.g., quinine, solanidin, strychinine), antibiotics (nonactin, boromycin, ferrioxams, and rifamycin), asymmetric organic and biological syntheses, and descriptor rules for stereoisomers. His many honors include the 1975 Nobel Prize in Chemistry for "his research into the stereochemistry of organic molecules and reactions" [112].

Many years ago, Prelog and Wieland reported the chromatographic separation of enantiomers of Troeger's base using starch as a stationary phase [113]. While Prelog performed brilliant research on stereoisomers over many later decades, as recognized by his receipt of the 1992 Chirality Award, he apparently did not study further the area of chiral chromatography [112].

W. H. Pirkle is the third Chiral Awardee who has advanced chiral chromatography [99]. He has described his many contributions [1, see Chap. 5].

These early initiatives by E. Gil-Av and W. H. Pirkle in chiral chromatography have been followed by many other investigators and their chiral research papers, so much in volume that 10 books on this specific theme were published during 1959–1998 [101–110]. Other advances in chiral chromatography have been described for GC, LC, preparative chromatography, capillary electrophoresis, and capillary electrochromatography in the thorough 1998 book by Beesley and Scott [110]. A 2000 symposium focused on many advances in chiral capillary electrophoresis [114] with 10 reviews, including two on amino acids and peptides, one on herbicides, plus many related original articles. This tidal wave of a major change in research direction is consistent with the "paradigm shift" hypothesis.

Additional evidence of the "paradigm shift" in the area of chiral chromatography includes the research of the deceased Awardees (G. E. Hesse and A. Zlatkis—see Ref. 1) and at least 10 living Awardees (see Ref. 1 and Author/Scientist Index for those with a code letter of "q").

What are the theoretical implications of chirality in metabolism and pharmaceuticals? The significance of stereochemistry in natural or synthetic drugs is not a new concept. Differences were found earlier in the enantiomers for their biological activities, potencies, toxicities, mechanisms of transport, and pathways of metabolism. For earlier years and at present, many drugs are sold as racemates after having been evaluated earlier for their risk/benefit and cost/benefit factors. Nearly 60% of the most frequently prescribed drugs in the United States have one or more asymmetric centers. Some drugs have been resolved by selective enzymatic and microbiological processes, or by fractional crystallization. Hence, the new approach in this area is chiral chromatography for both the analytical and preparative objectives.

What is the practical significance of chiral chromatography? Probably the most dramatic adverse effect of using a racemate drug is the thalidomide disaster in the early 1960s. Thalidomide was used as a sedative and antinausea preparation during human pregnancy, but caused the tragedy of serious birth defects in newborn babies from mothers receiving thalidomide prior to the development of chiral HPLC. Subsequent mouse studies suggest that $R(+)$ thalidomide was inactive, whereas the $S(-)$ thalidomide was teratogenic [102]. Propanolol and many other β-blocker drugs have been resolved by chiral chromatography. Beyond the present emphasis on chiral drugs for human patients is the extension of chiral chromatography for veterinary drugs, pesticides, food and drink additives (the R isomer of aspartame, the sugar substitute, is sweet, but the S isomer of asparagine is bitter, etc.). Further discussion is beyond the range of this book (see Ref. 110 and later references).

Biotechnology has led to a major thrust for new biopharmaceuticals [115]. Parallel to these advances is the growth of chiral drugs, with the worldwide sale of single-enantiomer drugs rising 21% [116]. Areas covered include drugs for diseases of the cardiovascular system, central nervous system, antiviral antiinflammatory mechanisms (e.g., ibuprofen), lung (albuterol), and cancer [117]. Several companies are now producing chiral separation phases; both academic and industrial laboratories have accelerated the asymmetric synthesis of organic compounds and enantioselective chemistry [118].

TABLE 4.4. Chiral Chromatography in Process Production

Period	Phase	Column	Production Rate
Late 1980s	Semipreparative scale	1, 2, or 5 cm i.d.[a]	2 g/day
Early 1990s	Preparative scale	5–20 cm i.d.	10 g–1 kg/week
Late 1990s	Pilot and production scale	6–16 columns, continuous flow	Kilograms to metric tons/year

[a]Inner diameter.

For instance, chiral stationary phases are now being coupled to HPLC developmental for faster screening of chiral pharmaceuticals [119–123]. Further discussion of this general area may be found in Chapter 8; see also Table 4.4.

To summarize, what was once an idea, or an obstacle to overcome, or a question in the minds of the chiral pioneers in the 1960s–1970s, has opened major new doors, leading to new insights, additional processes, and subsequent achievements. What was thought to be impossible (prior to Gil-Av in 1966) has become reality. Hence, the development of chiral chromatography is another example of a "paradigm shift" in chromatography.

4.G. SUPERCRITICAL-FLUID CHROMATOGRAPHY (SFC)

Supercritical fluids have been known for a long time (1822), have viscosities and densities between those of gases and liquids, and hence show efficiencies and analysis times between those of GC and LC [124,125]. Myers and Giddings [124] compared many different gases at high pressures (2000 atm) and found that N_2, He, CO, HCN, H_2S, $CClF_3$, and CH_3F were ineffective; dense CO_2 was best for nonpolar compounds, and dense NH_3 was suggested for polar compounds (some 70 compounds were tested). Pentane, CCl_2F_2, and $CHClF_2$ had some desirable qualities. While GC has high sensitivity and selectivity, it cannot be used with compounds having thermolability or low volatility. Although LC and SFC can separate much larger molecules than can GC, plus those that are reactive, thermolabile, and nonvolatile, SFC stands ahead of LC in that quantitation by flame ionization detection (FID) is feasible [125].

In the first report on SFC, Klesper and coworkers in 1962 [126] used supercritical Freons as the carrier for metal porphyrins moving through a chromatography column, but the work had some practical and theoretical limitations [125]. Giddings and colleagues employed high pressures (up to 2000 atm) and explored the theoretical aspects of what they called "dense gas chromatography" [124]. Gradually as pumps, injectors, and detectors were improved for LC in the 1980s, they were transferred for use with SFC (see details in Refs. 124c and 125). Other key early SFC papers were by T. H. Gouw and R. E. Jentoft (1972), E. Klesper (1978), U. v. Wasen et al. (1980), and M. L. Lee and K. E. Markides (see Ref. 125 for these reviews), and others (see Refs. 127–129).

One reason for the rapid growth in understanding of SFC was a symposia series initiated by Drs. Milton L. Lee and Karin Markides; the first one in 1988 was called the "Workshop on Supercritical-Fluid Chromatography." For the third meeting in 1991, the

TABLE 4.5. International Symposia on Supercritical-Fluid Chromatography, Extraction, and Processing

Date	Place	Organizers	Keynote Lecturers (Awardees)
1988	Park City, UT	M. L. Lee and K. Markides	E. Klesper
1989	Snowbird, UT	Same organizers	D. Ishii
1991	Park City, UT	Same organizers	M. V. Novotny
1992	Cincinnati, OH	J. W. King, L. T. Taylor, K. D. Bartle, and T. L. Chester	M. L. Lee
1994	Baltimore, MD	Same organizers	S. Hawthorne
1995	Uppsala, Sweden	K. Markides, G. Jonsall, K. D. Bartle, H. Engelhardt, and P. Sandra	(none awarded)
1996	Indianapolis, IN	J. W. King, L. T. Taylor, and T. L. Chester	K. D. Bartle
1998	St. Louis, MO	L. T. Taylor and T. L. Chester	J. W. King
2000	Munich, Germany	H. Engelhardt and T. L. Chester	
2001	Myrtle Beach, SC	T. L. Chester, L. T. Taylor, and J. W. King	K. Markides
2004	Pittsburg, PA	T. L. Chester, J. W. King, L. T. Taylor, T. Berger, and L. Chordic	E. J. Beckman

Source: This table is based on information supplied by Dr. Tom L. Chester.

name was changed to the present one below. The organizers select a recipient, who receives a plaque as an Award of Excellence for research in SFC and presents the keynote address. From the 1989 meeting, the attendees developed a book on SFC and SFE [130]. (See also Table 4.5.)

The abstracts of the abovementioned Symposia were published by the organizers and distributed to the participants. Some are now out-of-print; those that are still available can be ordered from Supercritical Conferences, 7122 Larchwood Dr., Cincinnati, OH 45241. Consistent with the growth of experimental SFC is the rapid increase in books devoted to this subject, the first one in 1990 written by M. L. Lee and K. Markides [130]. Then as problems were resolved and instrumentation improved, an increasing number of original papers and eight books appeared in the 1990s [131–133], supplemented by more recent reviews [134–137].

Other scientists have also made many research contributions to SFC. Some research groups have reported difficulties with SFC. E. Klesper reports that "the phase behavior in the supercritical region allows one to consecutively combine GC, SFC and HPLC in a single uninterrupted chromatogram" [1, Chap. 5]. M. L. Lee concludes that SFC was particularly applicable for "separations of low molecular weight polymers, polymer additives, petroleum fractions, agrochemicals and chiral chromatography" [1, Chap. 5].

To summarize, the initial exploration with SFC was a turning point in the overall flow of investigations; once recognized, there was a rapid flow of experiments, theses, research papers, reviews, and now many books. As the above investigators and others entered the field, a degree of contagious spirit, or some would say competitive style, has prevailed in the past decade. Thus SFC becomes another recent example of a "paradigm shift."

4.H. SIZE-EXCLUSION CHROMATOGRAPHY

Dextran, a glucose polymer made by *Leuconostoc mesenteroides*, was developed as a plasma substitute in World War II [138]. Then in the 1950s, dextran was modified to be the polymer for size-exclusion chromatography or "molecular sieving" or "steric exclusion chromatography"—a method for the differential separation of solutes from a chromatography column of porous material, based mainly on the molecular size of solutes and steric exclusion of the solutes from the pore volume. Although gel filtration was recognized in the 1950s, J. Porath and P. Flodin [139] performed the systematic studies in 1959 that led to Sephadex—an epichlorohydrin crosslinked dextran (see further details by J. Porath and P. Flodin in Ref. 1, Chap. 5). Hjertén and Mosbach [140] developed in 1962 a crosslinked [*N*,*N*'-methylene bis(acrylamide)] polyacrylamide, called "Bio Gel P." Hjertén [141] also separated the multibranched agarose from the straight-chain agar, giving the commercial product "Sepharose"; then the preparation of the soft gels in bead form by Flodin was a key step forward. For each, molecules move through the bed of porous gel and diffuse into the pores, while larger molecules do not enter and move through the column faster. Systematic variation of crosslinking of the gel provides a series of porous beads useful for the separation of different-molecular-weight analyte [142].

In succeeding decades, the SEC columns have had such functions as separation and purification of high-molecular-weight (MW) solutes, determination of MW, buffer exchange, concentration of solutes, solute diffusivity and shape, and pore dimensions. The objective of MW determinations requires pure, stable, and available polymers as standards, usually proteins, enzymes, or viruses that have been identified and are now commercially available [143].

These SEC gels have been supplemented by substituted methylacrylate, modified silicas, mixed polymers, and other materials. Their use in chromatography has been broadly examined and reviewed with respect to column materials, theoretical aspects, degree of crosslinking, particle size, characterization of analyte size, separating power (resolution), steric hindrance, band broadening, gel stability microcolumn SEC, fast size SEC, process SEC, and a wide variety of applications [142–144].

Since SEC is usually limited to hydrophilic gels and aqueous or polar solvents, a variation, introduced by J. C. Moore [145], is *gel permeation chromatography* (GPC) with less polar resins, such as crosslinked polystyrene beads, and used in hydrophobic chromatography. It has been extensively employed in the analysis of high-MW synthetic and natural polymers, the interactions of analyte association and aggregation, binding with ligands, solvation in mixed solvents, polymer–salt–solvent interactions, preparative-scale isolations, and other more specific properties [142]. This sketch may be supplemented by referring to several reviews [145–152].

4.I. HIGH-PERFORMANCE LIQUID CHROMATOGRAPHY

As supports and stationary phases, instrumentation, detectors, and other facets developed in the 1960s and as the need to improve the performance of LC to that of GC, the practice of "high-pressure liquid chromatography" was frequently used, but was soon replaced by the present term, *high-performance liquid chromatography* (HPLC). Csaba Horváth wrote 27 years ago that HPLC "is a method of unsurpassed versatility and a microanalytical tool *par excellence*. Like GC, HPLC is characterized by a linear elution mode and by the use of a sophisticated instrument, high-efficiency columns and sensitive detectors" [153]. More recent advances in instrumentation, column engineering, and theory have considerably broadened the field of application HPLC, which now finds use in virtually all branches of science and technology [153]. The many contributions of Csaba Horváth to HPLC have been summarized [154]; he is often called "the father of HPLC."

Related research papers are too numerous to mention here; many excellent reviews are available elsewhere, although the one by L. S. Ettre stands out [155]. HPLC has been included as a chapter in the many overall comprehensive books [156] and is the primary subject of many HPLC-focused books [157–8]. HPLC and GLC have become the predominant modes of chromatography [158]. For instance, the 23rd International Symposium on HPLC Separations and Related Techniques (Spain, May 30–June 4, 1999) required 1545 pages to present the 148 papers that were only a part of the presentations contributed [159]. A new large-scale version of HPLC, called "simulated moving-bed technology," is a continuous process used by companies on an industrial scale. By the 1990s, HPLC had grown so much that it was raised to first place in total analytical instrument sales, as well as scientific contributions [160]. This brief sketch is complemented by more recent reviews [160–169] and by the awardees who have made contributions to HPLC, that is, those with the code letter "h" (see Ref. 1, Author/Subject Index). Hence, this major subject of HPLC and its modifications can be brief for this Chapter 4.I.

The rise of *combinatorial chemistry*—the combination "of a small number of chemical reagents, in all combinations defined by a given reaction scheme to yield a large number of defined products" [170]—means a greater number of chromatographic purifications by GLC, HPLC, TLC, CEC, and other methods. Subsequent "screening" for defined properties, such as the search for new drugs, must be also performed on a repetitive or simultaneous basis. Hence developments in automation, microarrays, and the "lab-on-a-chip" concept arose in the 1990s and will probably continue well into the present century [170,171].

4.J. DETECTORS AND SCIENTIFIC INSTRUMENTS IN CHROMATOGRAPHY

For the 1940s and 1950s, the older generation of chromatographers remember the earlier methods of detection:

- Separation of visible plant pigments on chromatographic adsorption columns
- Identification of amino acids by spraying with ninhydrin to give a purplish color, or with pH indicator dyes for organic acids, in paper chromatography or TLC
- Examination of radioactive ^{14}C intermediates or those with ^{32}P, ^{35}S by radioautography (M. Calvin and others)

- Reliance of UV absorption to locate purines, pyrimidines, nucleosides, nucleotides, and many natural products in paper chromatography or TLC
- Detection of antibiotics by growth-inhibiting spots, or vitamins by growth-enhancing spots on paper chromatograms or TLC, and so on and so forth
- Cut out peaks of graphs in LC or GC to weigh and determine concentration of solutes
- Hand titration of organic acids from GC columns (e.g., A. J. P. Martin and A. T. James, in 1952)
- Measurement of amino acid–ninhyrin color, or nucleic acid bases in UV spectrophotometers or other compounds

With the passage of several decades and numerous inventions, detection is now instrument based—continuous, automated, recorded on charts, and analyzed by a computer. UV detectors for fixed wavelength and later variable wavelength have been employed for LC and later HPLC. Thermal conductivity detector followed by flame-ionization detectors (FIDs) were included for GC. Two other detectors were the argon ionization detector and electron capture detector designed by James E. Lovelock [1, Chap. 5]. The details of evolving sensitivity and efficiency for UV, IR, MS, NMR, fluorescence spectroscopy, interfermetric refractometry, Raman spectroscopy, light-scattering instruments, chiral detector to measure optical rotation, universal detectors for nonchromophoric components, and others have been reviewed elsewhere [172–178], and hence this section will be brief.

After the recent decades of automation, we need to take a step back for a moment. For earlier centuries and the present one, the advance of science depended on precise measurements, which in turn were built on new instruments. Some instruments date back to the early empires of the Mediterranean; others were conceived in the Middle Ages. Their range includes [179]

- Time-telling instruments (hourglasses, sundials, clocks, etc.)
- Navigational instruments (quadrant, sextant, compass, chronometer, etc.)
- Surveying instruments (barometer, pedometer, compass, chronometer, etc.)
- Drawing and calculating instruments (rulers, pantographs, abacus, slide rule, adding machines, etc.)
- Optical instruments (microscopes, telescopes, cameras, etc.)
- Physical instruments for mechanics (magnetism, electricity, heat, sound, temperature, light, etc.)
- Meteorological instruments (thermometers, barometers, etc.)
- Medical instruments (saws, catheters, forceps, spectacles, etc.)

The chemical sciences were based on the early inventions of the balance—a system of weights, and the measurement of volume (calibrated flasks, burettes, and pipettes), all of which underwent many refinements in the late 1800s and 1900s. Others have written to describe the pattern in which originally brilliant inventions have become commonplace laboratory tools: pH meter, spectrophotometer, mass spectrometer, X-ray crystallography, etc.; each has undergone considerable modifications with the passing decades. Such were the intellectual/scientific milestones into which chromatography was conceived by Mikhail Tswett and others in the early decades of the twentieth century (described in Ref. 1—see Chap. 1).

Subsequently, chromatography has expanded and flowered because of the insights, curiosity, and inventiveness of many key investigators in the 1920s–1990s. Soon after the earlier "pioneers in chromatography" (see Chapter 1), each form of chromatography was enhanced in their development by the subsequent "discoverers" (see Chapter 3)—sometimes in column details, sometimes in instrumentation, but frequently with a significant concept or unique research contribution. Since such a description would be quite lengthy, the author has concentrated on the subjects in the preceding sections 4.B to 4.I.

For the past several centuries, scientific discoveries have served as the precursors of invention and then subsequential technological processes. Sir Humphry Davy stated, "Nothing begets good science like the development of a good instrument." Both the theoretical and practical aspects of chromatography, usually published in the scientific literature, have led to (or is it driven to?) the invention of today's complex chromatography instruments, with more sensitive detectors, pumps, valves, collectors, supports, and stationary phases. Such developments have usually focused on needed organizations of scientists—namely, those in the scientific industries. These small and large companies have dedicated groups of chromatographers, biochemists, chemists, engineers, physicists, and others to develop the early research leads, the protection of patents, the needed capital and administrators, plus the usual marketing resources of the world economy [179]. Leslie S. Ettre has presented a summary of the evolution of instrumentation in GC and LC, including the design and construction of key components by early companies, such as the Perkin-Elmer Corporation [180]. Summaries of other scientific companies may be found in the articles sponsored by the Chemical Heritage Foundation (co-initiated by the American Chemical Society): Arnold O. Beckman (inventor of the pH meter, gas chromatographs, mass spectrometers, leak detector, portable gas analyzer, amino acid analyzer, automated data processing, etc. [181]), or Robert F. Finnigan (1927–1990) (developer of the first quadrupole mass spectrometer of GC [182,183]. Further information on these subjects may be found in the scientific news magazines: *Science*, *Nature*, *LC-GC*, *American Laboratories*, *Chemical and Engineering News*, *AOAC Management*, and similar magazines (see Chapter 11, Sections 11.A–11.E).

The mid-1950s ushered in the modern phase of instrumentation in chromatography that started first with GC for petroleum fractions, then in 1958 for amino acid analysis by IEC and followed in the 1960s for LC, GPC, HPLC, and the variants of CE. The last 40 years are presented elsewhere in this book:

- Research publications (see Chapter 5)
- International and national symposia (see Chapter 5)
- Chromatography—the bridge to other selected sciences [1]
- The accumulated and continuing advances in the instruments related to chromatography

The last several decades have witnessed a major swing as computers have assisted development of methods in chromatography, particularly HPLC, TLC, and IEC [174] and other areas subsequently [184]. Beginning in 1965, minicomputers and then after 1971 microcomputers and microprocessors were interfaced with a variety of analytical instruments, including GC and LC systems [185–188].

Major research advances with capillary gel electrophoresis and capillary electrochromatography have been made and may be found in Ref. 1. Another recent vigorous research area is the development of miniaturized, automated, separation devices, which have acquired the popular name "lab-on-a-chip" (see A. Guttman's contribution in Ref. 1, Chap. 5 for an excellent description) and are a present research tool in genomics and proteomics.

As noted in other areas of science, the penetration of instrumentation into chromatography practice has opened new doors for research possibilities and will continue to enhance our understanding of the scientific world in which we are immersed. With the rise of automation, shorter elution times, and the complexities of mass spectrometric data, the reliance on computers has become a necessity in chromatography [184–189]. Thus, the development of scientific instruments has both the elements of "normal growth" and at certain times and places, the features of major rapid changes—a "paradigm shift" (see Section 3.D). Beyond these characteristics are the human mind and the creative nature of scientists, which has also been explored in these chapters.

4.K. HYPHENATED/COUPLED/TANDEM TECHNIQUES IN CHROMATOGRAPHY

These closely related terms reflect the growing closeness and interdependence of the chromatographic separations with the detection mechanisms over the past several decades (see Fig. 4.1). Some possible pairs of techniques have advanced chromatographic solutions; others are incompatible and some pairs are only appropriate for specific solutes. Hence, we now see frequent use of GC-MS, GC-IR, LC-UV, LC-MS, LC-IR, LC-NMR, etc. and extended to capillary electrophoresis: CE-MS, CEC-MS, and MEKC-MS (micellar electrokinetic chromatography-mass spectrometry) [190–195]. This brief sketch serves as an introduction and invites attention to the papers in a key early symposium [196], several reviews [192–195,201], several pertinent books [197–200,202], plus the approximately 25 chromatographers, who earlier described these combined approaches [1] and to the present chromatographers in Chapter 5, below.

What is the frequency of papers relying on one of these instruments for detection? A MEDLINE book search (chromatography plus the instrument below in the title or abstract was conducted for the period 1998–2007 and provided the clear recognition of mass spectrometry shown in Table 4.6.

A follow-up search, also using MEDLINE, showed the rapid increase in recent years for citations for mass spectrometry (Table 4.7). MEDLINE would cover the biomedically related papers. By contrast, *Chemical Abstract* listed thousands and thousands more citations than did MEDLINE, presumably with a similar time-dependent increase (search by

Advances in Chromatography to Coupled Systems

Figure 4.1. Advances in chromatography to coupled systems.

TABLE 4.6. Instrumental Detection of Analyte in Chromatography (1998–2007)[a]

Instrument for Chromatography	Number of Citations per Decade
Mass spectrometer	24,658
Nuclear magnetic resonance	1,184
Flame ionization	613
Electron capture	774
Photometer	11
Refractive index	162
Polarimeter	23

[a]A Boolean search in MEDLINE for the combination of chromatography with the listed detection term in either the articles' title or abstract was conducted for the decade 1998–2007 on 9/18/08. Search by Kate Anderson, MU/HSL.

TABLE 4.7. Frequency of Chromatography Books, Research Papers, and Chromatography Papers[a] with Mass Spectrometry

	Number of Citations per Decade for Chromatography and Mass Spectrometry		
Decade	Chromatography (Books)	Chromatography (Journal Articles)	Chromatography and Mass Spectrometry (Journal Articles)
1938–1947	0 (WorldCat lists 3 books for 1938–1947)	11	0
1948–1957	2	1,371	0
1958–1967	35	4,598	23
1968–1977	119	16,305	1,009
1978–1987	231	49,765	3,748
1988–1997	225	60,907	7,249
1998–2007	186	78,804	23,395

[a]A Boolean search in MEDLINE on the indicated online source was performed for chromatography and mass spectrometry for the period 1998–2007.

Kate Anderson, University of Missouri Health Science Library). The hyphenated approach (LC-MS) has been reviewed earlier [202].

4.L. WOMEN SCIENTISTS IN CHROMATOGRAPHY

The reader may have noted the predominance of male scientists in these chapters. Although the Editors suggest that this is a reflection of cultural values and educational goals in the decades prior to 1960, women chromatographers have contributed significantly. Their precise contributions may be found in the indicated biographical references listed in column 4 of Table 4.8.

The early and distinctive contributions of Dr. Erika Cremer were recognized by being the first recipient of the M. S. Tswett Award (1974) [203]. Dr. Mary Ellen McNally, of E. I. Dupont de Nemours & Co., Wilmington, Delaware has been a major leader in the Chromatography Forum of the Delaware Valley, which administers the Stephen Dal Nogare Award (see Chapter 5) and has completed noteworthy research on supercritical fluids for extraction and chromatography, plus biosensors [203].

Marja-Liisa Riekkola is another leading chromatographer. She is Professor of Analytical Chemistry, Department of Chemistry, University of Helsinki, Finland. Earlier, Dr. Riekkola was a 1983 Visiting Scientist with Dr. K. Grob in Switzerland and a 1984 Visiting Researcher with Dr. R. Sievers at the University of Colorado, USA. For the 1994 year, she was a Visiting Professor, Toyohashi University of Technology, Japan (with Prof. K. Jinno). Her research interests—expressed in about 134 papers and two books [204,205]—include GC, LC, capillary electrophoresis, field-flow fractionation, sample preparation methods, and automation. She is at present

TABLE 4.8. Distinguished Women Chromatographers

Women Chromatographers	Organizational Affiliation	Award and Date[a]	Reference[b] for Additional Information
Phyllis R. Brown	Univ. Rhode Island, USA	2–5—1988 2–7—1989	Section S9E
Erika Cremer	Univ. Innsbruck, Austria	2–5—1974	Appendix 2
Marjorie G. Horning	Baylor Univ. College of Med., USA	2–5—1987	Appendix 2
Karin E. Markides	Uppsala Univ., Sweden	2–9—1992	Section S9E
Mary Ellen McNally	E. I. Dupont de Nemours Co., USA	—	See text, present chapter
Marja-Liisa Riekkola	Univ. Helsinki, Finland	—	See text, present chapter

[a]These award numbers for specific years (e.g., 2–5 in 1988 for Phyllis R. Brown) are described in detail in Chapter 2 of C. W. Gehrke, R. L. Wixom, and E. Bayer (Eds.), *Chromatography: A Century of Discovery (1900–2000)—the Bridge to the Sciences/Technology*, Vol. 64, Elsevier, Amsterdam, 2001.

[b]C. W. Gehrke, R. L. Wixom, and E. Bayer (Eds.), *Chromatography: A New Discipline of Science*, Elsevier, Amsterdam, 2001.

on the Editorial Boards of five chromatography-related journals, has served on many scientific boards in Finland, and is a current member of the IUPAC Commission on Separation Methods in Analytical Chemistry.

The reader should note that seven women have received the Nobel Prize in Science. Hopefully, both groups of women will become role models for other young women in the coming years.

4.M. SUMMARY

The reasons for the subtitle of this book, *The Bridge to the Sciences/Technology*, has been reflected in Chapters 1 and 3, and is now reinforced here in Chapter 4. The scientists in chromatography have described themselves also as

- Chemists (analytical, physical, biological, natural product, polymer, and other varieties),
- Biologists (botanists, zoologists, insect physiologists, microbiologists, pharmacologists, molecular biologists, etc.)
- Chemical engineers, food scientists, etc.

Others have built the bridge to connect and relate chromatography to parts of the medical sciences, agricultural sciences, food sciences, pharmaceutical sciences, environmental sciences, space sciences, petroleum industry, and so on; further amplification of these thoughts is found in Chapters 5–9. Thus, with the considerable variety, mingling, and blending within chromatography, chromatography in its growth phase has become a central directive in the overall structure of science. Furthermore, *chromatography has become a bridge*—a connection with each of the above scientific/technological subjects. Thus chromatographic science in many ways has moved through the "normal phase" of growth, but in at least nine areas (carotenoids, natural products, amino acids/peptides/ proteins, affinity chromatography, chiral chromatography, supercritical-fluid chromatography, size-exclusion chromatography, high-performance liquid chromatography, and chromatographic instrumentation), the characteristics of a "paradigm shift" have been identified. Each will probably continue as a "normal- or mainstream science" (see Chapter 3, particularly Fig. 3.8).

The near future in science depends on research in the recent past. At this millennium, new concepts are being expressed; new research tools have been brought into practice! *New paradigm shifts are underway*! Evidence for those statements is expressed daily in the fields of *genomics* (to probe the sequence in DNA, including their structure, functions, and relationships) and *proteomics* (to probe the sequence in proteins and their structures, functions, and relationships) [206–209], where chromatography has had a leading role to date, and has the potential for an even more significant future [206–213]. The recent groundwork of miniaturization and microfluidics has led to the present research thrust based on a "lab-on-a-chip" or microfluidics [214], robotics, automations, microarray technologies, and MS applications for the Human Genome Project [211]—see also Section 11.F, later. The recent considerable progress in combinatorial chemistry has already contributed to published research on protein–protein interaction, target-guided ligand assemblies, microencapsulated catalysts, nanoelectrospray MS in conjunction with LC, and other areas (see reviews in Refs. 212 and 215). As this chapter winds up, recognition should be indicated to the Editors of *LC-GC* for the experimental/instrumental aspects of chromatography by their regular columnists: Ronald E. Majors, John W. Dolan, John V. Hinshaw, Tim Wehr, and Leslie S. Ettre [216, 217]. The reader is encouraged to examine the pertinent reviews and books in the *Journal of Chromatography* (*A* and *B*), *Advances in Chromatography* (Vol. 45, 2007, Marcel Dekker, Publisher), *Analytical Chemistry*, and Chromatographic Science Series (Marcel Dekker and John Wiley & Sons, Publishers). Thus this chapter on the *trails of research in chromatography*, primarily from 1960 to 2000, will serve as an introduction for Chapters 5–9 and 11, which follow.

Paradigm shifts continue to lead to new discoveries and to create a broader diversity in chromatography in the following areas (and probably also in other sciences):

Selectivity!
    Detection!
        Sensitivity!
            Coupled (tandem) systems!
                Automation!
                    Computer-controlled systems!
                        Miniaturization!

Parallel column separation!

Chiral analyses and purification!

Molecular recognition in separation!

[See also Fig. 4.1.]

REFERENCES

1. C. W. Gehrke, R. L. Wixom, and E. Bayer (Eds.), *Chromatography: A Century of Discovery (1900–2000)—the Bridge to the Sciences/Technology*, Vol. 64, Elsevier, Amsterdam, 2001.

2. C. W. Gehrke, R. L. Wixom, and E. Bayer (Eds.), *Chromatography: A New Discipline of Science*, Elsevier, Amsterdam, 2001; Sections S-7–S-15 are online at Chem Web Preprint Server http://www.chemweb.com/preprint/.

3. Z. Deyl (Ed.), *Separation Methods*, Vol. 8 in *New Comprehensive Biochemistry*, Elsevier, Amsterdam, 1984.

4. T. G. Cooper, *The Tools of Biochemistry, a Wiley-Interscience Publication*, Wiley, New York, 1997.

5. L. S. Ettre and J. V. Hinshaw, chapters in the *Evolution of Chromatography*, Imperial College Press, London, 2008.

6. R. E. Majors, Glossary of HPLC and LC separation terms, *LC-GC (Liquid Crystal Gas Chromatography)* **26**(2), 119–168 (2008).

References on Carotenoids

7. P. Karrer and E. Jucker, *Carotinoide*, Birkhäuser, Basel, 1948; Engl. transl. *Carotenoids* (E. A. Braude, trans.), Elsevier, Amsterdam, 1950.

8. O. Isler, W. Huber, A. Ronco, and M. Kofler, Synthese des vitamin A, *Helv. Chim. Acta* **30**, 1911–1927 (1947).

9. P. Karrer and C. H. Eugster, Synthese von carotinoiden II. Total synthese des β-carotins I, *Helv. Chim. Acta* **33**, 1172–1174 (1950).

10. T. W. Goodwin, *The Biochemistry of the Carotenoids*, Vol. 1, *Plants*; Vol. 2, *Animals*, 2nd ed., Chapman & Hall, London, 1980, 1984.

11. T. W. Goodwin and G. Britton, Distribution and analysis of carotenoids, pp. 61–132; Biosynthesis, pp. 133–182; Functions, pp. 231–255; in T. W. Goodwin (Ed.), *Plant Pigments*, Academic Press, London, 1988.

12. R. Macrae, R. K. Robinson, and M. J. Sadler (Eds.), *Encyclopedia of Food Science and Technology*, 8 vols; Academic Press, New York, 1993; see Vol. 1, *Carotenoids*, pp. 707–718.

13. F. J. Francis (Ed.-in-Chief), *Encyclopedia of Food Science and Technology*, Wiley, New York, 1st ed., 1992, 4 vols.; 2nd ed., 2000, 4 vols.; see Vol. 1, *Carotenoids*, pp. 272–282.

14. G. Britton, S. Liaaen-Jensen, and H. Pflander (Eds.), *Carotenoids*, Vol. 1A, *Isolation and Analysis*, Chaps. 1, 2, 5, and 6; Vol. 1B, *Spectroscopy*, Vol. 2, *Synthesis*; Birkhäuser Verlag Bard, Basel, Switzerland, 1995–1996.

15. *Comprehensive Biochemistry*, Vol. 33A, 1979, Chap. 61, The isoprenoid pathway, by M. Florkin and E. H. Stotz, pp. 239–324.

References on Natural Products and Chromatography

16. L. Zechmeister (Founding Ed.), *Fortschritte der Chemie organischer Naturstoffe (Progress in the Chemistry of Organic Natural Products)*, Springer-Verlag, Vienna/New York, Vol. 1 (1938)–Vol. 29 (1969); W. Herz and three or four co-editors, Vol. 28 (1970)–Vol. 79 (2000).

17. R. H. F. Manske and H. L. Holmes (initial Eds.), A. Brossi and G. A. Cordell (current Eds.), *The Alkaloids, Chemistry and Pharmacology*, Academic Press, San Diego, CA, Vol. 1 (1950)–Vol. 53 (1999).

18. H. F. Linskens and J. F. Jackson (Eds.), *Modern Methods of Plant Analyses*, Springer-Verlag, Berlin, New Series, Vol. 1 (1985)–Vol. 18 (1995); Vol. 3 on GC-MS and Vol. 5 on HPLC.

19. J. ApSimon (Ed.), *The Total Synthesis of Natural Products*, Wiley-Interscience, New York, Vol. 1 (1973)–Vol. 9 (1992).

20. K. Nakanishi, T. Goto, S. Itô, S. Natori, and S. Nozoe (Eds.), *Natural Products Chemistry*, Kodanska, Tokyo; Academic Press, New York, NY, 1974–1983, 3 vols.

21. B. Josephson, Separation techniques, Chap. 1, pp. 2–93; in P. J. Scheuer (Ed.), *Marine Natural Products—Chemical and Biological Properties*, Academic Press, New York, 1980, 3 vols.

22. A. B. Svendsen and R. Verpoorte, *Chromatography of the Alkaloids, Part A: Thin Layer Chromatography, J. Chromatogr.* Library, 1983, Vol. 23A.

23. Royal Society of Chemistry, *Natural Products Reports, a Journal of Current Developments in Bio-Organic Chemistry*, Royal Chemical Society, London, UK, Vol. 1 (1984)–Vol. 25 (2000); result of a merger of earlier specialist periodical reports on alkaloids, biosynthesis, aliphatic and related natural product chemistry, terpenoids, and steroids.

24. A. U. Rahman (Ed.), *Studies in Natural Products Chemistry*, Vol. 1 (1988)–Vol. 29 (2003); for stereoselective synthesis, see Vols. 6, 8, 10, 11, 12, 14, 16, and 18; for structure and chemistry, see Vols. 7, 9, 15, 17, 19, and 20; for bioreactive natural products, see Vols. 13 and 21–29.

25. K. Torssell, *Natural Product Chemistry: A Mechanistic, Biosynthetic and Ecological Approach*, Swedish Pharmaceutical Society, Stockholm, Sweden, 1997.

26. P. B. Kaufman, L. J. Cseke, S. Warber, J. A. Duke, and H. L. Brielman, *Natural Products from Plants*, CRC Press, Boca Raton, FL, 1999; see Chap. 7 on bioseparatin of compounds.

27. D. Barton, K. Nakanishi, and O. Meth-Cohn (Eds.), *Comprehensive Natural Products Chemistry*, Elsevier, Amsterdam, 1999, 9 vols.

28. L. M. L. Nollet (Ed.), *Chromatographic Analysis of the Environment*, 3rd ed., Vol. 93 in the Chromatographic Science Series, CRC Press, Taylor & Francis Group, Boca Raton, FL, 2006.

References on Chromatography of Amino Acids, Peptides, and Proteins

29. J. P. Greenstein and M. Winitz, *Chemistry of the Amino Acids*, Wiley, New York, 3 vols., 1961 (see particularly Vol. 2, Chap. 15, pp. 1366–1511).

30. A. E. Garrod, Inborn errors of metabolism, *Lancet* **2**, 1, 73, 142, 214 (1908); see also *Inborn Errors of Metabolism*, Oxford Univ. Press, Oxford, UK, 1923.

31. C. R. Scriver, A. L. Beaudet, W. S. Sly, and D. Valle (Eds.), *The Metabolic and Molecular Basis of Inherited Disease*, 3 vols., 7th ed., McGraw-Hill, New York, 1995.

32. D. L. Rimoin, J. M. Connor, and R. E. Pyertz (Eds.), *Principles and Practice of Medical Genetics*, 3rd ed., Churchill-Livingston Publishers, New York, 1996, 2 vols.; see Chaps. 87–99 on metabolic disorders.

33. L. Pauling, H. A. Itano, S. J. Singer, and I. C. Wells, Sickle cell anemia, a molecular disease, *Science* **110**, 543–548 (1949).

34. *Nobel Lectures—Chemistry* (1942–1962), Vol. 3, pp. 423–439.

35. L. C. Pauling, *Nobel Laureates in Chemistry* (1993), pp. 368–379.

36. G. R. Lee, J. Foerster, J. Lukens, F. Paraskevas, J. P. Greer, and G. M. Rodgers (Eds.), *Wintrobe's Clinical Hematology*, 10th ed., Williams & Wilkins, Baltimore, MD, 1999, 2 vols.; see pp. 1329–1335.

37. Ref. 32 continued: see Chaps. 72–76 on hematologic disorders.

38. C. K. Larive and 11 co-authors, Separation and analysis of peptides and proteins, *Anal. Chem.* **71**(12), 389R–423R (1999).

39. See Ref. 2—Apps. 4–7 have many books on chromatography of amino acids, peptides and proteins that cover purification, and analysis and the determination of sequential structure; consideration of secondary and tertiary structure are only incidentally covered in these references.

References on Affinity Chromatography

40. R. Axén, J. Porath, and S. Ernback, Chemical coupling of peptides and proteins to poly-saccharides by means of cyanogen halides, *Nature* **214**, 1302–1304 (1967).

41. P. Cuatrecasas, M. Wilchek, and C. B. Anfinsen, Selective enzyme purification by affinity chromatography, *Proc. Natl. Acad. Sci. USA* **61**, 636–643 (1968).

42. J. Turková, *Affinity Chromatography*, in *J. Chromatogr.* Library Series, Vol. 12, Elsevier, Amsterdam, 1978.

43. H. Schott, *Affinity Chromatography: Template Chromatography of Nucleic Acids and Proteins*, Marcel Dekker, New York, 1984.

44. J. Turková, Bioaffinity chromatography, in Z. Deyl (Ed.), *Separation Methods*, Vol. 8 in A. Neuberger and L. L. M. van Deenen (General Eds.), *New Comprehensive Biochemistry*, Elsevier, Amsterdam, 1984, pp. 321–361.

45. P. Mohr and K. Pomerening, *Affinity Chromatography—Practical and Theoretical Aspects*, Chromatographic Science Series, Vol. 33, Marcel Dekker, New York, 1985.

46. T. Kline (Ed.), *Handbook of Affinity Chromatography*, Chromatographic Science Series, Vol. 63, Marcel Dekker, New York, 1993.

47. T. M. Phillips, Affinity chromatography, in E. Heftmann (Ed.), *Chromatography, Part A: Fundamentals and Techniques*, 5th ed., Elsevier, Amsterdam, 5, 1992, Chap. 7.

48. R. F. Service, Problem solved—sort of, *Science* **321**, 783–786 (2006).

49. E. A. Bayer and M. Wilchek, The use of the avidin-biotin complex as a tool in molecular biology, *Meth. Biochem. Anal.* **26**, 1–45 (1980); Avidin-biotin technology, *Meth. Enzymol.* **184** (1990).

50. I. M. Chaiken, M. Wilchek, and I. Parikh (Eds.), *Proc. 5th Int. Symp. Affinity Chromatography and Biological Recognition*, Academic Press, New York, 1983.

51. M. Wilchek (Ed.), *Proc. 8th Symp. Affinity Chromatography and Biological Recognition*, 1989; see *J. Chromatogr.* **510**, 1–375 (1990).

52. Y. Zhang and I. W. Wainer, On-line chromatographic analysis of drug-receptor interactions, *Am. Lab.* **31**, 17–20 (1999).

53. References for six more books on affinity chromatography are presented in Ref. 2, Apps. 4-G, 5-ABG.

54. D. S. Hage (Ed.), *Handbook of Affinity Chromatography*, 12th ed., Vol. 92, Chromatographic Sciences Series, Taylor & Francis, Boca Raton, FL, 2006.

References on Ubiquitin's Roles

55. A. Ciechanover, Y. Hod, and A. Hershko, A heat-stable polypeptide component of an ATP-dependent proteolytic system from reticulocytes, *Biochem. Biophys. Res. Commun.* **81**, 1100–1105 (1978).

56. A. Hershko, A. Ciechanover, H. Heller, A. L. Haas, and I. A. Rose, Proposed role of ATP in protein breakdown: Conjugation of proteins with multiple chains of polypeptide of ATP-dependent proteolysis, *Proc. Natl. Acad. Sci. USA* **77**, 1783–1786 (1980).

57. M. J. Schlesinger and A. Hershko (Eds.), *The Ubiquitin System*, Cold Spring Harbor Laboratory, Cold Spring Harbor, NY, USA, 1988.

58. A. Hershko and A. Ciechanover, The ubiquitin system for protein degradation, *Ann. Rev. Biochem.* **61**, 761–807 (1992) (216 references cited therein).

59. K. D. Wilkinson, Roles of ubiquitination in proteolysis and cellular regulation, *Ann. Rev. Nutr.* **15**, 161–189 (1995) (136 references).

60. D. Finley and V. Chau, Ubiquitination, *Ann. Rev. Cell Biol.* **7**, 25–69 (1991) (173 references).

61. D. A. Parsell and S. Lindquist, The function of heat-shock proteins in stress tolerance: Degradation and reactivation of damaged proteins, *Ann. Rev. Genet.* **27**, 437–496 (1993) (291 references).

62. M. Pagaon, Cell cycle regulation by the ubiquitin pathway, *FASEB J.* **11**, 1067–1075 (1997) (122 references).

63. A. L. Goldberg, ATP-dependent proteases in prokaryotic and eukaryotic cells, *Semin. Biol.* **1**(6), 423–432 (1990) (67 references).

64. A. L. Goldberg and K. L. Rock, Proteolysis, proteasomes and antigen presentation, *Nature* **357**(6377), 375–379 (1992) (92 references).

65. O. Coux, K. Tanaka, and A. L. Goldberg, Structure and function of 20S and 26S proteasomes, *Ann. Rev. Biochem.* **65**, 801–847 (1996) (287 references).

66. C. M. Pickart, Targeting of substrates to the 26S proteasome, *FASEB J.* **11**, 1055–1066 (1997) (99 references).

67. S. H. Lecker, V. Solomon, W. E. Mitch, and A. L. Goldberg, Muscle protein breakdown and the critical role of ubiquitin- proteasome pathway in normal and disease states, *J. Nutr.* **129** (Suppl.), 227S–237S (1999).

68. D. Voges, P. Zwickel, and W. Baumeister, The 26S proteasome: A molecular machine designed for controlled proteolysis, *Ann. Rev. Biochem.* **68**, 1015–1068 (1999) (414 references).

69. S. Jentsch, The ubiquitin system, *Ann. Rev. Genet.* **26**, 179–207 (1992) (158 references).

70. M. Hochstrasser, Ubiquitin-dependent protein degradation, *Ann. Rev. Genet.* **30**, 405–439 (1996) (115 references).

71. A. L. Schwartz and A. Ciechanover, The ubiquitin-proteasome pathway and human diseases, *Ann. Rev. Med.* **50**, 57–74 (1999) (79 references).

72. S. Borman, Intricacies of the proteasome, *Chem. Eng. News* **78**(12), 43–47 (2000).

73. A. J. Ciechanover and A. L. Schwartz, *Cellular Proteolytic Systems*, Wiley-Liss, New York, 1994.

74. R. J. Mayer and I. R. Brown (Eds.), *Heat Shock Proteins in the Nervous System*, Academic Press, London, 1994.

75. J.-M. Peters, J. R. Harris, and D. Finley (Eds.), *Ubiquitin and the Biology of the Cell*, Plenum Press, New York, 1998.

76. M. C. Rechsteiner, Ubiquitin-mediated proteolysis: An ideal pathway for systems biology analysis, in L. K. Opresko, J. M. Gephart, and M. B. Mann (Eds.), *Advances in Systems Biology*, Kluwer Academic/Plenum Publishers, New York, 2004, pp. 49–59.

77. D. C. Rubinsztein, The roles of intracellular protein-degradation pathology in neurodegeneration, *Nature* **497**(7113), 780–786 (2006).

References on Molecular Recognition and Affinity Chromatography

78. M. DeLaage (Ed.), *Molecular Recognition Mechanisms*, VCH Publishers, Weinheim, Germany, 1991.

79. E. Knobil and J. D. Neill (Eds.), *The Physiology of Reproduction*, 2nd ed., Raven Press, New York, 1994, 2 vols.; see particularly pp. 118–121, 1013–1031.

80. P. Cuatrecasas and M. F. Greaves (Eds.), *Receptors and Recognition*, Series A—Vol. 1 (1976)–Vol. 6 (1978); S. Jacobs and P. Cuatrecasas (Eds.), Series B—Vol. 1 (1977)–Vol. 17 (1984); see particularly Series B, Vol. 11, *Membrane Receptors—Methods for Purification and Characterization*; Chapman & Hall, New York.

81. A. D. Buckingham, A. C. Legon, and S. M. Roberts (Eds.), *Principles of Molecular Recognition*, Blackie Academic & Professional, London, 1993.

82. K. Jinno (Ed.), *Chromatographic Separations Based on Molecular Recognition*, Wiley-VCH, New York, 1997.

83. T. M. Phillips and B. F. Dickens, *Affinity and Immunoaffinity Purification Techniques*, Eaton Publishing, Natick, MA, 2000.

84. F. Diederich and H. Kunzer (Eds.), *Recent Trends in Molecular Recognition*, Springer, New York, 1998.

85. D. Kriz, O. Ramström, and K. Mosbach, Molecular imprinting: New possibilities for sensor technology, *Anal. Chem.* **69**(11), 345A–349A (1997).

86. T. Takeuchi and J. Haginaka, Separation and sensing based on molecular recognition using molecularly imprinted polymers (a review), *J. Chromatogr. B* **728**, 1–20 (1999).

87. V. T. Remcho and Z. J. Tan, MIPs as chromatographic stationary phases for molecular recognition, *Anal. Chem.* **71**(7), 243A–255A (1999).

88. N. Dontha, W. B. Nowall, and W. G. Kuhr, Generation of biotin/avidin/enzyme nanostructures with maskless photolithography, *Anal. Chem.* **69**(14), 2619–2625 (1997).

References on Chiral Chromatography

89. L. Pasteur, Sur les relations qui peuvent exister entre la forme cristalline, la composition chimique et le sens de la polarisation rotatoire, *Ann. Chim. Phys.* **24**, 442–459 (1848).

90. J. H. van't Hoff, Sur les formules de structure dans l'espace, *Arch. Neerl. Sci. Exactes. Natur.* **9**, 445–454 (1874).

91. J. A. Le Bel, Sur les relations qui existent entre les formules atomiques des corps organiques et le pouvoir rotatoire de leurs dissolutions, *Bull. Soc. Chim. Paris* **22**, 337–347 (1874).

92. E. Fischer, Preparation of ℓ-glyceric aldehyde, *Science* **88**, 108 (1938).

93. A. M. Rouhi, Tetrahedral carbon redux—symposium commemorates 125 year-old idea that evolved into stereochemistry, *Chem. Eng. News* **77**(36), 28–32 (1999).

94. E. Gil-Av, B. Feibush, and R. Charles-Sigler, Separation by gas–liquid chromatography with an optically active stationary phase, *Tetrahedron Lett.* **1966**, 1009–1115 (1966).

95. E. Gil-Av, R. Charles, and S.-C. Chang, Development of the chiral diamide phases for the resolution of α-amino acids, in C. W. Gehrke, K. C. T. Kuo, and R. W. Zumwalt (Eds.), *Amino Acid Analysis by Gas Chromatography*, Vol. II, CRC Press, Boca Raton, FL, 1987, pp. 2–34.

96. V. Schurig, Emanuel Gil-Av—in Memoriam (1916–1996), *J. High Resol. Chromatogr.* **19**, 462–463 (1996).

97. S. Sarel and S. Weinstein, Emanuel Gil-Av (1916–1996), *Enantionmer* **1**(4–6) iv–viii (1996); includes references for his 120 published papers.

98. H. Frank, G. J. Nicholson, and E. Bayer, Rapid gas chromatographic separation of amino acid enantiomers with a novel chiral stationary phase, *J. Chromatogr. Sci.* **15**, 174–180 (1974).

99. W. H. Pirkle and D. W. House, Chiral high-pressure liquid chromatography stationary phases. I, *J. Org. Chem.* **44**, 1957–1960 (1979).

100. D. W. Armstrong and W. DeMond, Cyclodextrin bonded phases for the liquid chromatographic separation of optical, geometrical and structural isomers, *J. Chromatogr. Sci.* **22**, 411–415 (1984).

101. S. Allenmark, *Chromatographic Enantioseparation: Methods and Applications*, Ellis Horwood Ltd., Chichester, UK, 1st ed., 1988; 2nd ed., 1991.

102. W. J. Lough (Ed.), *Chiral Liquid Chromatography*, Blackie, Glasgow, UK, 1989.

103. J. Jacques and H. B. Kagan, Séparation chromatographique des stéréoisoméres, in E. Lederer (Ed.), *Chromatographie en Chimie Organique et Biologique*, Masson, Paris, 1959, pp. 569–634.

104. R. W. Souter, *Chromatographic Separations of Stereoisomers*, CRC Press, Boca Raton, FL, 1985.

105. M. Zief and L. J. Crane (Eds.), *Chromatographic Chiral Separations*, Chromatographic Science Series, Vol. 40, Marcel Dekker, New York, 1988.

106. A. M. Krstulović (Ed.), *Chiral Separations by HPLC—Applications to Pharmaceutical Compounds*, Ellis Horwood, Chichester, UK, 1989.

107. D. Stevenson and I. D. Wilson (Eds.), *Recent Advances in Chiral Separations, A 1989 Chromatographic Society Symposium*, Plenum Press, New York, 1990.

108. S. Ahuja (Ed.), *Chiral Separations by Liquid Chromatography*, American Chemical Society, Washington, DC, 1991; S. Ahuja (Ed.), Chiral Separations: Applications and Technology, American Chemical Society, Washington, DC, 1997.

109. G. Subramanian (Ed.), *A Practical Approach to Chiral Separations by Liquid Chromatography*, VCH, Weinheim/New York, 1994.

110. T. E. Beesley and R. P. W. Scott, *Chiral Chromatography*, Wiley, New York, 1998.

111. Y.-S. Liu, T. Yu, and D. W. Armstrong, HPLC detection and evaluation of chiral compounds with a laser-based chiroptical detector, *LC GC North Am.* **17**(10), 946–957 (1999).

112. V. Prelog, *Nobel Lectures—Chemistry* (1971–1980), Vol. 5, pp. 200–216; *Nobel Laureates in Chemistry* (1993), pp. 578–583; V. Prelog, *My 132 Semesters of Studies of Chemistry*, American Chemical Society, Washington, DC, 1991.

113. V. Prelog and P. Wieland, Über die Spaltung der Tröger'schen Base in optische Antipoden, ein Beitrag zur Stereochemie des dreiwertigen Stickstoffs, *Helv. Chim. Acta* **27**, 1127–1134 (1944).

114. H. Nishi and S. Terabe (Eds.), Applications of chiral capillary electrophoresis—a symposium, *J. Chromatogr. A* **875**, 1–492 (2000).

115. A. M. Thayer, Great expectations in biopharmaceutical industry, *Chem. Eng. News* **76**(32), 19–31 (1998).

116. S. C. Stinson, Counting on chiral drugs, *Chem. Eng. News* **76**(38), 83–104 (1998).

117. S. C. Stinson, Chiral drug interactions, *Chem. Eng. News* **76**(41), 101–120 (1998).

118. S. C. Stinson, Fertile ferment in chiral chemistry, *Chem. Eng. News* **78**(19), 59–70 (2000).

119. S. C. Stinson, Chiral drugs, *Chem. Eng. News* **78**(43), 55–78 (2000).

120. H. Y. Aboul-Enein and I. Ali, Macrocyclic antibiotics as effective chiral selection for enantiomeric resolution by liquid chromatography and capillary electrophoresis, *Chromatographia* **52**, 679–691 (2000).

121. A. X. Wang, J. T. Lee, and T. E. Beesley, Coupling chiral stationary phases as a fast screening approach for HPLC method development, *LC GC North Am.* **18**(6), 626–639 (2000).

122. J. Ryan, A. Newman, and M. Jacobs (Eds.), *The Pharmaceutical Century—the Decades of Drug Discovery*, a supplement to ACS Publications, Aug. 2000; also C. M. Henry, The pharmaceutical century, *Chem. Eng. News* **78**(43), 5, 85–100 (2000).

123. H. Y. Aboul-Enein and A. Imran, *Chiral Separations by Liquid Chromatography and Related Techniques*, Vol. 90 in Chromatographic Science Series, Marcel Dekker, New York, 2003.

References on Supercritical-Fluid Chromatography

124. (a) M. N. Myers and J. C. Giddings, High column efficiency in gas liquid chromatography at inlet pressures to 2500 p.s.i., *Anal. Chem.* **37**, 1453–1457 (1965); (b) J. C. Giddings, Some aspects of pressure-induced equilibrium shifts in chromatography, *Separation Sci.* **1**, 73–80 (1966); (c) M. N. Meyers and J. C. Giddings, Ultra-high-pressure gas chromatography in micro columns to 2000 atmosphers, *Separation Sci.* **1**, 761–776 (1966); (d) J. C. Giddings et al., High pressure gas chromatography of non-volatile species, *Science* **162**, 67–73 (1968).

125. M. L. Lee and K. E. Markides, Chromatography with supercritical fluids, *Science* **235**, 1342–1347 (1987).

126. E. Klesper, A. H. Corwin, and D. A. Turner, High pressure gas chromatography above critical temperatures, *J. Org. Chem.* **27**, 700–701 (1962).

127. S. T. Sie, W.v. Beersum, and G. W. A. Rijinders, High-pressure gas chromatography and chromatography with supercritical fluids. I. The effect of pressure on partition coefficients in gas-liquid chromatography with carbon dioxide as a carrier gas, *Separation Sci.* **1**, 459–490 (1966); S. T. Sie and G. W. A. Rijnders, Chromatography with supercritical fluids, *Anal. Chem. Acta* **38**, 31–44 (1967); S. T. Sie and G. W. A. Rijnders, *Separation Sci.* **2**, 699–727, 729–753, 755–777 (1967).

128. M. Novotny, S. R. Springston, P. A. Peaden, J. C. Fjeldstad, and M. L. Lee, Capillary super-critical fluid chromatography, *Anal. Chem.* **53**(3), 407A–414A (1981).

129. D. R. Gere, R. Board, and D. McManigill, Supercritical-fluid chromatography with small particle diameter packed column, *Anal. Chem.* **54**(4), 736–740 (1982).

130. M. L. Lee and K. E. Markides, *Analytical Supercritical-Fluid Chromatography and Extraction*, Chromatography Conferences, Provo, UT, 1990.

131. P. J. Schoenmakers and L. G. M. Uunk, Supercritical-fluid chromatography, in E. Heftmann (Ed.), *Chromatography*, 5th ed., Elsevier, Amsterdam, 1992, Chap. 8, pp. A339–A391.

132. Seven additional books on supercritical-fluid chromatography (SFC) appear in Ref. 2, see Apps. 5A, 5B, 5D, and 5Gh.

133. J. R. Williams and T. Clifford (Eds.) *Supercritical Fluid Methods and Protocols*, Vol. 13 in Methods in Biotechnology series, Humana Press, Totawa, NJ, 2000.

134. B. Erickson, SFC in flux, *Anal. Chem.* **69**(21), 683A–686A (1997).

135. T. L. Chester, J. D. Pinkston, and E. E. Raynie, Supercritical fluid chromatography and extraction, *Anal. Chem.* **70**(12), 301R–319R (1998).

136. T. L. Chester and J. D. Pinkston, Supercritical fluid and unified chromatography, *Anal. Chem.* **72**(12), 129R–135R (2000).

137. J. R. Williams and A. A. Clifford (Eds.), *Supercritical Fluid Methods and Protocols*, Humana Press, Totowa, NJ, 2000.

References for Size-Exclusion Chromatography

138. T. C. Laurent, History of a theory (gel filtration), *J. Chromatogr.* **633**, 1–8 (1993).

139. J. Porath and P. Flodin, Gel filtration: A method for desalting group separation, *Nature (Lond.)* **183**, 1657–1659 (1959).

140. S. Hjertén and R. Mosbach, Molecular-sieve chromatography of proteins on columns of cross-linked polyacrylamide, *Anal. Biochem.* **3**, 109–118 (1962).

141. S. Hjertén, Chromatographic separation according to size of macromolecules and cell particles on columns of agarose suspensions, *Arch. Biochem. Biophys.* **99**, 466–475 (1962).

142. R. B. Bywater and N. V. B. Marsda, Gel chromatography, Part A, in E. Heftmann (Ed.), *Chromatography*, 4th ed., Elsevier, Amsterdam, 1983, Chap. 8, pp. A257–A330.

143. D. Berek and K. Marcinka, Gel chromatography, in Z. Deyl (Ed.), *Separation Methods*, Vol. 8, in A. Neuberger and L. L. M. van Deenen (General Eds.), *New Comprehensive Biochemistry*, Elsevier, Amsterdam, 1984, Sec. 4.6, pp. 271–361.

144. L. Hagel and J.-C. Janson, Size-exclusion chromatography, Part A, in E. Heftmann (Ed.), *Chromatography*, 5th ed., Elsevier, Amsterdam, 1992, Chap. 6, pp. A267–A307.

145. J. C. Moore, Gel-permeation chromatography. I. A new method for Mol. Wt. distribution of high polymers, *J. Polym. Sci. A* **2**(2), 835–843 (1964).

146. J.-C. Jansen, On the history of the development of Sephadex, *Chromatographia* **23**, 361–369 (1987).

147. J. Porath, From gel filtration to adsorptive size exclusion, *J. Protein Chem.* **16**(5), 463–468 (1997).

148. H. G. Barth, B. E. Boges, and C. Jackson, Size-exclusion chromatography and related separation techniques, *Anal. Chem.* **70**(12), 251R–278R (1998).

149. L. S. Ettre, Chromatography: The separation technique of the 20th century, *Chromatographia* **51**(1), 7–17 (2000).

150. Chi-San Wu (Ed.), *Handbook of Size-Exclusion Chromatography and Related Techniques*, 2nd ed., Chromatographic Science Series, Vol 91, Marcel Dekker, New York, 2004.

151. For additional SEC books, refer to Ref. 2—see Apps. 4CF, 5BD, and 7.

152. Refer to the contributions by J. O. Porath and S. Hjertén in Ref. 1, Chap. 5 and other chromatographers who have conducted research on SEC (see Author/Scientist Index for those with a code letter "g").

References on High-Performance Liquid Chromatography

153. C. Horváth (Ed.), *High-Performance Liquid Chromatography*, Academic Press, New York, 1980–1988, 5 vols.

154. A. Guttman, Editorial—happy birthday to C. Horváth, *Am. Lab.* **32**(13), 6–10 (2000).

155. L. S. Ettre, Evolution of liquid chromatography: A historical overview, Vol. 1, in C. Horváth, *High-Performance Liquid Chromatography*, Academic Press, New York, 1980–1988, 5 vols.; see Chap. 1, pp. 1–74.

156. Comprehensive chromatography books—see Ref. 2 and its Apps. 4A, 5G, and 7AB.

157. L. S. Ettre, The evolution of modern liquid chromatography, *LC North Am.* **1**, 408–410 (1983).

158. Books focusing on HPLC—see Ref. 2 for ∼14 books in App. 4C, ∼16 books in Apps. 5ABDE, ∼33 books in App. 5G, and ∼20 books in App. 7H.

159. E. Gelpi (Ed.), 23rd International symposium on high performance liquid phase separations and related techniques, *J. Chromatogr. A.* **869**, 1–529 (2000); **870**, 1–522 (2000); **871**, 1–494 (2000).

160. W. R. LaCourse and C. O. Dasenbrock, Column liquid chromatography: Equipment and instrumentation, *Anal. Chem.* **70**(12), 37R–52R (1998).

161. M. C. Ringo and C. E. Evans, Liquid chromatography as a measurement tool for chiral interactions, *Anal. Chem.* **70**(19), 315A–321A (1998) .

162. J. G. Dorsey, W. T. Cooper, B. A. Siles, J. P. Foley, and H. G. Barth, Liquid chromatography: Theory and methodology, *Anal. Chem.* **70**(12), 591R–644R (1998).

163. L. R. Synder, HPLC—Past and present, *Anal. Chem.* **72**(11), 412A–420A (2000).

164. R. E. Majors, HPLC columns—Q's and A's, *LC-GC* **24**(1), 16–28 (2006).

165. R. E. Majors, Technology and applications—highlights of HPLC 2007, *LC-GC* **25**(10), 1000–1012 (2007).

166. W. R. LaCourse, Column liquid chromatography: Equipment and instrumentation, *Anal. Chem.* **72**(12), 37R–51R (2000).

167. Refer to Author/Scientist Index in for Ref. 1 the names of approximately 63 chromatography awardees and contributors whose research advanced HPLC; they are identified with the code letter "h." Of course, there are many other scientists in this area.

168. K. Cabrera, D. Lubda, K. Sinz, C. Schäfer, and D. Cunningham, HPLC columns: The next great leap forward, Part 1, *Am. Lab. News Ed.* **33**(4), 40–41 (2001); Part 2, ibid. **33**(9), 25–26 (2001).

169. K. M. Gooding and F. E. Regnier (Eds.), *HPLC of Biological Macromolecules*, 2nd ed., Chromatogrpahic Science Series, Vol. 87, Marcel Dekker, New York, 2002.

170. A. W. Czarnik, Combinatorial chemistry, *Anal. Chem.* **70**(11), 378A–386A (1998).

171. J. N. Kyranos and J. C. Hogan, Jr., High-throughput characterization of combinatorial libraries generated by parallel synthesis, *Anal. Chem.* **70**(11), 389A–395A (1998).

References on Detectors and Scientific Instruments in Chromatography

172. H. M. McNair, Instrumentation in gas chromatography, in E. Heftmann (Ed.), *Chromatography—a Laboratory Handbook of Chromatographic and Electrophoretic Methods*, 3rd ed., Van Nostrand Reinhold, New York, 1975, Chap. 9, pp. 189–277.

173. S. H. Hansen, P. Helboe, and U. Lund, Liquid column chromatography—types, instrumentation, detection, in Z. Deyl (Ed.), *Separation Methods*, Vol. 8, in A. Neuberger and L. L. M. van Deenen (General Eds.), *New Comprehensive Biochemistry*, Elsevier, Amsterdam, 1984, Chap. 4, pp. 151–201.

174. G. Patonay (Ed.), *HPLC Detection—Newer Methods*, VCH, New York, 1992.

175. H. H. Hill and D. G. McMinn (Eds.), *Detectors for Capillary Chromatography*, Wiley, New York, 1992.

176. R. E. Sievers (Ed.), *Selective Detectors: Environmental, Industrial and Biomedical Applications*, Wiley, New York, 1995.

177. R. P. W. Scott, *Tandem Techniques*, Wiley, Chichester, UK, 1997.

178. For additional books with a focus on detection, see the many chapters of the more comprehensive books cited in Ref. 2, Apps. 4, 5, and 7.

179. L. E. Gerald and T. Turner, *Scientific Instruments—1500–1900*, Univ. California Press, Berkeley, 1998.

180. L. S. Ettre, American instrument companies and the early development of gas chromatography, *J. Chromatogr. Sci.* **15**, 90–110 (1977).

181. Anonymous, *Arnold O. Beckman—the Post-War Boom*; available online at http://www.chemheritage.org/explore/Beckman/chptr8.htm.

182. J. Poudrier and J. Moynihan, Instrumentation Hall of Fame: The luminaries of the analytical instrument business, *Todays Chemist at Work* **8**, 37–38 (1999).

183. R. P. Valleri (Ed.), The leading 50 separation science companies, *LC GC North Am.* (Aug. 1999 Suppl.).

184. J. L. Glaich and L. R. Snyder (Eds.), Computer-assisted method development in chromatography (symposium), *J. Chromatogr.* **485**, 1–675 (1989).

185. K. Jinno, *A Computer-Assisted Chromatographic System*, Huethig, Heidelberg, Germany, 1990.

186. P. C. Lu and G. W. Xu, *Expert System on Gas Chromatography*, Slasdong Press of Science and Technology, Jinan, People's Republic of China, 1994.

187. S. R. Crouch and T. V. Atkinson, The amazing evolution of computerized instruments, *Anal. Chem.* **72**(17), 596A–603A (2000).

188. W. M. A. Niessen, *Liquid Chromatography—Mass Spectrometry*, 3rd ed., Chromatographic Science Series, Vol. 97, CRC Press, Taylor & Francis Group, Boca Raton, FL, 2006.

189. N. Flanagan, Mass spec looms large in discovery, *Genet. Eng. Biotechnol. News* **28**(10), 1–27 (May 15, 2008).

References on Hyphenated/Coupled/Tandem Techniques

190. G. Bringmann, C. Günther, J. Schlauer, and M. Rückert, HPLC-NMR on-line coupling including the ROESY technique, *Anal. Chem.* **70**(15), 2805–2811 (1998).

191. J. L. Glajch and L. R. Snyder, Computer-assisted method development in chromatography, *J. Chromatogr.* **485**, 1–675 (1989).

192. P. Girörer, J. Schewitz, K. Puseker, and E. Bayer, On-line coupling of capillary separation techniques with H-NMR, *Anal. Chem. News Features* **4**, 315A–321A (1999).

193. C. Henry, Can MS really compete in the DNA world? *Anal. Chem.* **69**(7), 243A–246A (1997).

194. I. D. Wilson, J. C. Lindon, and J. K. Nicholson, Advancing hyphenated chromatographic systems, *Anal. Chem.* **72**(15), 534A–542A (2000).

195. P. Bócek, R. Vespalec, and R. W. Giese, Selectivity in CE, *Anal. Chem.* **72**(17), 586A–595A (2000).

196. M. A. Brown (Ed.), *Liquid Chromatography—Mass Spectrometry: Applications in Agricultural, Pharmaceutical, and Environmental Chemistry*, ACS Symposia Series 420, American Chemical Society, Washington, DC, 1990.

197. K. Jinno (Ed.), *Hyphenated Techniques in Supercritical-Fluid Chromatography and Extraction*, Elsevier, Amsterdam, 1992.

198. R. P. W. Scott, *Tandem Techniques*, Separation Science Series, Wiley, Chichester, UK, 1997.

199. M. Oehme, *Practical Introduction to GC-MS Analysis with Quadrupoles*, Hüthig, Heidelberg, Germany, 1998.

200. W. M. A. Niessen, *Liquid Chromatography—Mass Spectrometry*, 2nd ed., Marcel Dekker, New York; 1st ed., 1992; 2nd ed., 1999.

201. M. A. Grayson, Exploring the evolutionary relationship between computers and mass spectrometry, *Am. Soc. Inform. Sci. Technol.* 190–202 (2004).

202. R. E. Ardrey, *Liquid Chromatography—Mass Spectrometry—an Introduction*, Wiley, Hoboken, NJ, 2003.

References on Women Scientists in Chromatography

203. F. U. Bright and M. E. P. McNally (Eds.), *Supercritical Fluid Technology: Theoretical and Applied Approaches to Analytical Chemistry*, American Chemical Society, Washington, DC, ACS Symposium Series 488, 1992.

204. M.-L. Riekkola, *Capillary Gas Chromatography and Volatility Characteristics and Metal Dialkyldithiocarbamate Chelates*, Ph.D. thesis, Suomalaisen Tiedeakatemia, Helsingfors, Finland, 1983.

205. M.-L. Riekkola and T. Hyötyläinen (Eds.), *Column Chromatography and Capillary Electromigration Techniques* (in Finnish), Yliopistopaino, Finland, 2000.

References Related to Chapter 4 Summary

206. J. C. Giddings, *Unified Separation Science*, Wiley-Interscience, New York, 1991.

207. J. F. Parcher and T. L. Chester, *Unified Chromatography*, American Chemical Society, Washington, DC, distributed by Oxford Univ. Press, Oxford, UK, 2000.

208. T. L. Chester and J. F. Parcher, Blurring the boundaries, *Science* **291**, 502–503 (2001).

209. S. Borman, Proteomics—taking over where genomics leaves off, *Chem. Eng. News* **78**(31), 31–37 (2000).

210. A. Emmett, Scientists ponder the meaning of all 23 human chromosome pairs, *Scientist* **14**(15), 17–19 (July 24, 2000); see also 11 other papers that focus on genes and genomes cited therein.

211. W. Hancock, A. Apffel, J. Chakel, K. Hahnenberger, G. Choudhary, J. A. Traina, and E. Pungor, Integrated genomic/proteomic analysis, *Anal. Chem.* **71**(21), 742A–749A (1999).

212. E. K. Wilson, Gearing up for genomics protein avalanche, *Chem. Eng. News* **78**(39), 41–44 (2000).

213. J. S. Roach, What's in a genome? *Anal. Chem.* **72**(17), 609A–611A (2000).

214. D. Figeys and D. Pinto, Lab-on-a-chip: A revolution in biological and medical sciences, *Anal. Chem.* **72**(9), 320A–335A (2000).

215. S. Borman, Combinatorial chemistry—redefining the scientific method, *Chem. Eng. News* **78**(20), 53–65 (2000).

216. R. E. Majors, Glossary of HPLC and LC separation terms, *LC-GC* **26**(2), 118–168 (2008).

217. L. S. Ettre and J. V. Hinshaw (Eds.), chapters in the *Evolution of Chromatography*, Imperial College Press, London, 2008; a valuable organized collection of earlier papers (to 1973).

5

TODAY'S CHROMATOGRAPHERS AND THEIR DISCOVERIES (2000–2008)

Robert L. Wixom

Department of Biochemistry, University of Missouri, Columbia

Charles W. Gehrke

Department of Biochemistry and the Agricultural Experiment Station Chemical Laboratories, College of Agriculture, Food and Natural Resources, University of Missouri, Columbia

A discovery is finding or observing something new, something unknown or unnoticed before. It is noticing what was always there, but had been overlooked by all before. It is stretching out into untouched and uncharted regions. Discoveries open new horizons, provide new insights, and create vast fortunes. Discoveries advance human knowledge; they mark the progress of human civilizations.

—Hendell Haven, *100 Greatest Science Discoveries of All Times*, 2007

CHAPTER OUTLINE

Chromatography: A Science of Discovery. Edited by Robert L. Wixom and Charles W. Gehrke
Copyright © 2010 John Wiley & Sons, Inc.

5.A. INTRODUCTION

This chapter introduces the major national and international professional societies that present annual awards in chromatography. Of course, there are some societies that don't have such awards, as well as many local or regional chromatography societies that present awards; space prevents citing the latter here. For more comprehensive descriptions of chromatography awardees, see also Section 11.E (of Chapter 11).

This, the cardinal chapter of our book, has brief biographies of all awardees followed by their stories on their unique activities, their research, experience, the relations of their research to others in nearby fields, and pertinent photographs. These biographies, in their first-person accounts and individual writing styles, have included some unique personal associations, some humorous events, some man-to-man, woman-to-woman interactions at national or international professional meetings or major research conferences. The reader of Ref. 1 (see References list at end of Section 5.B) knows that the stories of chromatography exist in the undercurrents of sciences. Although the formal scientific journals usually omit such informality, it does exist and, if recognized, has a role in the maturation of graduate students and young professors.

This chapter is organized as follows:

1. The society or institution presenting the award is described. Please bear in mind that although many other awards and honors are presented each year; only the most significant national and international chromatography awards are presented here.

2. The 2000–2007 Awards and Awardees, followed by their research contributions, are described in this chapter. Since these awards have no chronological nor organizational order, the awardees are presented here in alphabetical order according to their surnames. To achieve a degree of order, each scientist has one or more letters in the text; each letter represents a *seminal concept* as defined earlier (see Chapter 1, Table 1.3). Some contributions by awardees are missing for various reasons.

Note: The annual chromatography awards for the 1900s through 1999 and the recognized earlier awardees are described in Reference 1, Chaps. 2, 3, and 5.

5.B. PROFESSIONAL SOCIETIES PRESENTING AWARDS

5.B.1. National Award in Chromatography of the American Chemical Society (2000–2007)

This award was established in 1959, underwent a name change in 1971, and recognizes outstanding contributions in the field of chromatography with particular attention to the development of new methods. The awardees for 1961–2000, a single scientist each year, were described in our 2001 book [1, Chaps. 2, 5]. The 2000–2007 awardees describe their scientific contributions later in this chapter. Several 2000 awardees who did not present their manuscripts in time for inclusion in our earlier book are now included. Several of the 2000–2007 awardees had also received awards earlier as cited in our 2001 book [1]. (See also Table 5.1.)

5.B.2. National Award in Separation Science and Technology of the American Chemical Society (2000–2007)

This award was initiated in 1982 by the Rohm and Haas Co., with the aim of recognizing outstanding accomplishments of scientists and engineers in fundamental or applied research devoted to separations science and technologies. The awardees for 1984–2001 were presented in our 2001 book [1]. The award has been sponsored by the ACS Division of Industrial and Engineering Chemistry and several other companies. The 2008 award was sponsored by the Waters Corporation. Separation science is broader than chromatography, and awardees have also contributed to chemistry, biology, engineering, medicine, ecology, and other areas. Table 5.2 lists the 2000–2007 awardees in chromatography.

TABLE 5.1. Recent Recipients of the National Award in Chromatography of the American Chemical Society

Year	Awardee, Affiliation, and Country
2000	Charles W. Gehrke, Univ. Missouri, Columbia, MO, USA [1]
2001	Ernest Bayer, Univ. Tubingen, Germany [1]
2002	Edward S. Yeung, Iowa State Univ., Ames, IA, USA [1]
2003	William S. Hancock, Thermo Finnigan, San Jose, CA, USA [1]
2004	Shigeru Terabe, Himeji Inst. Technology, Kamigori, Japan [1]
2005	Patrick J. F. Sandra, Ghent Univ., Belgium and Univ. Stellenbosch, South Africa [1]
2006	John G. Dorsey, Florida State Univ., FL, USA
2007	J. Michael Ramsey, Univ. North Carolina, Chapel Hill, NC, USA

TABLE 5.2. Recent Awardees of the National Award in Separation Science and Technology of the American Chemical Society

Year	Awardee, Affiliation, and Country
2000	Earl P. Horwitz, Eichron Industries, Inc., Darien, IL, USA
2001	Csaba Horwath, Yale Univ., New Haven, CT, USA, [1]
2002	Edward L. Cussler, Univ. Minnesota, Minneapolis, MN, USA
2003	Ralph T. Yang, Univ. Michigan, Ann Arbor, MI, USA
2004	William H. Pirkle, Univ. Illinois, Urbana—Champaign, IL, USA [1]
2005–2007	No award

5.B.3. A. J. P. Martin Award of the Chromatographic Society (2000–2007)

This award was established in 1978 by the Chromatographic Discussion Group, UK, to honor A. J. P. Martin, the founder of partition chromatography. It is now sponsored by the Chromatographic Society. Their aim is to recognize some special contributions for the advancement of chromatography, although not necessarily limited to its scientific aspects. Building on the annual awardees in 1978–2000 [1], the recent awardees are listed in Table 5.3.

TABLE 5.3. Recent Recipients of the A. J. P. Martin Award

Year	Awardee, Affiliation, and Country
2000	C. H. Mosback, Univ. Lund, Lund, Sweden
2000	W. S. Hancock, Agilent Technologies, Palo Alto, CA, USA
2001	J. Michael Ramsey, Oak Ridge Natl. Lab., now Univ. North Carolina, Chapel Hill, NC, USA
2002	Paul Haddad, Univ. Tasmania, Tasmania, Africa
2002	Werner Engewald, Univ. Leipzig, Leipzig, Germany
2003	Jack Henion, Cornell Univ., Ithaca, NY, USA
2004	Terry Berger, Berger Instruments (Mettler-Toledo AutoChem), Columbus, OH, USA
2005	Vadim Davankov, Russian Acad. Sciences, Moscow, Russia [1]
2006	Jim Waters, venture capitalist, founder of Waters Foundation, Waters Corp., USA
2007	Ronald E. Majors, Agilent Technologies, Wilmington, DE, USA
2007	Johan Roeraade, Anal. Chem. Div., Royal Inst. Technology, Stockholm, Sweden

5.B.4. M. J. E. Golay Award in Capillary Chromatography by the International Symposia on Capillary Chromatography (2000–2007)

The Golay Award in Capillary Chromatography was established in 1989 by the organization of the Symposia on Capillary Chromatography, and was continued after 1990 by the Perkin-Elmer Corporation. The earlier 1989–1999 awardees have been described in Ref. 1. The post-2000 awardees are listed in Table 5.4.

TABLE 5.4. Recent Recipients of the M. J. E. Golay Award

Year	Awardee, Affiliation, and Country
2000	John H. Knox, Univ. Edinburgh, UK, [1]
2000	Ernst Bayer, Univ. Tubingen, Tubingen, Germany
2001	Ben Burger, Stellenbosch Univ., Stellenbosch, South Africa
2001	Fred Regnier, Purdue Univ., Lafayette, IN, USA
2002	Keith Bartle, Leeds Univ., Leeds, West Yorkshire, UK
2002	Stellan Hjerten, Univ. Uppsala, Sweden [1]
2003	Andreas Manz, Imperial College, London, UK (Inst. for Analytical Sciences)
2003	D. Jed Harrison, Univ. Alberta, Edmonton, Canada
2003	J. Michael Ramsey, Oak Ridge National Lab., now Univ. North Carolina, Chapel Hill, NC, USA
2004	Wilfred A. Konig, Univ. Hamburg, dec. 11/2004
2004	Volker Schurig, Univ. Tubingen, Tubingen, Germany
2005	Frantisek Svec, Univ. California, Berkeley, CA, USA
2006	Edward Yeung, Iowa State Univ., Ames, IA, USA, [1]
2007	Nobuo Tanaka, Kyoto Inst. Technology, Kyoto, Japan

5.B.5. Stephen Dal Nogare Award in Chromatography of the Chromatographic Forum of Delaware Valley

The Chromatography Forum of the Delaware Valley encompasses states bordering the Delaware River–New Jersey, Pennsylvania, and Delaware, which house some of the important chemical industries of the United States. The Forum started in 1966 with Stephen Dal Nogare, E. I. Du Pont de Nemors and Co., one of the largest chemical companies in the world. Dr. Nogare served as the second President of the Forum, but died suddenly in 1968. In 1972, the group started this annual award in his memory. The award is presented at PITTCON (see Section 11.A) and consists of a honorarium and a plaque; see Ref. 1 for earlier (1972–1999) awards. (See also Table 5.5).

TABLE 5.5. Stephen Dal Nogare Award of the Chromatography Forum of the Delaware Valley

Year	Awardee, Affiliation, and Country
2000	William F. Pirkle, Univ. Illinois, Urbana, IL, USA
2001	Harold M. McNair, Virginia Polytech Inst., Blacksburg, VA, USA
2002	Walter G. Jennings, Univ. California, Davis, CA, USA
2003	William S. Hancock, Thermo Finnigan, Waltham, MA and Northeastern Univ., Boston, MA, USA
2004	Milos Novotny, Indiana Univ., Bloomington, IN, USA
2005	Daniel W. Armstrong, Iowa State Univ., Aimes, IA, USA
2006	Victoria L. McGuffin, Michigan State Univ., East Lansing, MI, USA
2007	John W. Dolan, LC Resources, Walnut Creek, CA, USA

5.B.6. Silver Jubilee Award of the Chromatographic Society (2000-2007)

The Chromatographic Society initiated this silver medal in 1982 at its Silver Jubilee to be awarded mainly to scientists in the middle of their careers to recognize their chromatography contribution. The 1982-1999 awardees may be found in our 2001 book [1, Chaps. 2 and 5]. (See also Table 5.6.)

5.B.7. Separation Sciences Award of the Eastern Analytical Symposium (2000-2007)

The Eastern Analytical Symposium (EAS) is a major American scientific conference and has met annually since 1959, usually in the New York City area. Their sessions cover all branches of analytical chemistry; one of its awards is for Achievements in Separation Science, which began in 1986. Usually one or two scientists are recognized and present

TABLE 5.6. Silver Jubilee Award of the Chromatography Society (UK)

Year	Awardee, Affiliation, and Country
2000	Klaus Albert, Univ. Tubingen, Tubingen, Germany
2000	Phillip Marriott, Royal Melbourne Inst. Tech., Melbourne, Australia
2001	Marie-Claire Hennion, Ecole Superieure de Physique et de Chemie de Paris, France
2001	David Perrett, Barts and the London, Queen Mary's School of Medicine and Technology, London, UK
2002	Christopher R. Lowe, Univ. Cambridge, Cambridge, UK
2002	Nobuo Tanaka, Kyoto Inst. Technology, Kyoto, Japan
2003	Tom Lynch, BP Chemicals, London, UK
2003	Ian M. Mutton, GlaxoSmithKline, Stevenage, UK
2004	Ernest Dawes, Scientific Glass Engineering, Austin, TX, USA
2004	Antoine-Michel Sioffi, Univ. Paul Cezanne, Aix-en-Provence, France
2005	Kevin Altria, GlaxoSmithKline, Univ. Liverpool, UK
2006	Peter Myers, Independent Consultant, Univ. Liverpool, UK
2006	Steven Wren, AstraZeneca, Macclesfield, UK
2007	Mel Euerby, AstraZeneca Co., Loughborough, UK

TABLE 5.7. Recent Recipients of the Eastern Analytical Symposia Award

Year	Awardee, Affiliation, and Country
2000	Peter W. Carr, Inst. Tech., Minneapolis, MN, USA [1]
2001	Georges Guiochon, Univ. Tennessee, Knoxville, TN, USA [1]
2002	Karel Cramers, Eindhoven Tech. Univ., Belgium [1]
2003	Edwards S. Yeung, Iowa State Univ., Ames, IA, USA [1]
2005	Frantisek Svec, Univ. Calif., Berkeley, CA, USA
2006	Phyllis Brown, Univ. Rhode Island, Providence, RI, USA [1]
2007	Peter C. Uden, Univ. Massachusetts, Amherst, MA, USA

TABLE 5.8. Recent Recipients of the Colacro Medal

Year	Awardee, Affiliation and Country
2000	Milos V. Novotny, Indiana Univ., Bloomington, IN, USA [1]
2000	R. Saelzer, Univ. Conception, Conception, Chile
2000	M. Vega, Univ. Conception, Conception, Chile
2002	Janusz Pawliszyn, Dept. Chemistry, Univ. Waterloo, Waterloo, Canada

an award address. Currently it is sponsored by the Waters Corporation. The earlier (1986–1999) awardees are described in Ref. 1. (See also Table 5.7).

5.B.8. COLACRO Medal of the Congressio Latino Americano de Chromatografia

The biannual meeting of COLACRO (Congresso Latino Americano de Chromatografia) has had presentations to scientists, who have made a significant contribution to chromatography, usually in Latin America. The 1986–2000 awardees are listed in our 2001 book [1, Chap. 5]. The more recent awardees are listed in Table 5.8.

5.B.9. Other Awards and Awardees

Because of their close similarities in subject matter, two awardees were placed in Chapter 4. Table 4.3 lists the recipients of the Chirality Medal Awards (1991–2007). Table 4.5 lists the awardees of the Supercritical-Fluid Chromatography, Extraction and Processing Symposia (1988–2004).

REFERENCES

1. C. W. Gehrke, R. L. Wixom, and E. Bayer (dec.) (Eds.), *Chromatography: A Century of Discovery (1900–2000)—the Bridge to the Sciences/Technology*, Elsevier Science, Amsterdam, 2001, Seven chapters (709 pp). Hereafter in this chapter this book will be cited in abbreviated form, namely Gehrke et al., *Chromatography* (2001).

2. C. W. Gehrke and R. L. Wixom (Eds.), *Chromatography: A Science of Discovery*, Wiley, Hoboken, NJ, 2010 (present volume), Chap. 4.

3. C. W. Gehrke, R. L. Wixom, and E. Bayer (dec.) (Eds.), (2001s) *Chromatography: A Century of Discovery (1900–2000)*, Sections S7–S15 and seven appendixes—these nine chapters are available on the Internet at Chem. Web Preprint Service (http://www.chemweb.com/preprint).

4. L. S. Ettre, *Milestones in the Evolution of Chromatography*, ChromSource, Franklin, TN, 2002.

5. L. S. Ettre and J. V. Hinshaw, chapters in the *Evolution of Chromatography*, Imperial College Press, London, 2008.

6. Many other earlier references could be cited. Please refer to Ref. 3 above and its Appendixes.

5.C. PROMINENT CHROMATOGRAPHERS (AWARDEES)

The biographic sketches that follow are presented in alphabetic order according to surname. Affiliations are omitted from the headings because they are presented in detail in the Contributors list at the beginning of this book. Each biographic section contains a list of references authored by or related to that particular awardee; there is no composite end-of-chapter list (in this chapter or in Chapters 6–9 and 11).

TERRY BERGER

Figure 5.1. Terry Berger.

Terry Berger (Fig. 5.1) received a Ph.D. in Analytical Chemistry and a Diploma of Imperial College (DIC) from Imperial College, London in 1976. He worked in Research for Hewlett-Packard for 18 years, where he defined the HP-SFC. With another HP employee, he bought this product line in 1995 and formed Berger Instruments (BI), where he variously acted as Vice President, President, and Chief Technical Officer. With minimal outside funding, BI achieved a cumulative growth rate of >85% per year from 1995 to 2001, grew to over 40 employees, and made a profit! BI was sold to Mettler-Toledo (on the NYSE) in Fall 2000.

Terry is considered to be the "father" of modern supercritical-fluid chromatography (SFC), for which he was awarded the 2004 Martin Gold Medal by the Chromatographic Society of Great Britain. Berger Instruments received an IR&D 100 award for one of the 100 most important innovations in R&D in 2000, for the development of a semiprep(aratory) SFC. In 2008, he received an award as the outstanding alumni for achievement from Carroll College in Waukesha, Wisconsin.

He has written approximately 60 papers, 25 patents, 10 book chapters, six encyclopedia entries, dozens of application notes, and >300 technical talks. He wrote the book *Packed Column SFC*, published in 1995 by the Royal Society for Chemistry. He is an editor for the journal *Chromatographia*, and section editor of the *Encyclopedia of Separation Science*.

With three colleagues he founded AccelaPure Corporation in August 2004. In 2006, AccelaPure founded an Institute of SFC Excellence within the Chemistry Department of Princeton University. In 2007, he co-founded Aurora SFC to manufacture fourth-generation SFC instrumentation. Terry lives in Florida, with his wife of 36 years, and works with his surviving daughter.

See Chapter 1, Table 1.3, a, k, o.

SFC Research Conducted by Terry Berger, 1979–2004. In 1979, at Hewlett Packard (HP), I was "the next bench" to Dennis Gere, who started modern interest in supercritical-fluid chromatography (SFC), with a series of papers delivered at the 1979 Pittsburgh conference. Our research group later commercialized a kit converting a HP

1084 liquid chromatograph into a packed-column SFC in 1981. In the same year, capillary, or open tubular SFC was introduced by Milton capillary columns (mostly for gas chromatography), making that advance possible.

The general perception was that capillary SFC was easier, and cheaper with higher performance. One of the perceived advantages of capillary SFC was the use of pure carbon dioxide (CO_2) from a pump with pressure control, while packed columns required two high-pressure pumps with flow control. Most workers abandoned packed columns. After the obsolescence of the 1084 in 1983, HP management dropped SFC.

I spent 1985 just thinking, traveling, and experimenting in SFC. I visited every academic and industrial user of SFC that I could find, including a 2-week whirlwind tour of Europe (Germany, Netherlands, France, Switzerland, Norway, and England) visiting several users per day. We worked all day and traveled at night to the point of exhaustion. (We spent 5 nonconsecutive nights in different rooms in the same hotel in Düsseldorf, which was extremely disorienting.)

I bought, and built equipment and performed experiments. After a year, I came to the conclusion that SFC had value, and demonstrated separations that piqued the interest of management. I was allowed to continue my research.

A number of issues had emerged, which I attempted to systematically overcome, scientifically. The general consensus among SFC users was that carbon dioxide (CO_2) was much more polar than it actually is. Modifiers were presumed to change retention, not by changing mobile-phase polarity but by changing its density. This misconception can be traced back to an elutrophic series published in *Science* in 1968 by J. Calvin Giddings. Giddings was probably the most influential chromatographer of the 1960s and had helped change it from an empirical method, to a science, bounded by theory. Never the less, Giddings overestimated Hildebrandt solubility parameters for CO_2 and wrongly calculated it to have a polarity similar to isopropanol. This suggested, in general, that programming pressure could change the polarity of CO_2 from hydrocarbon-like, to alcohol-like, which could elute a wide range of solute polarity. (Giddings' towering reputation meant nobody challenged his elutrophic series for many years.)

Jerome (Jerry) Deye was a strong believer in SFC when he retired from DuPont in 1988. I hired him to work part time. I borrowed a high pressure densitometer from the University of Delaware, and we measured the density of methanol/CO_2 mixtures (not available in the literature). We performed chromatographic separations at constant density. The results [1] disproved the main objections to the use of modifiers. Modifiers changed retention by mostly changing polarity, not density. Still people clung to the idea that CO_2 was as polar as an alcohol.

I read a paper by John Dorsey on solvatochromic shifts, where the spectra of certain dyes shifted in wavelength depending on the polarity of the solvent sheath around the dye. Experiments with such dyes produced our own elutrophic series, which placed CO_2 near hydrocarbons in polarity.

We also measured the solvent strength of mixtures of modifiers in CO_2, which showed that small concentrations of polar modifier had a very nonlinear effect on solvent strength [2]. However, there are still people who believe the Giddings' estimates.

Around 1990, several competing theories emerged claiming that pressure drops across the column could not exceed 20 bar without serious band broadening. Thus, it was claimed that packed-column SFC was limited to no more than 20,000 plates, whereas capillary SFC was virtually unlimited in efficiency (due to very low pressure drops). I connected 11–20-cm-long columns packed with 5-μm particles, in series, and published the separations, demonstrating 220,000 plates [3]. Clearly, the theories were wrong! It is probably that the theories were generated using an instrumental configuration where the column head pressure, not the column outlet pressure was controlled.

This work demonstrated that packed columns generated high efficiencies faster than capillary columns (5 μm particles vs. 50 μm tube i.d.). Through the 1990s, capillary SFC slowly disappeared.

Curiously, in the early 1900s, many working with pharmaceuticals had concluded that SFC was too *nonpolar* for small drug-like molecules, opposite to many of the older theories about solvent strength. This was actually a backlash against the overselling of capillary SFC. Very polar compounds often did not elute or eluted with severe tailing. We performed a series of experiments with very polar additives (strong acids or bases) dissolved in the modifier. A number of papers followed, including the first use of additives in SFC [4]. A subset of this controversy was the concern that the CO_2 would react with amines, forming an insoluble precipitate. In one paper, a series of benzylamines and other primary amines were separated with high efficiency and symmetric peaks. Another paper beautifully separated carbamate pesticides, similar to the presumed reaction product. Now, amines are routinely used as additives and virtually all other amines elute with symmetric peaks.

There was another widespread misconception that small drug-like molecules are too water-soluble to be amenable to SFC [5]. However, the problem that medicinal chemists are usually trying to overcome is extreme hydrophobicity. I published a plot of log P (the partition coefficient of drugs between octanol and water) of 5000 commercial drugs. The average value is 3.5, meaning there are 3600 molecules in the aqueous phase. SFC has now been demonstrated to cover the log P range from at least -1 to $+7$, making it the ideal technique for small drug-like molecules.

Finally, our success in separating a wide range of compounds at the analytical scale, brought pressure to extend SFC to the semipreparative scale. At Berger Instruments, we developed hardware to separate the modifier from CO_2 without making aerosols. We further developed gas delivery systems to supply high flows of clean CO_2, at very low cost. This equipment changed the methods for semipreparative separation in the laboratory. Since it is used mostly for chiral separations, CO_2 mostly replaces heptanes, dramatically improving safety and allowing the separations in regular laboratories, not a bunker. The CO_2 is considered "green", since it is recycled. Most of the mobile phase vaporizes when the pressure is dropped, leaving polar solutes in a small volume of an organic solvent. Additionally, diffusion coefficients of solutes in the mobile phase are 3–10 times higher than in high-performance liquid chromatography (HPLC), making SFC 3–10 times faster than the more traditional HPLC. Low viscosity allows use of smaller particles at the semipreparative scale.

Some Important Publications

1. T. A. Berger and J. F. Deye, Composition and density effects using methanol/carbon dioxide in packed column SFC, *Anal. Chem.* **62**, 1181 (1990).
2. J. F. Deye, T. A. Berger, and A. G. Anderson, Nile Red as a solvatochromic dye for measuring solvent strength in normal liquids and mixtures of normal liquids with supercritical and near critical fluids, *Anal. Chem.* **62**, 615 (1990).

3. T. A. Berger and W. H. Wilson, Packed column SFC with 220,000 plates, *Anal. Chem.* **65**, 1451 (1993).

4. M. Ashraf-Korassani, M. G. Fessahaie, L. T. Taylor, T. A. Berger, and J. F. Deye, Rapid and efficient separation of PTH-amino acids employing supercritical CO_2 and an ion pairing agent, *J. High Resol. Chromatogr.* **11**, 352 (1988).

5. T. A. Berger, Analytical and semi-preparatory SFC in the pharmaceutical industry, in P. York, U. B. Kompella, and B. Y. Shekunov (Eds.), *Supercritical Fluid Technology for Drug Product Development*, Drugs and the Pharmaceutical Sciences, Series, Vol. 138, Marcel Dekker, New York, 2004, Chap. 12.

The Future of SFC Chromatography. I will limit my comments about the future of chromatography to SFC. Most developments in HPLC are toward the use of very small particles to increase speed of analysis. SFC achieves the same objectives without the excess pressure drops caused by small particle size, since the solute binary diffusion coefficients are 3–10 times larger in SFC compared to HPLC. Thus, SFC is inherently faster. Further, the use of very small particles in HPLC is not scalable, and semipreparative SFC should be the primary means of delivering gram-scale purifications in the foreseeable future. SFC is far safer, faster, cheaper for semipreparative separations. Thus, SFC is likely to both replace HPLC for most semipreparative applications and significantly increase the number and economic importance of such applications.

PHYLLIS R. BROWN

Phyllis R. Brown (Fig. 5.2), a pioneer in the application of HPLC and CE to biochemical problems, attended Simmons College and received her B.S. in Chemistry from George Washington University and her Ph.D. in Chemistry from Brown University. She did postdoctoral work in the pharmacology section of Brown and in 1973 joined the faculty of the University of Rhode Island, where she was promoted to Full Professor in 1980 and now is Professor Emeriti. She is an authority on the analysis of purine and pyrimidine compounds by both HPLC and CE and has applied these assays to every phase of nucleic acid metabolism. She is the author of over 200 articles and has written and/or edited five books on HPLC and/or CE. For 25 years she was an editor of the Advances in Chromatography Series. Dr. Brown has

Figure 5.2. Phyllis R. Brown.

been on the editorial boards of *Analytical Chemistry, Journal of Chromatography*, and *Biomedical Applications*. She is member of Phi Beta Kappa, Sigma Xi, and Phi Kappa Phi. She received a Fulbright Fellowship, the Tswett Medal in Chromatography, the Dal Nogare Award in Chromatography, the Governor's Medal for Contributions to Science and Technology in the State of Rhode Island, the Scholarly Achievement

Award for Excellence in Research in the Field of Chemistry, the Csaba Horvath Medal for Separation Science, and the 2006 Award by the Eastern Analytical Symposium for Achievements in Separation Science. For the past four decades she has been an activist in supporting and promoting women in chemistry.

See Chapter 1, Table 1.3, d, h, t.

My career in chromatography was serendipitous. My major in my Ph.D. work was physical organic chemistry, and I took the last course in analytical chemistry that Brown University offered. Therefore the last thing I expected to be doing was research in separation science. However, my first published paper from my thesis was a short note in the *Journal of Chromatography* on "The TLC of sulfur containing compounds" [1]. Maybe it was a prediction of things to come. My first job was in the Section of Pharmacology at Brown, synthesizing purine analogs to be used as chemotherapeutic drugs in the treatment of cancer, but assaying the drugs for purity was difficult. I had blithely thought that I could use gas chromatography. That was impossible because these compounds were thermally labile. The traditional method for analyzing nucleotides or nucleosides was to perform open-column chromatography in a cold room, collect the fractions, and then take the UV spectra of all the fractions. The reproducibility was poor, and a complete nucleotide profile (the mono-, di-, and triphosphate nucleotides of CTUAG) could take up to a week. So, when the head of our section came home from a conference and announced that he had leased a "nucleotide analyzer," I jumped at the chance to try it out. No one else wanted anything to do with it. The "nucleotide analyzer" turned out to be one of the early HPLCs developed by Horvath and Lipsky, adapted for the analysis of nucleotides. I was given 3 months in which to make it work. If I did not get results, it was going back. The rest is history. I did get results in the required time and wrote a paper that was rapidly published in the *Journal of Chromatography* on "The rapid separation of nucleotides in cell extracts using high performance liquid chromatography" [2]. It was one of the first papers in which HPLC was applied to actual biological samples. Until this time mainly standards were separated by HPLC and most of the previous work had been on the theory of separation and the development of instrumentation. My paper turned out to be a classic in the field and was widely read and quoted. My graduate students and I then evaluated microparticle, chemically bonded packings for separating biologically active compounds [3] and then turned our attention to the factors affecting the anion exchange of nucleotides. We found that many factors affected the analyses; among them, the procedures for extracting the nucleotides from biological matrices, detection of the compounds, and peak identification. We were among the first to investigate the use of the compounds and peak identification. We were among them first to investigate the use of the reversed-phase mode and applied it to the separation of nucleosides and their bases in serum or plasma [4]. We went on to establish a solid body of work on the analysis of nucleic acid constituents and showed that HPLC could be applied to many problems in biomedical research, such as studies of cancer and heart disease. We used HPLC in the determination of enzyme activities [5], kinetic studies [6], in investigations of DNA synthesis [7], and in studies of other biologically active moieties such as amino acids, vitamins, and glicerophospholipids in cells and physiological samples. We were also pioneers in the application of CE to both biological problems [8] and in the determination of inorganic ions in marine matrices [9]. Some of the assays we developed are still being used in biochemical, medical, clinical, or environmental laboratories and were essential in many medical breakthroughs. We also miniaturized HPLC by using microbore and very short columns and investigated scaling up to preparative scale. We designed HPLC systems for large clinical studies and formulated structure–retention relationships in the RPLC of purines and pyrimidines. Some of our later work involved the use of tunable coated capillaries for CE that show much promise.

I am happy to say that my prediction, made in the early 1970s that "HPLC will open new horizons in the analysis of nonvolatile, thermally labile compounds that are so important in biochemistry and medical research" [10], has come true and paved the way for the biotechnology era. Without HPLC and subsequent CE, the Human Genome Project would have taken many more years to complete and would not have been as useful in medical research, studies of genetics, and forensic science. I dreamed big dreams, but never did I think that some of the techniques developed in my laboratory could help revolutionize the life sciences and that I would play a role, albeit a small one, in the birth of biotechnology.

REFERENCES

1. P. R. Brown and J. O. Edwards, Quick TLC test for the detection of mercaptan groups in the presence of other types of sulfur functional groups, *J. Chromatogr.* **38**, 543 (1968).

2. P. R. Brown, The rapid separation of nucleotides in cell extracts using high pressure liquid chromatography, *J. Chromatogr.* **521**, 252–272 (1970).

3. R. A. Hartwick and P. R. Brown, Evaluation of microparticle chemically bonded anion exchange packings in the analysis of nucleotides, *J. Chromatogr.* **112**, 651–661 (1975).

4. R. A. Hartwick and P. R. Brown, Evaluation of microparticle, chemically bonded reverse phase packings in the high pressure liquid chromatography analysis of nucleosides and their bases, *J. Chromatogr.* **126**, 679–691 (1976).

5. A. P. Halfpenny and P. R. Brown, Simultaneous assay of the activities of HGPRTase and PNPase by HPLC, *J. Chromatogr. Biomed. Appl.* **345**(1), 125–133 (1985).

6. J. M. Miksic and P. R. Brown, The reaction of reduced nicotinamide dinucleoticde in acid: Studies by reversed phase high pressure liquid chromatography, *Biochemistry* **17**, 2234–2238 (1978).

7. S. J. Friedman, P. J. Skahan, M. L. Polan, A. Fausto-Stering, and P. R. Brown, DNA synthesis during early development of Drosophila melanogaster, *Insect Biochem.* **4**, 381–384 (1974).

8. S. E. Geldart and P. R. Brown, Optimization for the separations of ribonucleotides by capillary electrophoresis at high pH, *J. Chromatogr.* **792**, 62–73 (1997).

9. G. Zuang, Z. Yi, R. A. Duce, and P. R. Brown, Link between iron and sulfur cycles suggested by Fe(II) in remote marine aerosol, *Nature* **355**, 537–539 (1992).

10. P. R. Brown, *High Pressure Liquid Chromatography: Biochemical and Biomedical Applications*, Academic Press, New York, 1973.

BAREND (BEN) VICTOR BURGER

Figure 5.3. Barend (Ben) Victor Burger.

Barend (Ben) Victor Burger (Fig. 5.3) was born in Robertson, South Africa, on June 4, 1937.

In 1940 we moved to Riversdale, South Africa where I completed my schooling at the Langenhoven High School in 1954. I am married to Wina Burger and we have three children.

I commenced my university studies in 1955 and received the degrees B.Sc. (1975), M.Sc. cum laude (1960, thesis *The Structure of Ammonium Phosphomolibdate*) and D.Sc. (1966, dissertation *Dimers of Vitamin A*) from the University of Stellenbosch. In 1962 I was offered a position as Lecturer in General Chemistry at my alma mater. Subsequently, I was promoted to Lecturer in Organic Chemistry (1966), Senior Lecturer in Organic Chemistry (1969), Associate Professor of Organic Chemistry (1976), and Professor of Organic Chemistry (1995). These promotions were all merit promotions. From 1996 to 1998 I served an obligatory term as Chairman of the Department of Chemistry. Over the years I have lectured and presented practical courses to students from their first to their fourth (Hons-B.Sc.) years on subjects including basic systematic organic chemistry, the mechanisms of organic reactions, radical chemistry, and advanced stereochemistry. Since 1990, my annual Kurt Grob Commemorative Courses on Practical Gas Chromatography have attracted participants from all over South Africa.

My doctoral studies culminated in the determination of the structure of kitol, a natural dimer of vitamin A. After carrying out my first GC analysis in 1962, during my doctoral studies, I continued using the technique with great success to monitor the progress of organic syntheses. During 1967 the focus of my research began to shift from synthetic organic chemistry to the characterization of natural products, which in later years crystallized into two main topics, *viz.* chemical characterization of the volatile constituents of the secretions and excretions of living organisms, and the development of analytical methods for the determination of these and other volatile organic compounds. I commenced my pheromone research on the identification of the sex attractant of the pine emperor moth, *Nudaurelia cytherea*, in 1967, and in 1968 I spent a sabbatical year as holder of an Alexander von Humboldt Stipend at the University of Cologne with Prof. Leonhard Birkofer, one of the pioneers in the development of trimethylsilyl (TMS) derivatization agents. In 1975 I spent another Von Humboldt year at the University of Heidelberg with Prof. Hermann Schildknecht, where I continued working on one of my own projects, namely, the dorsal secretion of the springbok. The available facilities were inadequate for the envisaged research, but the applications laboratory of Varian MAT in Bremen made a GC-MS system available to me, one day per month. I therefore had ample time available to pursue other interests, which included making glass capillary columns. This fortunate coincidence made it possible for me to eventually build up a large supply of tailor-made capillary columns for my research (currently more than 500). In 1980 and 1986 Prof. Kurt

Grob spent two periods, the last one lasting 4 months, in my laboratory. I owe the quantum jump in the quality of the laboratory's capillaries in the mid-1980s to his willingness to share his unpublished expertise with me.

In March–April 1990 I gave a series of lectures as invited Tamkang Chair Lecturer at the University of Tamkang, Taiwan, and at the National Taiwan University, on selected aspects of our research. The contents of these lectures were subsequently published by Tamkang University Press (Tamkang Chair Lecture Series No. 100, 239 pp., 1991). (Prof. Carl Djerassi was a previous Tamkang Chair Lecturer.) In 2004 I was invited to give a further series of lectures as a Visiting Professor at Tamkang University.

See Chapter 1, Table 1.3, d, p, t, w.

Educational Background and Research Interests. For more than 20 years I have been carrying out intermittent research, together with my students and postdoctoral associates, into the semiochemicals of dung beetles, mainly focusing on the abdominal sex-attracting secretion of male dung beetles belonging to the genus *Kheper*. Of all the projects concerned with the identification of insect semiochemicals, I enjoyed our work on the semiochemicals of dung beetles the most. This research made it possible in some cases to discover unique phenomena and to elucidate the chemical basis of the phenomena. While carrying out behavioral tests with *Kheper* beetles, we were surprised to find that another species, *Pachylomerus femoralis*, in addition to rolling dung, also rolls the fruit-flesh-covered seeds of the yellow monkey orange tree, *Strychnos madagascariensis*. GC-FID/EAD and GC-MS analyses of the volatile constituents of the aroma of yellow monkey oranges trapped on open tubular capillary traps revealed that the ball-rolling behavior was induced by 1-butanol and a series of esters that are responsible for the fruit's pleasant aroma [1]. The individual esters also attracted *P. femoralis*. This dung beetle does not prepare dung balls, but it does roll suitably formed dung fragments it comes across. The beetle's size and its aggressive behavior also enable it to take dung balls away from smaller dung beetles, especially from *K. lamarcki*, which often has the same habitat. In this regard it is interesting that some of the esters present in the monkey orange are also present in the abdominal secretion of *K. lamarcki* [2]. Thus, although the esters do not elicit any endocrine activity disruption (EAD) activity in the antennae of this dung beetle, and thus are apparently not components of this insect's pheromone, they are utilized as a feeding kairomone by *P. femoralis*. Although the seeds are utterly useless to *P. femoralis*, this dung beetle is tricked into spreading the seeds of the tree.

Since 1972, most of our projects have been devoted to the identification of the constituents of the glandular secretions of mammals. Using GC-MS, and ancillary techniques such as enantioselective GC-MS, we have analyzed the following exocrine secretions of several South African antelope species: the interdigital secretions of bontebok, *Damaliscus dorcas dorcas*, blesbok, *D. d. phillipsi*, red hartebeest, *Alcelaphus buselaphus*, and gnu, *Connochaetes gnou*; the dorsal secretion of springbok, *Antidorcas marsupialis*; and the preorbital secretions of bontebok, blesbok, grysbok, *Raphicerus melanotis*, steenbok, *R. campestris*; blue duiker, *Cephalophus monticola*; red duiker, *C. natalensis*; gray duiker, *Sylvicapra grimmia*; oribi, *Ourebia ourebi*; suni, *Neotragus moschatus*; and klipspringer, *Oreotragus oreotragus* [3]. We have identified more than 1060 constituents of mammalian secretions. Many of these compounds are present in more than one of the

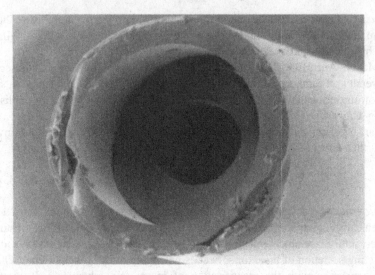

Figure 5.4. Section through an ultra-thick film trap (d_f 145 μm) made by the insertion of a polydimethylsiloxane (PDMS) rubber tube into a 0.52-mm-i.d. fused-silica capillary.

secretions, and some of them are unique or were not previously found in mammals. In the process we have produced practically all the information currently available on the chemical composition of the exocrine secretions of South African antelope.

Throughout my research career I have attempted to be not only a user of available technology, but also to improve or develop techniques for our research. For example, the open tubular traps coated with activated charcoal, and the thick films of stationary phases [4] or containing a lining of polydimethylsiloxane (PDMS) rubber (Fig. 5.4) [5] that we developed between 1986 and 1993 were used very successfully in our research, for example, on the dung beetle *P. femoralis*. In later years PDMS rubber became the sorptive medium of choice in some convenient and highly efficient sample enrichment techniques for GC and GC-MS analyses.

The more recently developed high-capacity sample enrichment probe (SEP) [6] is another example of the development of a tailor-made technique for our research. Using the SEP technique it is possible to use larger volumes of PDMS as sorptive medium than the relatively small volumes that are used in SPME, without having to cryofocus the analytes on the column.

Outputs. I have authored or co-authored two books two chapters in books, on semiochemistry and mass spectrometry, respectively, six patents, 86 publications in refereed journals, 70 contributions at international symposia, and 55 contributions at symposia held in South Africa. Since 1975 I have given 20 lectures on my research at universities and other research establishments in Europe. GC and GC-MS have played a key role in all my research projects.

Memberships. I am a member of ChromSA (Member of Executive Committee, Western Province Branch), the SA (South Africa) Association for Mass Spectrometry

(Manager, Western Province Branch), the Gesellschaft Deutscher Chemiker, and the International Society of Chemical Ecology (since 1979, member of the Board from 2002 to 2004). I was a member of the scientific organizing committees of the International Symposium on Chromatography and Mass Spectrometry in Environmental Analysis, St. Petersburg, in 1994; the Second International Symposium on Chromatography and Spectroscopy in Environmental Analysis and Toxicology, St. Petersburg, in 1996; the organizing committees of Analitika 1994 in Stellenbosch; and the First International Symposium on Chromatography, Analitika, 2002, in Stellenbosch. I was a member of the Board of the International Foundation for Environmental Assistance for Russia while that organization existed. I was also been a member of the Editorial Advisory Board of the *Journal of High Resolution Chromatography* from 1989 to 2001.

Awards and Medals. I received Alexander von Humboldt Stipends in 1968 and 1975, the Havenga Gold Medal for Chemistry from the South African Academy for Art and Science in 1989, the ILSA-Shimadzu Floating Trophy for outstanding contributions in chromatography from ChromSA in 1991, and the Gold Medal from the SA Chemical Institute in 2004. In 1997 I was elected Fellow of the Royal Society of South Africa. I am most honored to have received the M. J. E. Golay Medal in 2001 for pioneering work in the development of capillary chromatography.

Future Perspectives. To compensate for inadequate training of personnel, and to ensure the integrity of chromatographic data, equipment will in future have to be practically fully automated. However, this will also make life difficult for researchers who are interested in the development of novel analytical techniques. For research purposes I myself would prefer to use instruments that allow me to override certain settings embedded in the instruments' software.

It can be foreseen that in future all GCs on the market could have comprehensive two-dimensional capabilities. The required hardware is relatively simple and inexpensive. However, the problem of the prohibitively high cost of the cooling liquids used for cryomodulation still has to be solved.

REFERENCES

1. B. V. Burger and W. G. B. Petersen, Semiochemicals of the Scarabaeinae III: Identification of the attractant for the dung beetle Pachylomerus femoralis in the fruit of the spineless monkey orange tree, Strychnos madagascariensis, *Z. Naturforsch.* **46c**, 1073–1079 (1991).
2. B. V. Burger, W. G. B. Petersen, and G. D. Tribe, Pheromones of the Scarabaeinae IV: Identification of an attractant for the dung beetle Pachylomerus femoralis in the abdominal secretion of the dung beetle Kheper lamarcki, *Z. Naturforsch.* **50c**, 675–680 (1995).
3. B. V. Burger, Mammalian Semiochemicals, *Topics Curr. Chem.* **240**, 231–278 (2005) and references cited therein.
4. B. V. Burger, M. le Roux, Z. M. Munro, and M. E. Wilken, Production and use of capillary traps for headspace gas chromatography of airborne volatile organic compounds, *J. Chromatogr.* **552**, 137–151 (1991).

5. B. V. Burger, M. le Roux, and W. J. G. Burger, Headspace analysis : a novel method for the production of capillary traps with ultra-thick stationary phase layers, *J. High Resol. Chromatogr.* **13**, 777–779 (1990).

6. B. V. Burger, B. Marx, M. le Roux, and W. J. G. Burger. Simplified analysis of organic compounds in headspace and aqueous samples by high-capacity sample enrichment probe, *J. Chromatogr. A* **1121**, 259–267 (2006).

EDWARD L. CUSSLER

Edward Cussler (Fig. 5.5) holds the following degrees (all in chemical engineering): a B.E. from Yale University (cum laude; 1961), an M.S. and a Ph.D. from the University of Wisconsin (1963, 1965), a D.Sc. from Lund University (honoris causa; 2002), and a Ph.D. from Nancy Universite (honoris causa; 2007).

He was Assistant Professor, Associate Professor, and Full Professor at Carnegie-Mellon University during 1967–1980; Professor at the University of Minnesota during 1980–1995; and has served as a Distinguished Institute Professor at the University of Minnesota since 1996.

Figure 5.5. Edward L. Cussler.

Books Published

E. L. Cussler, *Diffusion*, 3rd ed., Cambridge Univ. Press, 2008.

E. L. Cussler and G. D. Moggridge, *Chemical Product Design*, Cambridge Univ. Press, 2001.

R. W. Baker, E. L. Cussler, W. Eykamp, W. J. Koros, R. L. Riley, and H. Strathmann, *Membrane Separation Systems*, Noyes Data, NJ, 1991.

P. Belter, E. L. Cussler, and W. S. Hu, *Bioseparations*, Wiley, New York, 1988.

E. L. Cussler, *Multicomponent Diffusion*, Elsevier, Amsterdam, 1976.

Principal Awards and Additional Publications

Alan P. Colburn Award, AIChE, 1975.

Minnesota Teaching Awards (12 times)

W. K. Lewis Award, AIChE, 2001

American Chemical Society Separations Science Award, ACS, 2002

National Academy of Engineering, 2002

Merryfield Design Award, American Society of Engineering Education, 2005

C. G. Gerhold Award, AIChE, 2006

Over 220 publications, including the following

Q. Liu and E. L. Cussler, Barrier membranes made with lithographically printed flakes, *J. Membrane Sci.* **285**, 56–67 (2006).

D. Yang, R. S. Barbero, D. J. Delvin, E. L. Cussler, C. W. Colling, and M. E. Carrera, Hollow fibers as structured packing for olefin/paraffin separation, *J. Membrane Sci.* **279**, 61–69 (2006).

T. Shimotori, W. A. Arnold, and E. L. Cussler, Fate of mobile products generated within reactive membranes, *J. Membrane Sci.* **291**, 111–119 (2007).

L. Chen, W. A. Phillip, E. L. Cussler, and M. A. Hillmyer, Robust nanoporous membranes templated by a doubly reactive block copolymer, *J. Am. Chem. Soc.* **129**, 13786 (2007).

A. Shrivastava, E. L. Cussler, and S. Kumar, Mass transfer enhancement due to a soft elastic boundary, *Chem. Eng. Sci.* **63**, 4302–4305 (2008).

See Chapter 1, Table 1.3, h, n.

Chemical Separations on Membrane Affiliations. For over 40 years, my research has been on chemical separations. While some of this has been on chromatography, most of it has been on membranes. I started using membranes to improve the efficiency of absorption and distillation. I studied the use of membranes to extract proteins and to separate ammonia from mixtures of hydrogen and nitrogen. I have looked at ways to use membranes to recover bromine from the sea and to adjust the humidity inside houses during a Minnesota winter. I have even used hollow-fiber membranes to try to improve the efficiency of preparative chromatography.

My current research on membranes is now based on block copolymers. In some cases, these membranes form lamellar structures of potential value in barrier packaging. In others, they provide ways to effect separations of gas mixtures, for example, by removing carbon dioxide from methane. In still others, they provide monodisperse, nanoporous membranes suitable for removing viruses from drinking water.

500 nm ▬▬▬▬▬▬

Figure 5.6. Detail of membrane surface.

Block Copolymer: Pore Diameter 13.5 nm.

Solute	Permeance (Experimental)	Permeance (Knudsen)	Permeance (Kinetic Theory)
He	265.7	260.5	5815.7
Ar	84.2	82.2	633.9
N_2	100.4	98.3	702.0
O_2	89.4	92.0	713.4

Track Etched: Pore Diameter 29.5 nm.

Solute	Permeance (Experimental)	Permeance (Knudsen)	Permeance (Kinetic Theory)
He	1.54	1.48	13.4
Ar	0.48	0.47	1.46

Figure 5.7. Pores showing Knudsen diffusion.

It is these nanoporous membranes that have unexplored potential for preparative chromatography. As shown in Figure 5.6, the membranes' pores characteristically are 10–50 nm in diameter but can be several millimeters long. They show Knudsen diffusion, so that gas molecules in the pores collide only with the pore walls (Fig. 5.7). They are monodisperse so that liquid flowing through them obeys the Hagen–Poiseuille law. They have reactive groups on the pore walls that could be modified to serve as selective sites. Most important, they are dense occupying 30% of the solid polymer matrix. Thus they have the potential to allow HPLC at higher productivity. To date, this potential has not been explored at all.

VADIM DAVANKOV

Figure 5.8. Vadim Davankov.

Vadim Davankov (Fig. 5.8) is a major contributor to the field of chromatography. In modern societies, there probably are only few families where all members exercise the same profession. My (Fig. 5.8) family was of that kind: my father Alexander, my mother Nadezhda and my two sisters Diana and Lyudmila worked or studied chemistry at the same Mendeleev Institute of Chemical Technology in Moscow. As a child, I loved to visit my father's laboratory at the Institute and make some chemical experiments, such as adsorption of dyes on polystyrene-type ion-exchange resins. During my study at the Mendeleev Institute (1954–1957) and later at the Technische Hochschule, Dresden (1957–1962) in the former German Democratic Republic, I was thinking of developing selective ion-exchange resins which could solve most difficult separation tasks, such as separation of enantiomers of organic compounds. Having this in mind, I selected polystyrene as the topic of my diploma and examined the anionic polymerization of styrene under the supervision of Dr. Gottfried Gloeckner. I happened to be his first student to write a university thesis, and we worked in close contact with each other. Notably, both he and I kept polystyrene as the main research subject for the whole of our careers.

I returned to Moscow from Dresden in 1962, together with my wife Evtichia and a 9-month-old son Andrei (1961). Evtichia, daughter of a Greek emigrant's family, continued her study at the Lomonosov University, while I joined the Institute of Element-Organic Compounds (INEOS) of the Russian Academy of Sciences. There, I gradually moved from a position of a technician through several grades of scientific research fellow to Head of Department and Deputy Director (1998–2004) and then again to Head of Department. This Department on Stereochemistry of Adsorption Processes was opened in 1975, when it became evident that the new chromatographic technique that I suggested in 1968, namely, ligand exchange chromatography of enantiomers deserved more close attention and further development. From the Nesmeyanov Institute INEOS, I received my Ph.D. degree (1966) and Doctor of Science degree (1975) as well as the position of Full Professor (1980). For more than a decade, I served as Chairman of the Scientific Council on Chromatography of the Russian Academy of Sciences, then as Vice Chairman of the Scientific Council on Adsorption and Chromatography.

Up to the present, my wife has taught Greek and German languages at the Lomonosov University, while Andrei, in the middle of his successful career as fighter pilot, had to change his profession after the hasty disarmament of Russia and switch to construction activity. My younger son Alexander graduated from the Chemistry Department of the Lomonosov University in 1973 and is rather satisfied with his work at the Russian cosmetics company Faberlic. I am a happy grandfather of five grandchildren, ranging in age from 24 years to less than one month.

Thanks to my knowledge of German and English, as well as to interesting results of my research, I was able to participate in numerous international meetings, mostly in the area of polymer research, chromatography, ion exchange, stereochemistry, and biomedical applications of polymers. This gave me the unique possibility to travel around the world and acquire many good friends.

For my scientific activity I received several awards, the most important of which are

- 1978—M. Tswett Medal for Chromatography, USSR Acad. Sci.
- 1992—Diploma for the scientific discovery of "Participation of achiral molecular structures in the chiral recognition of enantiomers" (Russia)
- 1996—State Award of Russian Federation for Science and Technology
- 1999—"Chirality Medal 1999"—international gold medal for stereochemistry
- 2005—"Martin Gold Medal," issued by the Chromatography Society (UK) for contributions to the field of chromatography
- 2005—Distinguished Scientist of the Russian Federation

See Chapter 1, Table 1.3, c, e, b, i, q.

Chiral Ion Exchangers, Hypercrosslinked Sorbents, and General Stereochemistry. The Technical School (Technische Hochschule) of Dresden gave me a very broad chemical education, so that in my scientific career I dared to start projects in different branches of chemistry. Herewith, in all initial trials I preferred going after my intuition, rather than studying carefully at the very beginning all existing literature on the subject. Probably, this helped me to approach the goal from rather original positions, and not spend efforts on improving the existing protocols.

CHIRAL ION EXCHANGERS. Thus, not knowing of manifold previous attempts, only partially successful, to separate chromatographically enantiomeric ("left" and "right") organic molecules, I started enthusiastically synthesizing chiral ion exchangers on the basis of slightly crosslinked styrene–divinylbenzene copolymers. After chloromethylation of the latter, I enhanced the reactivity of the polymer by exchanging chlorine atoms with iodine (by simply heating the material with an acetone solution of NaI) and then allowed natural chiral amino acids to be *N*-alkylated by the iodomethyl groups. The chiral ion exchangers thus obtained were in position to partially resolve racemic amino acids into constituent enantiomers, but the enantioselectivity of the separation was far from being impressive. A real success came in 1968 when I decided to switch from the ion-exchange mechanism to a new process that was named *chiral ligand-exchange chromatography*. It was distinguished by introducing a transition metal cation, mostly copper (II), into the chromatographic system. The resin was first saturated with copper that formed a stable chelate with the carboxy and amino groups of the resin-bonded chiral selector. The copper ion, having the coordination number of 4, still exhibited the ability to additionally coordinate an amino acid molecule from the aqueous solution. In the ternary mixed-ligand complex thus formed, the two amino acids came into dense interaction with each other and easily recognized the configuration of the partner. This resulted in an unprecedented enantioselectivity of the ligand exchange process. Thus, the L-isomer of racemic proline easily eluted with water from the column packed with the L-proline incorporating polymer, while the D-isomer of proline required replacing the mixed-ligand copper complex with ammonia. Figure 5.9 presents this first example of a complete resolution of a racemate into two enantiomers by using liquid chromatography. Actually, the selectivity and loadability of the column corresponds to those of the best modern preparative chiral stationary phases. It can be mentioned here that the first complete resolution of a racemate by using capillary gas chromatography was reported by Gil-Av et al. just 2 years earlier, in 1966.

Ligand-exchange chromatography proved to be a general approach to separating enantiomers that are capable of forming chelates with transition metal ions, i.e., α- and β-amino acids, hydroxy acids, amino alcohols, diamine, but also some monodentate ligands with single functional groups [1].

In order to elucidate the mechanism of chiral recognition in the chromatographic systems, the structure and properties of chiral model copper(II) aminoacidato complexes have been examined in detail, and for the first time, thermodynamic enantioselectivity effects have been found in copper complexes with amino acids, diamine, and hydroxy-ketones, as well as in the mixed-ligand structures with these ligands. A unique example of a kinetic enantioselectivity in the formation of copper(II)–diamine complexes was

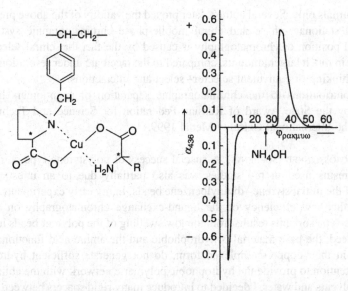

Figure 5.9. Resolution of 0.5 g of D,L-proline on a column (475 × 9 mm) containing 11 g of copper(II)-saturated polystyrene-type resin with L-proline (2.3 mmol/g) as the chiral selector. Particle size 30–50 μm; flow rate 7.5 mL/h; polarimetric detection. Abscissa, number of fractions. [*Source*: V. A. Davankov and S. V. Rogozhin, *Dokl. Akad. Nauk SSSR* **193**, 94 (1970).]

also discovered. It was further found that chiral structures, like the flat surface of a chromatographic packing or simply a solvent molecule, can actively participate in the process of chiral recognition of the enantiomers by the chiral selector. This kind of cooperation of the hydrophobic surface of a reversed-phase (RP) silica gel in the discrimination of enantiomers takes place in the ligand-exchange chromatography on chiral stationary phases prepared by a dynamic coating of a RP phase with hydrophobic derivatives of chiral amino acids. Combined with two coordination bonds to copper(II), the interaction of the solute with the achiral surface completes the number of interactions with the chiral selector to three, which are generally required for the chiral recognition of any enantiomeric pair. The dynamically coated chiral ligand-exchanging phases proved to be especially simple and efficient in the resolution of chelating racemates. The phases are still popular and available in the form of "Davankov columns" (Regis Technologies) and chiral thin of layer plates (Macherey Nagel).

An important theoretical question was raised concerning the enantiomeric purity of the chiral selector in the chromatographic system. I was first to state that chromatography permits obtaining quantitatively both enantiomers in enantiomerically pure state even in the case that the chiral selector contains admixtures of the opposite enantiomer. However, the journal *Chemical Communications* refused to publish this statement in my first publication on chiral ligand-exchange chromatography, since E. Eliel in the popular book *Stereochemistry of Carbon Compounds* stated the opposite, namely, that the insufficient enantiomeric purity of the chiral selector unavoidably results in the lower enantiomeric purity, or the yield, of the resolved enantiomers. My statement can be found in

Russian journals only. Several authors later proved the validity of the above prediction in both chiral stationary-phase and chiral mobile-phase chromatographic systems. The exceptional position of chromatography is caused by the fact that chiral selectors here are present in much larger amounts, compared to the racemate under resolution, whereas Eliel was thinking of equivalent selector–select and interactions.

My contributions to the chromatographic separation of enantiomers have been rewarded by the State Award of Russian Federation for Science and Technology in 1996 and the International Chirality Medal 1999.

HYPERCROSSLINKED SORBENTS. A special success of polystyrene-type chiral ligand-exchange resins used in my studies was also partially due to an unusual internal structure of the initial styrene–divinylbenzene beads. In my early experiments, I noticed that the rather low efficiency of the ligand-exchange chromatography on the chiral polystyrene-type sorbents results from the low swelling of the polymer beads in aqueous media. Indeed, the base material is hydrophobic and the amino acid functional groups, especially in their copper-complexed form, do not generate sufficient hydrophilicity. With an intention to provide the hydrophobic polymeric network with the ability to take up polar solvents and water, I decided to introduce many rigid spacers between the initial polystyrene chains taken in dissolved or highly swollen state. Indeed, this can be easily done by reacting polystyrene solution or swollen with ethylene dichloride styrene–divinylbenzene beads with bifunctional crosslinking agents, like p-bis(chloromethyl)benzene in the presence of a Friedel–Crafts catalyst. Actually, this basic idea contradicted the generally accepted notion that any additional crosslinks must reduce the swelling ability of any polymeric network. Still, my hope was based on the peculiarity of the new process, namely, that the many rigid spacers were introduced between the highly solvated polymer chains, which should result in the formation of an open-network rigid construction that should preserve its transparent structure even in bad solvating media, including water. Indeed, the idea worked out and resulted in obtaining polymeric networks with unusual structure and properties. They were named *hypercrosslinked networks*. Polymers of that class are distinguished by an extremely developed porous structure with the apparent inner surface area amounting up to 1000–$2000 \, m^2/g$, ability to strongly swell in any liquid media, regardless of their thermodynamic affinity to the polymer, and mechanical flexibility even at low temperatures down to $-100°C$. Hypercrosslinked polystyrene proved to present excellent adsorbing materials. They easily absorb large amounts of any organic compounds from aqueous or gaseous media [2]. Large-scale production of a whole series of the new class of neutral polymeric resins was organized in cooperation with Purolite International (Wales, UK) under the trade name "Hypersol Macronet." They have found wide application in removing and recovering organic contaminants from industrial waste streams, in decolorizing sugar syrups, removing odors from industrial exhausts, etc. They have a significant advantage over the previously used activated carbons in that they can be easily regenerated by either steam or organic solvents.

Recently, we found that the nanoporous hypercrosslinked sorbents of the NanoNet series (Purolite) effectively separate inorganic electrolytes according to the size-exclusion mechanism that capitalizes on the difference in the size of hydrated mineral ions. Any pair of electrolytes, salts, acids, bases, or any mixtures of them can be separated on a preparative scale, provided that their constituent hydrated ions differ in their diameter. In these dynamic systems, protons and hydroxyl ions do not experience exclusion effects, so that acids and bases always elute behind their salts. Since the polymers contain no functional groups and the ions are not retained by the sorbents, elution of the electrolytes with water automatically regenerates the column. Thus, the process is reagentless, generates no mineralized waste streams, and presents interest for hydrometallurgy and processing other concentrated and aggressive electrolyte solutions. Most unusual in the new process is that it results in obtaining the two separated components in concentrations significantly exceeding their concentration in the initial feed solution. This contradicts the general experience in chromatography, where solutes are retained by the stationary phase and require additional portions of the eluent to be transported through the column, which unavoidably results in the dilution of the solutes. To explain the self-concentration effect in the ion size-exclusion process, a new notion on "ideal separation process" had to be formulated, which denotes any separation process that does not introduce new matter into separated components. Being forced into fractions of the initial sample volume, the separated components experience an automatic self-concentration in such ideal separation process.

The size exclusion mechanism of the abovementioned separations was unambiguously confirmed by the spontaneous (partial) resolution of a mineral salt into constituent parent acid and base. Thus, being sent through a column with the nanoporous polystyrene, a portion of K_2SO_4 solution elutes in the form of acidic and basic zones, since the larger SO_4^{2-} anions tend to move along the column faster than do smaller K^+ cations. This remarkable effect can find an important practical application in the case of processing solutions that contain polyvalent metal ions, as, for example, a solution of titanium in concentrated mixed acids $HF + HNO_3$. Being separated from excess acids, the Ti(IV) salt hydrolyses giving an almost neutral fraction of oxo and hydroxy complexes of titanium (the latter easily precipitates as TiO_2, one target product) and an acidic fraction that contains more acids then existed in the form of free acids in the starting feed solution [3,4].

Beside the MacroNet and NanoNet resin series, smaller particles of hypercrosslinked polystyrene appeared on the market as "Purosep" (Purolite Int.), "Isolut" (IST), and "LiChrolut EN" (Merck). These are excellent adsorption materials for concentrating trace organic compounds from complex matrixes by the solid-phase extraction (SPE) technique. They are distinguished by an extremely high adsorption potential, combined with easy and complete recovery of adsorbed and thus concentrated target analytes. Importantly, these resins can be successfully employed not only for the determination of nonpolar analytes in aqueous or other polar media, where conventional SPE materials of the RP type can also be used but also for the preconcentration of polar analytes from both aqueous and nonpolar media, as well. Thus, polar cellulose degradation products in transformer oils or polyaromatic hydrocarbons in sunflower oils can be determined after their solid-phase extraction on hypercrosslinked polystyrene.

Another potential field of using hypercrosslinked polystyrene in analytical chemistry is high-performance liquid chromatography, where packing the HPLC columns with monosized microbeads of the polymer offers the possibility of using any mobile phase and thus operating the same column in many different types of chromatography procedures, such as reversed-phase or quasi-normal-phase modes or size exclusion. This is possible due to chemical resistance of hypercrosslinked polystyrene and its compatibility with any liquid media. Especially attractive is the unique selectivity of the material, caused by the tendency of aromatic and polar analytes to form $\pi-\pi$ and $n-\pi$ interactions with the π-electron systems of exposed aromatic rings of the packing. Although the HPLC-type microbeads are still under development, their potentials have been convincingly shown in a series of publications.

The principle of obtaining hypercrosslinked networks by the above postcrosslinking of solvated polymeric chains can be also carried out within individual polystyrene coils in a dilute polystyrene solution. The process results in compacting the coils into rigid "nanosponges" of about 15 nm in diameter at a molecular weight of about 330 kDa. These rigid spherical macromolecular species differ fundamentally from known globular macromolecules, in particular, globular proteins, by their hypercrosslinked open-network structure. The nanosponges, when dissolved in typical polystyrene solvents, display extremely low viscosity, but high diffusion and sedimentation coefficients. Most interesting is that the monodisperse nanosponges self-assemble in solution to regular clusters of several generations. The cluster of the first generation consists of 12 nanosponges, assembled around one central, 13th species. The second shell of the next generation cluster incorporates 42 additional species. To the best of my knowledge, there exist no other examples of formation of regular spherical clusters by self-assembling macromolecules of a homopolymer.

Experience collected in synthesizing various porous polymers proved to be extremely helpful in solving an important biomedical problem, namely, developing a hemosorbent capable of selectively sorbing the so-called middle molecular toxic proteins. One of them, β_2-microglobulin causes major problems to patients with renal failure. The protein is too large, about 12 kDa, to be effectively removed through the membrane during the conventional hemodialysis procedure. The toxin builds up in all physiological fluids and deposits in joints and blood vessels, causing painful inflammations. We were able to suggest a poly(styrene–co-divinylbenzene) sorbent that efficiently removed this toxin from the whole blood with only a minimal removal of albumin, which molecule is larger in diameter by a factor of only 1.5. The high selectivity is due to the size-exclusion phenomenon, combined with the higher hydrophobicity of β_2-microglobulin. The material was provided with good hemocompatibility by very small amounts of vinylpyrrolidone grafted on the outer surface of the beads. The material successfully passed clinical trials in Italy and in the USA on more than 100 patients, showing an average removal of over 400 mg β_2-microglobulin per hemodialysis/hemosorption session. The polymer could also be used in treating sepsis, acute intoxications, as well as during cardiopulmonary surgery, organ transplantations, etc. It is a pity, that practical implementation of perspective medical technologies is strongly retarded by many legal manufacturing and approval issues.

GENERAL STEREOCHEMISTRY. While my activity in polymer chemistry developed parallel to the chromatographic separations of enantiomers, the latter slowly evolved into routine determination of enantiomeric excess (ee) of many organic and metal–organic compounds. Command of this technique gave us the unique (for Russia) possibility to start intensive research in the field of enantioselective catalysis. We concentrated on developing new phosphite- and phosphamide-type chiral ligands, which often proved to be even more effective than the well known phosphine-type ligands. In addition, these compounds are much more resistant to air and water and are easily accessible. In rhodium- and palladium-catalyzed allylic substitution and hydrogenation reactions, enantioselectivity exceeding 99% and fast and complete conversion of initial substartes have become a frequent result achieved with the new ligands, especially when ionic liquids or supercritical carbon dioxide are employed as reaction medium.

My interest in general stereochemistry extended beyond separating enantiomers and synthesizing useful chiral compounds to the problem of the origin of homochirality of bio-molecules. I made a bold statement that, because of parity violation of weak interactions, all primary species of our matter, such as electrons, protons, and neutrons, build up a homochiral pool, while all photons and other electromagnetic energy quantums are also chiral, but exist in both enantiomeric forms. This leads to the inherent homochirality of all atoms, as well as a slight energetic nonequivalence of conventional enantiomeric mol-ecules. From this perspective, any compound must have a true enantiomer buildup of anti-matter species. Thus, the true enantiomer of Na^+Cl^- is a salt Na^-Cl^+ whose atoms are composed of antiprotons, antineutrons, and positrons. In cooperation with specialists in physics and astrophysics, I plan to conduct experiments on self-assembling complex molecules from primary particles, followed by testing their possible enantiomeric enrich-ment by using both chiral chromatography and the catalytic asymmetry induction tech-nique [5].

Finally, I would like to mention the story with the discussion on the physical mean-ing of the James and Martin's gas compressibility correction factor, which I started in *Chromatographia* in 1996. At that time, I was asked by Russian colleagues to prepare a *Compendium of Terms and Concepts in Chromatography*. As usual, I started formulat-ing definitions from my own understanding of their physical sense, rather than taking them from textbooks. However, my colleagues pointed out that my formulation of such a fundamental term as corrected retention volume in gas chromatography contradicts the generally accepted formula. I finally discovered that Martin and James presented in 1952 a correct equation for calculating this parameter, but did not explain the physical sense of the key component of these calculation, coefficient j_3^2. The name of this coeffi-cient, the *compressibility correction factor*, does not reveal its meaning, either. With the thermodynamic sense of the corrected retention volumes remaining obscure, Littlewood suggested reporting these values at normal temperature, by multiplying with the tempera-ture ratio $273/T_c$, where T_c is column temperature. For over 50 years this suggestion was recommended by all textbooks on chromatography, and no one objected to the obvious nonsense of this last operation; by going down from the column temperature T_c to 273 K, the retention volume of any analyte must significantly increase, whereas multiplication with $273/T_c$ diminishes this value! By deriving the j factor in a more logical way, I was able to show that it is merely the ratio of gas ambient pressure at the column exit to the pressure inside the column averaged over the length of the column. From this, it becomes evident that Martin and James suggested the formula for calculating the volume of the gas mobile phase under real chromatographic conditions, i.e., under column temperature and averaged pressure in the column, which makes the thermodyn-amic sense of the corrected retention volume crystal clear. Recalculation of this value to standard temperature by using a totally inappropriate (for the thermodynamic retention parameter) coefficient, $273/T_c$, was fundamentally wrong. This longstanding mistake tre-mendously delayed the use of chromatography in physicochemical studies. Although IUPAC finally eliminated the error from its official documents, the wrong tradition will stay in gas chromatography textbooks for many years to come, unfortunately [6].

The Chromatography Discussion Group awarded me the A. J. P. Martin Gold Medal in 2005. Beside my achievements in chiral ligand exchange chromatography and hyper-crosslinked polystyrene resins, the above rediscovery of the thermodynamic sense of the Martin corrected retention volume was also a reason for the high award.

V. A. Davankov—Perspectives. In the twenty-first century, I believe, chromatography will strengthen its positions as a preparative and industrial method of obtaining valuable compounds, new continuous procedures with minimized consumption of reagents and solvents will be developed, that meet requirements of "green" technology. Here, size exclusion chromatography can show it's thus far unrecognized potentials. Analytical applications will further develop toward miniaturization and automation.

REFERENCES

1. V. A. Davankov, J. D. Navratil, and H. F. Walton, *Ligand Exchange Chromatography*, CRC Press, Boca Raton, FL, 1988.
2. M. P. Tsyurupa and V. A. Davankov, Porous structure of hypercrosslinked polystyrene: State-of-the-art mini-review, *React. Funct. Polym.* **66**(7), 768–779 (2006).
3. V. A. Davankov, M. P. Tsyurupa, and N. N. Alexienko, Selectivity in preparative separations of inorganic electrolytes by size-exclusion chromatography on hypercrosslinked polystyrene and microporous carbons, *J. Chromatogr. A* **1100**(1), 32–39 (2005).
4. V. A. Davankov and M. P. Tsyurupa, Chromatographic resolution of a salt into its parent acid and base constituents, *J. Chromatogr. A* **1136**, 118–122 (2006).
5. V. A. Davankov, Chirality as an inherent general property of matter, *Chirality* **18**(7), 459–461 (2006).
6. V. A. Davankov, Critical reconsideration of the physical meaning and the use of fundamental retention parameters in gas chromatography. New IUPAC recommendations, *Chromatographia Suppl.* **57**, 195–198 (2003).

PAUL R. HADDAD

Figure 5.10. Paul R. Haddad.

Paul R. Haddad (Fig. 5.10) obtained the degrees of B.Sc., Ph.D., and D.Sc. from the University of New South Wales, Sydney, Australia, after graduating from the Royal Military College Duntroon in 1969. The first half of his academic career was spent at the Australian National University (Canberra) and the University of New South Wales. He then moved to the University of Tasmania (Hobart) in 1992 as Professor of Chemistry, and served as Dean of the Faculty of Science and Engineering for 8 years, before being awarded an Australian Research Council (ARC) Professorial Fellowship in 2003 and an ARC Federation Fellowship in 2006. He is

Director of the Australian Centre for Research on Separation Science and Director of the Pfizer Analytical Research Centre.

He is a Fellow of the Australian Academy of Science, a Fellow of the Australian Academy of Technological Sciences and Engineering, a Fellow of the Royal Australian Chemical Institute, a Fellow of the Royal Society of Chemistry, and a Fellow of the Federation of Asian Chemical Societies.

Paul Haddad's research interests cover a broad area of separation science, with particular emphasis on the separation of inorganic species. His research covers the fundamentals and applications of liquid chromatography, ion chromatography, capillary electrophoresis, capillary electrochromatography, and chemometrics. He is the author or co-author of over 420 scientific publications and patents, including two books, and he has presented in excess of 350 papers at local and international scientific meetings.

He is an editor of the *Journal of Chromatography A*, an Associate Editor of *Encyclopedia of Analytical Chemistry*, and a contributing editor for *Trends in Analytical Chemistry*. He recently completed a 6-year term as Editor of *Analytica Chimica Acta*. He is currently a member of the editorial boards of 10 other journals of analytical chemistry or separation science. Paul Haddad has been the recipient of several awards, including the HG Smith medal from the Royal Australian Chemical Institute, the Federation of Asian Chemical Societies Foundation Lectureship Award (2001), the Royal Society of Chemistry Australasian Lectureship for 2001, the 2002 A. J. P. Martin Gold Medal from the Chromatographic Society, and the 2003 Royal Society of Chemistry Award for Analytical Separation Methods.

See Chapter 1, Table 1.3, a, h, e, m.

Fundamentals on Chromatographic Separation of Inorganic Ions

OVERVIEW OF RESEARCH INTERESTS. Research has been conducted in the general field of separation science in liquid phases, with particular emphasis on the separation and quantification of ionic species. Studies have been undertaken using high-performance liquid chromatography (HPLC), ion chromatography (IC), capillary electrophoresis (CE), and capillary electrochromatography (CEC). With each technique, the general aim of the research has been to study separation mechanisms and methods of detection, with a view to improving fundamental understanding of these aspects and to apply this understanding to the development of new chromatographic and electrophoretic methods of analysis. The general approach taken in the study of separation mechanisms has been to make detailed measurements of the retention or migration of analytes in the desired system, to derive mathematical models that describe these observations, and to use the mathematical models to devise strategies for the computer-assisted optimization of separations for particular applications. A recurring theme has been the investigation and manipulation of separation selectivity. In the study of detection methods, emphasis has been placed on potentiometric detection using reactive metallic electrodes and the theory and application of indirect methods of spectrophotometric detection. The separation of complexed metal ions and sample handling methods have comprised further major research themes.

Major Research Achievements

MODELING AND OPTIMIZATION OF SEPARATION MECHANISM. A substantial research program has been directed toward the study of the fundamentals of separation mechanisms in various chromatographic and electrophoretic systems and the use of these mechanistic models for computer-assisted selection of optimal separation conditions. These studies have been applied in HPLC, CE, ion chromatography (IC) (including ion-exclusion chromatography), and CEC and have led to new approaches for selection of optimal separation conditions. Important achievements include the development of new software, which enables rapid and accurate simulation and optimization of IC separations in isocratic and gradient elution conditions, including elution profiles comprising individual isocratic and gradient segments, using only isocratic retention data as input [1].

DEVELOPMENT OF DETECTION METHODS IN IC AND CE. A large body of work has been undertaken on the theory and application of a unique potentiometric detector based on the use of a metallic copper sensing electrode. Another major field of study has been the theory of indirect spectrophotometric detection, especially in CE, which extends detection to analytes which are not UV-absorbing. Significant achievements in these areas are the design and elucidation of electrode response theory for a metallic copper potentiometric detector for IC and CE, elucidation of the source of "system peaks" in indirect detection in IC and CE, derivation of fundamental theory of response of indirect detection in CE, and recognition of the key role of the transfer ratio, and use of dyes as probes for very sensitive detection in CE leading to the achievement of the lowest reported detection limits for inorganic anions.

CHROMATOGRAPHY OF METAL COMPLEXES. Fundamental and applied research on the separation of metal complexes has been undertaken. Emphasis has been on the separation of dithiocarbamates and 8-hydroxyquinolinates, metallocyanides, Th(IV), U(VI), Nb(V), Ta(V), and V(V), the use of chelation IC, and separation of anions and metal ions as their complexes using CE. Major achievements include the first reported separation of all stable metallocyanide complexes, thiocyanate and free cyanide. This has been extended to include the simultaneous determination of free cyanide, cyanate, thiosulfate, and the Cu:CN ratio [2], leading to a commercial cyanide analyser for use on gold mines for the complete speciation of cyanide and its degradation products in process liquors and tailings dams.

CAPILLARY ELECTROCHROMATOGRAPHY (CEC). An intensive research program in CEC using ion-exchange stationary phases has been directed toward the design of separation systems showing controllable separation selectivity and enhanced sensitivity. Major achievements include development of a new wall-coated CEC system based on electrostatic binding of nanometre sized functionalised latex particles to the silica wall, development of an online sample preconcentration method for CEC using a latex-coated precapillary and a transient isotachophoretic gradient [3], and theory and applications of pseudo-stationary-phase CEC systems having variable and predictable selectivity. The selectivity of each of these systems can be varied between the extremes of ion-exchange and electrophoresis; detection limits are more than 1000 times lower than those obtained by other workers.

ELECTROSTATIC ION CHROMATOGRAPHY (EIC). In collaboration with Dr. Wenzhi Hu, the fundamentals and applications of the new chromatographic technique of electrostatic ion chromatography (EIC) have been studied. In this technique the stationary phase comprises zwitterionic functionalities and separation of analyte ions occurs via simultaneous electrostatic attraction and repulsion effects. Major achievements are large improvements in sensitivity of EIC through the use of water or dilute electrolytes compatible with IC suppressors as eluents. Haddad demonstrated the utility of EIC for the determination of inorganic and organic ions in sample matrices of high ionic strength, and elucidated a new mechanism for EIC and established the fundamental theory.

MONOLITHIC STATIONARY PHASES FOR ION EXCHANGE. More recent studies have centered around the development of new ion-exchange stationary phases based on the use of a monolithic polymeric scaffold, onto which is coated a layer of nanometer sized functionalized ion-exchange particles [4]. The resultant material combines the desirable flow properties exhibited by monolithic materials with high separation efficiency generated by the excellent mass transfer characteristics of nanoparticles. These materials have been patented and are being utilized extensively in a range of research projects, including capillary IC separations, leading to portable instrumentation, high speed ion-exchange separations in which the separation is performed in 1 minute or less, and to counterterrorism applications for the identification and detection of improvised explosives used commonly in terrorist attacks. This work involves postblast identification of explosives using a fingerprint analysis of inorganic residues (both anions and cations), as well as preblast detection of improvised explosives for airport screening implementation.

REFERENCE TEXTS. A major reference text, *Ion Chromatography: Principles and Applications*, co-authored with P. E. Jackson, has been published by Elsevier Scientific Publishers. The book was the most comprehensive treatment of IC yet published and was awarded the 1991 Royal Australian Chemical Istitute Archibald Olle Prize for the best work published by a member of the Institute and the top HPLC/IC product award for 1991 judged by readers of *Laboratory Equipment*. The book has also been included in the *Essential Reference Books* compilation published by the journal *LC-GC*, in which only the most important reference works in each field of chromatography are listed. Many parts of this text constitute original research results and have not been published elsewhere. The text was reprinted in 1993.

A successful undergraduate textbook, *Principles and Practice of Modern Chromatographic Methods*, has also been published (by Academic Press) in collaboration with K. Robards and P. E. Jackson. This text covers all forms of chromatography and presents a unified approach to different chromatographic techniques.

Establishment of the Australian Centre for Research on Separation Science (ACROSS). A lead role has been taken in creating a coordinated approach to research in separation science in Australia by the establishment of a cross-institutional research center, called *The Australian Centre for Research On Separation Science (ACROSS)*. This center comprises the separation science research groups at the University of Tasmania, RMIT University, and University of Western Sydney (Professor Philip Marriott). These groups are well-established and highly performing in both the national and international contexts, and the proposed center brings the groups into close, formal collaboration. The major objective of ACROSS is to enhance fundamental (discovery-based) and applied (linkage-based) research in separation science in Australia by the creation of an organized, coordinated structure in which research is focused into defined programs.

Perspective. The field of separation science of inorganic species continues to develop at a rapid pace. Unlike the broader field of HPLC, where the newest developments are based primarily in the use of small-diameter particles monolithic materials as stationary phases, and miniaturization on microchips, developments in ion chromatography fall more in the areas of capillary separations, software for rapid optimization of complex eluent compositions, and faster separations. Ion chromatography is also expanding into more diverse and challenging applications. Electrodriven separations of inorganic species have hitherto been limited by deficiencies in detection sensitivity, but these deficiencies are now being addressed through new approaches to online sample enrichment and also new modes of detection. Of particular relevance is contactless conductivity detection, which will soon become the dominant detection method for inorganic ions. Finally, both ion chromatography and capillary electrophoresis are filling important roles in counterterrorism research as effective means of preblast and postblast detection of improvised explosive devices used widely in terrorist attacks.

REFERENCES

1. R. A. Shellie, B. K. Ng, G. W. Dicinoski, S. D. H. Poynter, J. W. O'Reilly, C. A. Pohl, and P. R Haddad, Prediction of analyte retention for ion chromatography separations performed using elution profiles comprising multiple isocratic and gradient steps, *Anal. Chem.* **80**, 2474–2482 (2008).
2. P. A. Fagan, P. R. Haddad, I. Mitchell, and R. Dunne. Monitoring the cyanidation of gold-copper ores with ion chromatography: Determination of the CN:Cu mole ratio, *J. Chromatogr. A* **804**, 249–264 (1998).
3. M. C. Breadmore, M. Macka, N. Avdalovic, and P. R. Haddad, On-capillary ion-exchange preconcentration of inorganic anions in open-tubular capillary electrochromatography with elution using transient-isotachophoretic gradients. II. Characterisation of the isotachophoretic gradient, *Anal. Chem.* **73**, 820–828 (2001).
4. J. P. Hutchinson, P. Zakaria, M. Macka, N. Avdalovic, and P. R. Haddad, Latex-coated polymeric monolithic ion-exchange stationary phases. I. Anion-exchange capillary electrochromatography and in-line sample preconcentration in capillary electrophoresis, *Anal. Chem.* **77**, 407–416 (2005).

WILLIAM S. HANCOCK

Figure 5.11. William S. Hancock.

William S. Hancock (Fig. 5.11) is Professor of Chemistry and holds the Bradstreet Chair in Bioanalytical Chemistry within the Barnett Institute of Chemical and Biological Analysis at Northeastern University, Boston, MA. Dr. Hancock received his B.Sc. in Chemistry and Biochemistry in 1966, his Ph.D. in Chemistry in 1970, and a D.Sc. in 1993 from the University of Adelaide, South Australia. He began his career in biochemistry as a postdoctoral fellow at Washington University. He then served on the academic staff at Massey University, New Zealand and was a Visiting Scientist at the Bureau of Drugs of the Food & Drug Administration in Washington, DC. In 1985, he joined the Medicinal and Analytical Chemistry unit of Genentech in San Francisco, CA, where he rose to the position of Staff Scientist. He has held a Principal Scientist position at Hewlett-Packard Laboratories and a Visiting Professorship in Chemical Engineering at Yale University. Prior to joining Northeastern University in 2002, he had served as Vice President of Proteomics at Thermo Finnigan Corp. in San Jose, CA. Dr. Hancock served as President of the California Separation Science Society (CASSS), based in San Francisco, CA for 10 years (1994–2004). He is the Editor-in-Chief of the new ACS publication *Journal of Proteomic Research*. He has published over 190 scientific publications, including seven books, holds 15 patents, and has organized and lectured in numerous international meetings. He has received many honors, including the Martin Gold Medal in Separation Science (British Chromatographic Society, 2000), Stephen Dal Nogare Memorial Award in Chromatography (2003), and the ACS Award in Chromatography (2003). Dr. Hancock's research program has focused on the application of separation science to biological problems. His work has advanced the state of the art of HPLC as a required tool in the analysis of protein therapeutics. At the Barnett Institute, his research focuses on cancer proteomics and the plasma proteome and is a reference laboratory for HUPO and a co-chair of the National Cancer Institute (NCI) Alliance of Glycan markers for the early detection of cancer. Dr. Hancock is currently serving as President of US-HUPO (human proteome organization) and also serves on the International Council.

See Chapter 1, Table 1.3, h, t, y, z.

HPLC Contributions. Dr. Hancock's work in the field of HPLC began in 1975 with his first publication of analysis of proteins by HPLC and then in 1978 with the first study on peptide mapping. Subsequently these methods, in conjunction with those developed by other research groups, became the gold standard for the industry for NDA, enabling protein characterization studies. At a consensus forming meeting with industry scientists in 1983, he advised the FDA on the process for using HPLC as a major analytical method for approval of the first biotechnology product, insulin, and then in 1984 as a visiting scientist, he worked with the agency for the development of testing methods for rDNA produced insulin.

Dr. Hancock's work in protein separations by HPLC developed from his postdoctoral studies on the chemical synthesis of acyl carrier protein. In 1975, continuing his research on improved peptide synthesis and monitoring methods, Dr. Hancock published his first studies on the reversed-phase HPLC of peptides and proteins and then peptide mapping in 1978, in collaboration with Milton Hearn. At that time columns for liquid chromatography did not produce reasonable peak shapes, retention times or recoveries for the chromatography of small peptides (let alone proteins), due to silanol group interactions. However, the work at Massey University in New Zealand discovered an analytical system that included the addition of phosphoric acid to the mobile phase and resulted in excellent chromatography of peptides and even proteins. The work then explored the role of both hydrophilic and hydrophobic ion-pairing agents in polypeptide analysis. These studies were extended to cationic as well as anionic ion-pairing reagents, e.g., alkanoic acids such as trifluoroacetic acid. The reagents had the advantage of volatility, and selectivity differences from other ion-pairing reagents and have had broad use today by chromatographers that are separating peptides and proteins.

Dr. Hancock then joined Genentech to lead the development of HPLC assays for several biotechnology products. His research group elucidated the degradation pathways for protein pharmaceuticals and prepared the characterization studies required for FDA approval of growth hormone, tissue plasminogen activator and γ-interferon. Subsequently, assays were also developed to aid manufacturing (fermentation and process recovery), formulation development, product stability, and quality control. Other applications include elucidation of the degradation pathways for protein pharmaceuticals, assays for fermentation control, process recovery, formulation development, product stability, and quality control. Dr. Hancock's group was among the first to publish innovative studies on the application of capillary electrophoresis, fast-atom bombardment (FAB) and electrospray mass spectrometry, matrix-assisted laser desorption time-of-flight mass spectrometry (MALDI/TOF), hydrophobic interaction chromatography (HIC), displacement chromatography, NMR, and low-angle laser light scattering (LALLS). Of particular interest was the use of LC hyphenated with mass spectrometry for the characterization of complex biotechnology drugs, such as the glycoprotein, tissue plasminogen activator, and a humanized monoclonal antibody.

Dr. Hancock then moved to the central research laboratories of Hewlett-Packard and applied new analytical instrumentation to the analysis of complex biological mixtures. Highlights of this work include the development of multidimensional analyses for glycoproteins, capillary electrochromatography analysis of clinical samples and HPLC-MS methods for DNA mutation analysis. In a project analogous to the early days of biotechnology, Dr. Hancock has collaborated with Erno Pungor at Berlex Bioscience to develop analytical protocols for adenoviral vectors for gene therapy. In this case, the drug substance is a particle of mass 300 million daltons and requires integrated analysis of both the genome and the corresponding protein products. In the late 1990s a new opportunity for HPLC was in the emerging field of proteomics, Dr. Hancock at Hewlett-Packard and then Thermo Electron developed with his coworkers improved HPLC separations coupled to mass spectrometry, such as 2D HPLC with ion trap mass spectrometry. This instrumental platform has been used in many proteomic laboratories as the improved separation enabled more in-depth analyses of complex human tissue and blood samples.

With the rapid transition of proteomics into the clinical arena and the development of patient-focused drug development processes in the pharmaceutical industry, Dr. Hancock decided to return to the academic arena at the Barnett Institute, Northeastern University, Boston. The focus of his new laboratory is in the area of clinical proteomics and has involved extensive collaborations with members of the Human Proteome Organization (HUPO). Important aspects of this work on patient plasma samples have involved affinity chromatographic separations to remove abundant plasma portions with antibody columns as well as fractionation of glycoproteins with mixtures of lectins (sugar-binding proteins). The fractionated proteins are then digested with trypsin and separated by 1D HPLC and identified by electrospray tandem mass spectrometry. With this powerful technology it has been possible to identify many hundred glycoproteins per clinical study and has been applied to cancer (breast, colon, kidney, lung), autoimmune disease (psoriasis, rheumatoid arthritis and multiple sclerosis), and diabetes. More recently, the NCI has funded a new alliance to study glycans as candidate markers for the early detection of cancer; these HPLC approaches are an important part of the initiative.

Perspectives. The area of HPLC is as fresh as the mid-1970s, when availability of (then) modern instrumentation and packing materials enabled breakthrough studies of peptide and protein separations. At the time, most HPLC applications were an extension of earlier studies from GC and involved small-molecule analysis. The study of polar, high-MW separations was, however, an area awaiting a powerful new approach. Furthermore, this area of analysis has continued to grow from these early applications in protein biochemistry, to the emerging biotechnology industry, and then to clinical proteomics. In the initial transition of GC into HPLC, one often found that separations theory migrated in a reasonably linear manner from the gas to liquid phases. The challenge was different in the world of biopolymer analysis! For example, the effect of temperature on retention of protein molecules in reversed phase LC is driven by conformational changes in the 3D structure. In the area of clinical proteomics, the analysis of the plasma proteome is a huge challenge because of the complexity of the sample as well as the need to detect the presence of low-level disease biomarkers. Not only is HPLC married to high-information-content detectors such as mass spectrometry, but sample preparation is as important as in an earlier era of developing sensitive environmental applications. In conclusion, HPLC of polypeptides continues to be an important field with fresh applications and challenges for those fortunate enough to be working in the area.

REFERENCES

1. W. S. Hancock, C. A. Bishop, R. L. Prestidge, D. R. K. Harding, and M. T. W. Hearn, Rapid analysis and purification of peptides and proteins. The use of reversed phase high pressure liquid chromatography with ion pairing reagents, *Science* **20**, 1168–1170 (1978).

2. W. S. Hancock, C. A. Bishop, and M. T. W. Hearn, The use of pressure liquid chromatography for the peptide mapping of proteins, *Anal Biochem.* **89**, 203–212 (1978).

3. W. S. Hancock (Ed.), *CRC Handbook of HPLC for the Separation of Amino Acids, Peptides and Proteins*, CRC Press, Boca Raton, FL, Vols. I and II, 1984.

4. V. Ling, A. Guzzetta, E. Canova-Davis, J. T. Stults, W. S. Hancock, T. R. Covey, and B. Sushan, Characterization of the tryptic map of recombinant DNA derived tissue plasminogen activator by high-performance liquid chromatography-electrospray ionisation mass spectrometry, *Anal. Chem.* **63**, 2909–2915 (1991).

5. Z. Yang, L. E. Harris, D. E. Palmer-Toy, and W. S. Hancock, Characterization of multiple glycoprotein biomarker candidates in serum from breast cancer patients using multi-lectin affinity chromatography (M-LAC), *Clin. Chem.* **52**, 1897–1905 (2006).

MARIE-CLAIRE HENNION

Marie-Claire Hennion (Fig 5.12) is professor of Analytical Sciences at ESPCI in Paris. Born in 1950, French nationality, I was student at Ecole Normale Supérieure in 1971 and obtained an M.S. in Physical Chemistry in 1973 (Paris University). In 1975, I was recruited by the Institute of Physics and Chemistry of Paris (ESPCI) in the laboratory of analytical chemistry for teaching and research. My research area was then liquid chromatography (LC) with the synthesis of chemically bonded *n*-alkylsilicas and their retention mechanisms. I received my Ph.D. degree in 1982 from the University of Paris. I followed my university career and was nominated as Professor in 1997, with teaching in both analytical sciences and environmental sciences.

Figure 5.12. Marie-Claire Hennion.

I have been heading the Environment and Analytical Chemistry Laboratory since 1997 (with 20–25 permanent staff in research). Actually, I am part of several direction committees at the French National Research Centre, Paris University, and ESPCI.

A large part of my research work has been mainly devoted to liquid chromatography with emphasis on retention mechanisms, especially with the first evidence of high retention provided by porous graphitic carbon for very polar analytes. During 1985–1995, many studies were dedicated to trace analysis using solid-phase extraction with emphasis on the extraction of very polar and water-soluble trace compounds and the introduction of more selective sorbents such as immunosorbents. Actual trends are miniaturization in LC and totally automated analytical systems using hyphenated techniques, including bioanalytical concepts.

My research has been attested to up to now by the publication of more than 150 reviews and original papers in international journals, three books, 13 book chapters, 80 invited lectures in international symposia (including 20 plenary ones), and 220 oral or poster lectures. I received the Silver Jubilee Award in Chromatography in 2001.

I have chaired several symposia in environmental analytical chemistry and chromatography, among them the recent 25th International Symposium on Chromatography in Paris, in 2004 with 800 participants.

My research activity has been devoted mainly to separation sciences, especially chromatography, and the development of new methods for environmental, pharmaceutical, and biological areas.

See Chapter 1, Table 1.3, d, e, h, v, w.

Chromatography Research in Development of New Methods for Environmental, Pharmaceutical, and Biological Separation Mechanisms: n-alkylsilicas, styrene divinylbenzene copolymers, and Porous Graphitic Carbons. I started with studies in reversed-phase chromatography in 1975. At that time n-octadecyl silica columns were just appearing on the market and it was impossible to predict retention from analyte structure and alkylsilica characteristics. My work started with the synthesis of several chemically bonded silicas with various lengths of the n-alkyl moieties and various bonding densities or surface coverages. The results were that both chain length and density were the main parameters governing the chromatographic retention and selectivity and that these parameters could be represented by the hydrocarbonaceous surface of the alkyl chains. The highest this surface, the higher the selectivity was. A high surface could theoretically be obtained by increasing both density and chain length, but there was a limit because the maximum surface coverage was obtained with n-octadecyl chains. For longer chains there was a sharp decrease in surface coverage due to the diffusion process within the pores during the synthesis. Studies on the effect of the mobile phases have shown a close analogy between the analyte solubility in various organic–water solvents and analyte retention. Therefore the theory of an enriched layer of organic solvent at the chain surface in equilibrium with organic–water mobile phase was confirmed. On analyte injection this equilibrium was disrupted and accompanied by a displacement of organic solvent molecules from the enriched layer at the alkyl chain surface toward the mobile phase. The variation of the refractive index of methanol–water mixtures allowed identification of the first so-called injection peak or system peak as a peak that corresponded to a band containing a higher organic solvent concentration than the mobile phase. From the retention measurement of these bands, it was possible to calculate adsorption isotherms of the organic solvent at the chain surface and to calculate the void volume. It was then possible to predict retention factors from solubility parameters of analytes, water, and organic solvent depending on the mobile-phase composition.

In the mid-1980s, I collaborated in the field of solid-phase extraction with the late Prof. Roland Frei. One day he gave me a column and a few grams of graphitized porous carbon (PGC), asking me to test this new sorbent that was given to him by his friend Prof. John Knox. At that time, PGC was not yet commercialized, and initial studies showed that retention of apolar analytes was much higher than that provided by n-alkylsilicas and that its flat structure induced some specific interactions with some aromatic isomers. PGC was intended to be an ideal reversed-phase sorbent without all the drawbacks of residual silanols of C18 silicas. Surprisingly, our first experiments indicated a strong retention for some polar analytes and a retention order depending on analyte polarity that was opposite to that obtained with conventional RP sorbents. As an example, the retention factor of a very polar analyte such as phloroglucinol (1,3,5-trihydroxyben-zene) was 1000 with water as mobile phase on PGC, whereas it was 3 on PRP-1 and not at all retained on C18 silica. Extensive experiments then consisted in comparing retention properties obtained for several analytes (mainly mono-, di-, and trisubstituted aromatic compounds) using different reversed-phase packings: conventional n-akylsilicas, styrene divinylbenzene (SDB) PRP-1 copolymer, and PGC. I was among the first chromato-graphists to explain the extraordinary potential of PGC for retaining very polar analytes. That was at the origin of numerous wonderful and fruitful discussions with Prof. John Knox. Several Ph.D. candidates under my direction worked on that topic during the early 1990s; we have shown that if retention of apolar analytes was due essentially to hydrophobic interactions, that of polar analytes could be explained by the PGC-specific structure. The large bands of delocalized π electrons were able to generate dipole-induced-dipole interactions and consequently, the retentions of polar analytes could be related to the local dipoles of the analyte molecule. Because of the strong polarizability of PGC that was perpendicular to the bands, some polar planar analytes could strongly interact, and the higher the number of polar groups they possessed, the greater was their retention. Prediction could be done from these polar dipoles. We also showed that the eluotropic series of organic solvent was different from that obtained with classical revered-phase sorbents.

Analytical Extraction for Trace Analysis. Trace analysis of organic compounds from real samples requires an extraction–enrichment step. Extensive theoretical and experimental works have been achieved in the area of solid-phase extraction (SPE) with emphasis to the extraction of polar analytes from aqueous samples. SPE was mod-elled from both frontal and classical elution chromatography. It was then possible to pre-dict the SPE parameters from chromatographic and kinetics parameters. In the early 1990s, our joint work with the group at the Free University of Amsterdam (Roland Frei and Udo Brinkman) expressed the need for more retentive extraction sorbents that the existing SDB sorbents and were at the origin of the commercialization of new copo-lymers with high specific areas that appeared only in mid-1990s.

Several works were made for automated systems by online coupling SPE to LC with emphasis on quantitative analysis and to the compatibility between the extraction sorbent and the analytical columns. Using PGC in the extraction precolumn and in the analytical columns allowed us to extract very polar compounds from water. Original applications allowed the first evidence of very polar metabolites of pesticides that have leached in groundwater after degradation in soil.

My really relevant contribution to trace analysis was in the introduction of more selective sorbents in the extraction step using antibody-antigen interactions specifically designed for small molecules in the early 1990s. Immunoaffinity chromatography was at this period devoted only to protein purification, and I remember that my first paper was refused in a high-impact journal because it was not fair to elute analytes that had interacted with antibodies using mixtures with high content of an organic solvent. Theoretical studies have been done in order to optimize the extraction parameters in

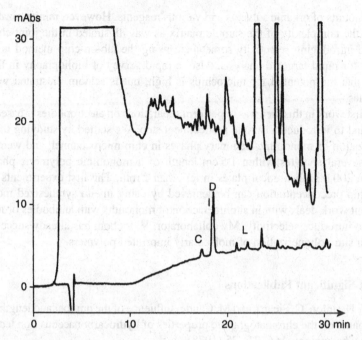

<u>Figure 5.13.</u> Online extraction and LC-UV analysis of 5 mL of River Seine sample (Paris city) using a C18 analytical column (250 × 1 mm i.d., flow rate 50 µL/min, water–acetonitrile gradient), and precolumns: on top a 5 × 1-mm-i.d. precolumn packed with C_{18} silica and on bottom a 10 × 1-mm-i.d. precolumm packed with an immunosorbant antiphenylurea. UV detection shown at 220 nm. The following pesticides were detected and confirmed by the DAD detection: C = Chlortoluron, 50 ng/L; I = Isoproturon, 220 ng/L; D = Diuron, 200 ng/L; L = Linuron, 40 ng/L. [*Source:* E. Schoenzeter, V. Pichon, D. Thiebaut, A. Fernadez-Alba, and M. C. Hennion, Rapid sample handling in microliquid chromatography using selective on-line immunoaffinity extraction, *J. Microcol. Separations* 12, 316–322 (2000).]

terms of extraction capacity and consequences for the quantitative results. Several studies have shown the potential of these immunosorbents for analysis of pesticides in water or soil, contaminants such as amines in industrial wastes, and toxins produced by microalgae.

Using silica as the solid matrix for the chemical bonding of antibodies, it was possible to optimize a microscale LC online system, as shown in the lower chromatogram in Figure 5.13. It illustrates how selective interactions are a key point to obtain rapid analyses from only 5 mL of sample and the comparison with use of a nonselective sorbent shows the complexity of the real sample (Fig. 5.13).

Miniaturization. Today there are increasing demands from society for very rapid and low-cost analyses that are possible with very low amounts of sample, easy to use in the field or in vivo and environmentally friendly. Miniaturization of the whole analytical system has been shown to be a solution for more rapid analysis while consuming far smaller amounts of organic solvents and various reagents. However, miniaturization will not solve the complexity of the sample matrix as was illustrated in the joint chromatogram, and introducing selectivity somewhere using the lab-on-chip method is still the key point for rapid targeted analysis. Also, a rapid survey of bibliography in that field indicates that the potential for microchips is high, but is seldom illustrated with real-life samples.

My first work in this area was to perform separations on electrophoresis-based microsystems and to introduce a sample pretreatment step. We started by studying the in situ polymerization of monolithic stationary phases in chip microchannel, and were able to separate several analytes within 1.5 cm length of a monolithic polymeric phase with more than 200,000 theoretical plates in less than 2 min. The first experiments showed that on-chip preconcentration can be achieved by using in situ synthesized monoliths. Our current work deals with in situ interaction of monoliths with antibodies or receptors in order to introduce selectivity. My collaborator, V. Pichon, has already succeeded for the first in situ polymerization of molecularly imprinted polymers.

Most Significant Publications

M. C. Hennion, C. Picard, and M. Caude, Influence of the number and length of alkyl chains on the chromatographic properties of hydrocarbonaceous bonded phases, *J. Chromatogr.* **166**, 21–35 (1978).

V. Pichon, L. Chen, M. C. Hennion, R. Daniel, A. Martel, F. Le Goffic, J. Abian, and D. Barcelo, Preparation and evaluation of immunosorbents for selective tare enrichment of phenylurea and triazine herbicides in environmental waters, *Anal. Chem.* **67**, 2451–2460 (1995).

M. C. Hennion, V. Coquart, and S. Guenu, Retention behaviour of polar compounds using porous graphitic carbon with water-rich mobile phases, *J. Chromatogr. A* **712**, 287–301 (1995).

M. C. Hennion, Solid-phase extraction: method development, sorbents, and coupling with liquid chromatography, *J. Chromatogr. A* **856**, 3–54 (1999).

Perspective of Chromatography for the Twenty-First Century. The recent developments in chromatography and its coupling to mass spectrometry were mainly to provide analytical techniques allowing higher resolution to solve complex samples and lower detection limits. In that area, progress will certainly continue, but one should not forget that one driving force of the twenty-first century is sustainable environment. This is particularly important for a field as chromatography, which has a key role to ensure better health, a better environment, and more safe food and other products. When looking to the routine analyses that are made at present, just a simple lifecycle analysis of our routine protocol to monitor our environment is very poor. There is a strong need for methods that consume far less organic solvent and energy and are robust enough for routine analysis. Miniaturization is a part of the answer, but will not involve only a simple miniaturization of chromatography. The microchip concept allows many combinations of chemical, physical, and biological processes. Chromatography and electrophoresis are part of them, and several examples suggest that their combination with bioassays on the same microchip, for instance, can be a good answer to rapid and efficient analysis for some toxic environmental micropollutants.

CHRISTOPHER R. LOWE

Figure 5.14. Christopher R. Lowe.

Professor Christopher R. Lowe (Fig. 5.14) is Director of the Institute of Biotechnology at the University of Cambridge. He is a Fellow of Trinity College, the Royal Academy of Engineering, the Institute of Physics, and the Royal Society of Chemistry. The principal focus of his biotechnology research programme over the last 35 years has been the high-value–low-volume sectors of pharmaceuticals, fine chemicals, and diagnostics. The work is characterized by not only being highly inter- and multidisciplinary and covering aspects of biochemistry, microbiology, chemistry, electrochemistry, physics, electronics, medicine, and chemical engineering but also covering the entire range from pure science to strategic applied science, some of which has significant commercial applications. The principal foci of his biotechnology research programme over the last 35 years have been (1) affinity chromatography, downstream processing of biopharmaceuticals proteomics; (2) biosensors/bioelectronics and nanobiotechnology; and (3) enzyme, protein, and microbial technology.

He has over 320 publications, including eight books and monographs, 60 patents, and has many collaborations worldwide, is an editorial board member of many academic journals, is a member of research council, grant-awarding and government committees, and is active in various legal and entrepreneurial roles. He has supervised over 60 Ph.D. students and won a number of national and international prizes. He has been the driving force for the establishment of eight spinout companies, including ProMetic BioSciences Inc., Purely Proteins Ltd., Affinity Sensors Ltd., Cambridge Sensors Ltd., Smart Holograms Ltd., Psynova Neurotech Ltd., Paramata Ltd., and Rebha Ltd.

Awards After 2000

2002	Silver Jubilee Medal, Chromatographic Society
2002	Russian Academy of Medical Sciences, Elected Fellow
2003	Henry Dale Medal and Prize; The Royal Institution (London)
2003	Institute of Physics; Elected Fellow (F.InstP)
2005	Royal Academy of Engineering (FREng), Elected Fellow
2005	Royal Society of Chemistry Award in Sensors: Medal and Prize
2005	*Journal of Chromatography* Top Cited Article Award 2000–2004

2005	Royal Society of Chemistry (FRSC), Elected Fellow
2006	Dade–Behring Award for Clinical Chemistry
2006	Albert Franks Memorial Lecture
2006	"Most Entrepreneurial Scientist of the UK" Award, UKSEC
2007	Queen's Anniversary Prize for Higher and Further Education

See Chapter 1, Table 1.3, h, i, n, t, v, z.

Major Research Program: Affinity Chromatography—New Horizons for Science and Technology.

A new armory of protein purification tools is essential to service the rapid advances in high-throughput proteomics required for the discovery, characterization, purification, and manufacture of the next generation of biopharmaceutical proteins. As much as 50–80% of the total cost of manufacturing a therapeutic product is incurred during downstream processing, purification, and polishing. There is intense pressure to revamp existing production processes to improve efficiency and yields and thereby reduce costs to the healthcare provider. Changes in the regulatory climate have shifted the focus of regulation from production processes per se to the concept of the "well-characterized biologic"; although process changes will still be regulated to some extent, manufacturers will be allowed to improve protocols for certain specified well-characterized products. The final protein should not only have defined purity, efficacy, potency, stability, pharmacokinetics, pharmacodynamics, toxicity, and immunogenicity but also be analyzed for contaminants such as nucleic acids, viruses, pyrogens, isoforms, residual host cell proteins, cell culture media, and leachates from the separation media. One way to address these economic and regulatory issues is to substitute conventional purification protocols with highly selective and sophisticated strategies based on affinity chromatography. This technique provides a rational basis for purification that simulates and exploits natural molecular recognition for the selective purification of the target protein. The key goal of my research over the past three decades has been to devise new techniques to identify highly selective affinity ligands that bind to the putative target biopharmaceuticals and allow scaleup into industrial processes.

My group has pioneered the development of selective separation technologies based on affinity chromatography since introduction of the technique in 1967. At that time, affinity chromatography exploited only highly specific inhibitors or substrate analogs, and it was apparent that the technique had to be made more generic to be universally accepted by the research community. My group pioneered the immobilization chemistry, introduced the concept of immobilized adenine nucleotide coenzymes in 1971, and coined the phrase "group-specific adsorbents" in 1972. This was followed by a series of six consecutive pioneering papers on novel chemistries, the interaction of these selective adsorbents with oxidoreductases and kinases, and identification of the key operational parameters required for successful chromatography, including the effects of immobilized ligand concentration, column geometry and dynamics, pH, temperature, cosolvent mixtures, and kinetics. As a follow-on, and in the laboratory of Professor Mosbach in Sweden, we synthesized the first catalytically active $NADP^+$ via a Dimroth rearrangement and immobilized the active coenzyme on soluble dextran and beaded agarose for applications in enzyme technology. However, despite a number of prominent papers in this sector, it was becoming clear to me by the end of the 1970s that natural coenzymes were too expensive to immobilize and not durable enough to be exploited in the newly burgeoning biotechnology industries and that "biomimetic" ligands might offer an alternative. My group pioneered the introduction of affinity chromatography on immobilized triazine dyes in 1980, based on the observation, at that time unexplained, that some adenine coenzyme enzymes interacted with the blue dye used as a void volume indicator in gel filtration. We contacted the manufacturer of these dyes, ICI plc, gained access to their expertise and facility in Manchester and embarked on an extensive program of joint research aimed at exploiting a wide range of these textile dyes in biotechnology. Our aim was to understand how they functioned and this warranted synthesis of ultra-pure and fully characterized dyes, a fundamental structure–activity study with the major classes of dyestuffs and binding proteins such as the oxidoreductases, kinases and blood proteins, inhibition kinetics, spectrophotometric behavior, and their use as reactive affinity labels. In a paper on the affinity labeling of enzymes with triazine dyes, we proved that the dichlorotriazinyl dye, Procion Blue MX-R, really was a mimic of the natural coenzyme NAD^+, by isolating and sequencing a Cys^{174}-modified pentapeptide unambiguously identified from the catalytic domain of horse liver alcohol dehydrogenase. We applied similar logic to provide unequivocal evidence that these biomimetic dyes selectively interacted with other proteins including the ricin A chain, hexokinase, and the metal-ion-mediated interaction of the red azo dye, Procion Red MX-8B, with the therapeutically important enzyme carboxypeptidase G2.

We were also the first to use these dyes and other ligands in high-performance liquid affinity chromatography (1981), charge transfer, and metal chelate chromatography (1982) on silica beads; introduce affinity precipitation on bifunctional dyes for the ultra-rapid purification of enzymes (1986); and use the dyes in novel acrylic two-phase polymer partition systems (1988) and process-scale high-performance liquid affinity chromatography (1986). Furthermore, we pioneered the immobilization chemistry and application of a range of novel perfluorocarbon, perfluoromethacrylate, and perfluorocarbon emulsions in separations technology (1989–1991). At this point, almost all the ligands used in these studies were conventional, albeit highly purified, dyes available commercially from ICI plc. It was becoming obvious that whilst these had many desirable features, and in some cases were exceptionally specific for the target proteins, they were designed as textile dyes, not as biological reagents. Consequently, in the late 1980s, we embarked on an extensive series of studies aimed at improving the specificity and affinity of textile dyes with specially designed dyes for target proteins and coined the term "designer dyes." For example, and as a model system, we examined in detail the interaction of the blue anthraquinone dye, CI (Color Index) Reactive Blue 2, with the known X-ray crystallographic structure of horse liver alcohol dehydrogenase. We undertook a comprehensive and systematic study synthesizing and characterizing terminal ring (1988) and anthraquinone (1990) analogs of the dye and related their kinetic, inhibition, and chromatographic behavior to the known structure of horse liver alcohol dehydrogenase. X-ray crystallography and affinity labeling studies showed that the dye binds to the coenzyme binding domain of the enzyme with the anthraquinone, diaminobenzene sulfonate, and triazine rings adopting similar positions as the adenine, adenosine ribose, and pyrophosphate groups of NAD^+. It was revealed that the terminal aminobenzene sulfonate ring of the dye was bound to the side of the main NAD^+-binding site in a crevice adjacent to the side chains of Arg and His residues.

Thus, the synthesis, characterization, and assessment of a number of terminal ring analogs of the dye confirmed the preference for a small, anionic *o*- or *m*-substituted functionality and substantially improved the affinity and selectivity of the dye for the protein. These conclusions have been confirmed with more recent studies with a range of new analogs and demonstrate how the use of modern design techniques can greatly improve the selectivity of biomimetic ligands. A similar study was conducted to design biomimetic dyes for the purification of calf intestinal alkaline phosphatase with phenylboronate-containing terminal ring analogs of CI Reactive Blue 2 (1988) and to introduce entirely novel cationic dyes for the purification of proteases such as trypsin, kallikrein, thrombin, and urokinase (1987).

The ability to combine knowledge of X-ray crystallographic, NMR, or homology structures with defined chemical synthesis and advanced computational tools has made the rational design of affinity ligands more feasible, more logical, and faster. The target site could be an catalytically active site, a solvent-exposed region, or a motif on the protein surface or a site known to be involved in binding a natural complementary ligand. Three distinct approaches to design can be distinguished: (1) investigation of the structure of a natural protein–ligand interaction and the use of the partner as a template on which to model a biomimetic ligand; (2) ligand design by construction of a molecule that displays complementarity to exposed residues in the target site, and (3) direct mimicking of natural biological recognition interactions.

My group pioneered the introduction of these selective design techniques with the very first approach to the design, synthesis, and characterization of ligands differentiating closely related serine proteases. Peptidal templates comprising two or three amino acids have been used to design highly selective affinity ligands for kallikrein (Arg–Phe) (1992) and elastase (Leu–Glu–Tyr) (2000) and were synthesized by substituting a triazine scaffold with appropriate analogs of the amino acids. For example, a durable, nonhydrolysable, low-molecular-weight affinity ligand, which mimics key features of the interaction between the naturally occurring turkey ovomucoid inhibitor and elastase, has been developed. A limited library of 12 ligands were designed and synthesized by mimicking various combinations of di- and tripeptides modeled on a known heptapeptide (–Pro–Ala–Cys–Thr–Leu–Glu–Tyr–) from turkey ovomucoid inhibitor. The ligands were composed of various substituents on the triazine nucleus, bridging secondary amines and the spacer assembly. One of these ligands was used to purify elastase from a crude porcine pancreatic powder with a 19.4-fold purification and 90% yield (2000), while a variant of this ligand was found to purify elastase from cod pyloric caeca with a 34.4-fold increase in specific activity and a yield in excess of 100%.

A further development of this technology involved the design, characterization, and evaluation of an artificial protein A, using a $Phe^{132}–Tyr^{133}$ dipeptide template identified from the interaction of the protein A from *Staphylococcus aureus* (SpA) with the CH_2–CH_3 hinge region of Immunoglobulin G (1998). This seminal work was based on identifying the key interactions in the X-ray crystallographic complex between the B domain (Fb) of SpA and the Fc fragment of IgG, modeling the Phe–Tyr dipeptide and creating, synthesizing, and characterizing a mimic using the triazine scaffold. An agarose-immobilized ligand comprising 3-aminophenol and 4-amino-1-naphthol moieties substituted on the triazine scaffold was found to display a high capacity for IgG from human plasma and was able to purify the protein in high yield (>95%) and purity (>99%).

A follow-up study employing the lead ligand synthesized in solution and immobilized on agarose indicated an affinity constant (K_A) for the immobilized ligand and human IgG of $1.4 \times 10^5 \, M^{-1}$ and a theoretical maximum capacity of 152 mg IgG/g moist-weight gel. This adsorbent showed selectivity similar to that of immobilized protein A and bound IgG from a number of species. The adsorbent showed an apparent capacity of 51.9 mg IgG/g moist-weight gel under the specified adsorption conditions and eluted IgG in ~68% yield with a purity of 98–99%. The immobilized artificial protein A was used to purify IgG from human plasma, murine IgG from ascites fluid, and isolate bovine IgG from foetal calf serum. Furthermore, the adsorbent was able to withstand cleaning-in-place procedures with 1 M NaOH for 1 week without loss of binding capacity for IgG. This study suggested that the immobilized ligand retained many of the advantages of selectivity and capacity of immobilized protein A, whilst obviating some of the disadvantages such as leakage and toxicity.

The selection of an appropriate target site and the design, synthesis, and evaluation of a complementary affinity ligand is, at best, only a semirational process, since numerous unknown factors are introduced during immobilization of the ligand on to beaded agarose. For example, the accessibility and affinity of the immobilized ligand for the complementary protein is determined partly by the characteristics of the ligand per se and partly by the matrix, activation, and coupling chemistry. Studies in free solution with soluble ligands do not fairly reflect the chemical, geometric, and steric constraints imposed by the complex three-dimensional matrix environment. These uncertainties involved in modeling solid-phase interactions led my group to introduce for the first time the concept of combinatorial chemistry into affinity chromatography. We refined the first-generation biomimetic ligands with the aid of a limited combinatorial library and introduced the concept of "intelligent" combinatorial libraries for the purification of biopharmaceuticals (1998). This paper won an award for a top cited paper in the *Journal of Chromatography* from 2000 to 2005.

Where there is inadequate or insufficient structural data on the formation of complexes between the target protein and a complementary ligand, designing molecules de novo with functionality matched to the target exposed residues of a specified active or surface site offers an alternative design route. This approach has been exemplified in the design, synthesis, and evaluation of an affinity ligand for a recombinant insulin precursor (MI3) expressed in *Saccharomyces cerevisiae* (1998). A characteristic feature of MI3 was a hydrophobic area comprising several aromatic residues, including three tyrosine residues (B:16-Tyr, A:19-Tyr, B:26-Tyr) and two proximal phenylalanine residues (B:24-Phe, B:25-Phe) that form a broad swathe across the waist of the globular protein and residing on both A and B chains. Inspection of this site revealed that B:16-Tyr and B:24-Phe were relatively exposed to solvent, while B:25-Phe was especially exposed. The accessibility of these aromatic residues suggested that a suitably designed affinity ligand could bind to these residues by aromatic $\pi-\pi$ stacking interactions. Preliminary computer-aided molecular modeling showed that a lead ligand comprising a triazine scaffold substituted with aniline and tyramine, showed significant $\pi-\pi$ overlap with the aromatic side chains of B:16-Tyr and B:24-Phe, and was thus used as a guide to construct a solid-phase combinatorial library. However, while the B:16-Tyr/B:24-Phe constellation was selected as the putative target for the directed library, other adjacent aromatic residues in the same target region were considered targets with equal potential and amenability, and thus a limited library of 64 members was synthesized from 26 amino derivatives of bi-, tri-, and heterocyclic aromatics, aliphatic alcohols, fluorenes, and acridines substituted with various functional groups.

The solid-phase library was screened for MI3 binding by application of a sample at pH 5.0, washing and eluting with 2 M acetic acid, with fractions from each column being analyzed by reversed-phase HPLC referenced to the known elution behavior of an authentic sample of MI3. Analysis of the adsorption and elution behavior of MI3 to the library revealed that the most effective ligands appeared to be bisymmetric compounds substituted with aminonaphthols or aminonaphthoic acids, with very high levels of discrimination noted with various substituents. For example, ligands constructed with 1-aminonaphthalene bearing hydroxylic functions in positions 5, 6, or 7 of the bicyclic nucleus proved very effective adsorbents, while those bearing substituents at positions 2 and 4 were ineffective. A similar trend was noted with hydroxylic isomers in the 2-aminonaphthol series. These differences in binding behavior were reconciled when modeling studies showed that bisymmetric bicyclic ring ligands display more complete $\pi-\pi$ overlap with the side chain of residues B:16-Tyr and B:24-Phe, than do the single-ring substituents of the original lead compound used for library synthesis. However, these screening data on the immobilized ligands serve to highlight the difficulties associated with the rational design process. Despite the value of computer modeling in visualizing putative interactions, the complexity of the three-dimensional matrix environments suggests that the combination of rational design with combinatorial chemistry constitutes an effective strategy to develop affinity ligands. Nevertheless, despite these reservations, a lead symmetric ligand was synthesized de novo in solution, characterized and immobilized to agarose beads, whence chromatography of a crude clarified yeast expression system revealed that MI3 was purified on this adsorbent with a purity of $>95\%$ and a yield of $\sim90\%$.

This study showed that a defined structural template is not required and that a limited combinatorial library of ligands together with the use of parallel screening protocols allows selective affinity ligands to be obtained for target proteins. This approach has been widely adopted in my group for the design of novel affinity ligands for the purification of human recombinant clotting factor VII (2002), prion proteins (2004), and clotting factor VIII (work in progress). The factor VII ligand was designed to interact with the γ-carboxyglutamate (Gla) residues of the Gla domain of the protease in order to create an adsorbent that bound recombinant factor VII only in the presence of Ca^{2+} ions and that could be released under mild nondenaturing conditions on removal or chelation of the ions (Fig. 5.15).

More recent work has resulted in the first synthetic mimic of protein L from *Peptostreptococcus magnus* (PpL), which normally binds the light chains belonging to the κ_1, κ_3, and κ_4 subgroups, but not to the κ_2 and λ subgroups of immunoglobulins.

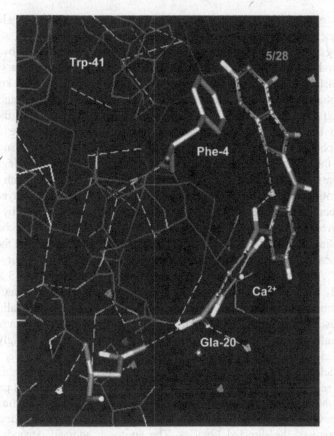

Figure 5.15. Factor VII–ligand complex. An illustration of the molecular model of the putative complex between triazine ligand 5/28 docked on the Gla domain of human coagulation factor VII using Quanta 97 software. The ligand comprising 3-aminobenzoic acid and 2-aminobenzimidazole substituents on a triazine nucleus interacts with the side chain of Phe-4 and forms a coordination complex between the side chain of Gla-20, a Ca^{2+} ion, and the carboxylic function of the 3-aminobenzoic acid substituent.

Our mimetic ligand behaved in a fashion similar to that of native PpL except that it was able to bind to both κ and λ light-chain subgroups and was able to isolate Fab fragments from papain digests of human IgG to a final purity of 97%. Recent work has also lead to the development of new cleavable crosslinkers for affinity chromatography (2005) and novel scaffold chemistries based on multicomponent reactions. Current work is also devoted to the exploitation of these design and synthesis techniques to develop highly selective, stable, and sterilizable adsorbents for the removal of contaminants such as endotoxins and retroviruses and to the ultrahigh resolution of misfolded forms, glyco-forms, and other posttranslationally modified proteins. In this respect, the artificial lectin (2000) was designed, synthesized, and evaluated in order to bind the glyco moiety of pharmaceutical proteins, which are sometimes present in relatively low concentrations by aberrant glycosylation in the host cell expression systems.

A highly selective mannose-binding affinity ligand has been designed by mimicking the way natural proteins, enzymes, lectins and antibodies, interact with sugars in aqueous solution. A detailed assessment of protein–carbohydrate interactions from known X-ray crystallographic structures was used to identify key residues that determine monosaccharide specificity and that were subsequently exploited as the basis for the synthesis of a limited library of glycoprotein-binding ligands. The ligands were synthesized using solid-phase combinatorial chemistry and were assessed for their sugar-binding ability with the glycoenzyme, glucose oxidase. Partial and completely deglycosylated enzymes were used as controls. A triazine-based ligand, bis-substituted with 5-aminoindan, was identified as a putative glycoprotein binding ligand, since it displayed particular affinity for core mannoside moieties. These findings were substantiated by interaction analysis between the ligand and mannoside moieties by NMR. ^{1}H NMR studies and molecular modeling suggested involvement of the hydroxyls on the mannoside moiety at C2, C3, and C4 positions. Partition coefficient experiments allowed the elucidation of binding constants (K_{ax}) for the enzyme and mannose, glucose and fucose to be $4.3 \times 10^5 \, \text{M}^{-1}$, $1.9 \times 10^4 \, \text{M}^{-1}$, and $1.2 \times 10^4 \, \text{M}^{-1}$, respectively. These affinities were remarkably similar to the affinities and specificities displayed by the plant lectin, concanavalin A, toward neutral hexoses. Experiments with partially and fully deglycosylated control proteins, specific eluants, and retardation in the presence of competing sugars strongly suggested that the ligand bound to the carbohydrate moiety rather than the protein per se.

This article describes techniques that we have developed to rationalize the design and synthesis of selective affinity ligands for the purification of pharmaceutical proteins. The strategy is straightforward and may involve either screening for target binding to combinatorial libraries of synthetic ligands, or, preferably, introduction of a design step to reduce the size of the directed libraries. The approach adopted depends to a large extent on what information is available at the outset; if structural data are available, the design approach is possible, while in the absence of such information, which may be the case in proteomics applications, a random screen of a combinatorial library would be the only plausible route. The author prefers the "intelligent" approach, since it drastically reduces the chemistry and screening necessary to identify a lead ligand. Nevertheless, combinatorial screening is still required to obviate many of the unknowns involved in the interaction of the protein with solid-phase immobilized ligands. The design strategy comprises several sequential steps: (1) selection of a target site or investigation of a known biological interaction, modeling of the site, and/or the interaction and preliminary design of a complementary ligand using the biological interaction as a template; (2) synthesis of a limited combinatorial library of first-generation ligands on a solid phase and their evaluation by chromatography; (3) solution phase synthesis and characterization of lead ligands; and (4) immobilization, optimization, and chromatographic evaluation of the final adsorbent with realistic feedstocks. Early work on the requirement for a defined structural template has been superseded by the synthesis of limited combinatorial libraries generated by the use of rationally designed ligands complementary to the target site or designed to mimic natural biological interactions. A key aspect of this system is that the chromatographic adsorption and elution protocols can be built into the total package at the screening stage and therefore lead to very rapid conversion of a hit ligand into a working adsorbent.

The use of synthetic ligands with few or no fissile bonds offers a number of advantages for the purification of pharmaceutical proteins: (1) the adsorbents are inexpensive, scaleable, durable, and reusable over multiple cycles; (2) the provision of a ligand with defined chemistry and toxicity satisfies the regulatory authorities; and (3) the the exceptional stability of synthetic adsorbents allows harsh elution, cleaning-in-place, and sterilization-in-place protocols to be used. These considerations remove the potential risk of prion or virus contamination, which may arise when immunoadsorbents originating from animal sources are used. Other types of affinity ligand based on peptide, oligonucleotide, or small protein libraries are likely to be less durable under operational conditions.

Perspective on Affinity Chromatography for the Twenty-First Century. Four decades after the term "affinity chromatography" was coined, this highly selective mode of chromatography remains an essential tool in the armory of separation techniques being used in research-intensive industries. The technique continues to be favored owing to its high selectivity and predictability, speed, ease of use, and yield, although widespread application of affinity chromatography in industry is tempered by the current high cost of the adsorbents. So the question is: What is the future of this technique in the twenty-first century? It is clear that the technique is likely to have a continuing impact on the biopharmaceutical industry to improve production processes for the "well-characterized biologic" in order to make them cost-efficient for the more stringent economic demands of healthcare providers. Development of new highly selective approaches to the resolution of posttranslationally modified proteins and their isoforms, elimination of leachates, and the removal of prion or viral contaminants from therapeutic preparations will require further attention. It is conceivable that ultra-high-throughput synthesis, selection, and optimization of affinity adsorbents may be conducted on-chip in the future. Furthermore, to implement the "omics" revolution, avid, high-resolution, and selective affinity techniques will be required to separate and analyze a range of low-abundance proteins found in biological samples.

Most Significant Publications in Chromatography (2001)

1. C. R. Lowe, Combinatorial approaches to affinity chromatography, *Curr. Opini. Chem. Biol.* **3**, 248–256 (2001).
2. P. R. Morrill, G. Gupta, K. Sproule, D. Winzor, J. Christiansen, I. Mollerup, and C. R. Lowe, Rational combinatorial chemistry-based selection, synthesis and evaluation of an affinity adsorbent for recombinant human clotting factor VII, *J. Chromatogr. B* **774**, 1–15 (2002).
3. P. R. Morrill, R. B. Millington, and C. R. Lowe, An imaging surface plasmon resonance system for screening affinity ligands, *J. Chromatogr. B.* **793** 229–251 (2003).
4. E. N. Soto Renou, G. Gupta, D. S. Young, D. V. Dear, and C. R. Lowe, The design, synthesis and evaluation of affinity ligands for prion proteins, *J. Mol. Recog.* **17**, 248–261 (2004).
5. G. Gupta and C. R. Lowe, An artificial receptor for glycoproteins, *J. Mol. Recog.* **17**, 218–235 (2004).
6. A. C. A. Roque, M. A. Taipa, and C. R. Lowe, Artificial protein L for the purification of immunoglobulins and fab-fragments by affinity chromatography, *J. Chromatogr. A* **1064**, 157–167 (2005).

7. A. C. A. Roque, M. A. Taipa, and C. R. Lowe, Synthesis and screening of a rationally-designed combinatorial library of affinity ligands mimicking protein L from peptostreptococcus magnus, *J. Mol. Recog.* **18**, 213–224 (2005).

RONALD E. MAJORS

Ron E. Majors (Fig. 5.16) presents an in-depth overview on HPLC column technology—state of the art to 2008 in *LC-GC* [1]. He enlisted 10 experts and pioneers in the field from industry and academia to contribute their technical knowledge.

HPLC Column Technology—State of the Art. It has often been stated (or maybe overstated) that the column is the heart of the chromatograph.

Figure 5.16. Ronald E. Majors.

Without the proper choice of column and appropriate operating conditions, method of development and optimization of the high-performance liquid chromatographic (HPLC) separation can be frustrating and unrewarding experiences. Since the beginning of modern liquid chromatography, column technology has been a driving force in moving separations forward. Today, the driving forces for new column configurations and phases are the increased need for high-throughput applications, for high-sensitivity assays, and to characterize complex samples such as peptide digests and natural products.

Advances are still being made in column technology with smaller porous particles (1–2 μm in diameter), ultra-high-pressure HPLC, high-temperature (≤200°C) columns, nanocolumns with diameters under 100 μm, and rapid separation columns enabling high-resolution separations in seconds. LC-on-a-chip experimentation is now driving columns to smaller and smaller dimensions but making LC-MS interfacing even easier. Polymeric and silica-based monoliths have seen major improvements with better reproducibility, a variety of stationary phases, and commercial availability. New particle designs such as superficially porous particles for high-speed applications have come on the scene. Improvements in application-specific columns such as those for chiral separations, sensitive biological samples, and very polar compounds are being shown every year. The area of multidimensional LC and comprehensive LC × LC has become a reality in the tackling of complex samples.

In time for the HPLC 2008 Symposium held in Baltimore this year, I have assembled a special edition of *LC-GC North America* to highlight the state of the art in HPLC column technology. Experts and pioneers in the field of HPLC column technology from industry and academia were asked to contribute their technical knowledge. In this issue, we have presented an overview of column advances in the last two years (Majors), followed by a look at high-throughput in high pressure LC (Rozing), polymeric monolithic columns (Svec and Krenkova), silica-based monolithic columns (Cabrera), high-temperature HPLC (Yang), chiral chromatography columns (Beesley), and enhanced stability stationary phases (Silva and Collins), rounded out with a treatise on hydrophilic interaction chromatography (McCalley). The contributors were asked to provide an update on the phase and column technology in their respective areas with a focus on advances made in recent years. Focus was directed primarily to the most recent advances. These papers are an excellent source of information on the recent developments in LC column technology.

REFERENCE

1. R. E. Majors, Recent developments in LC column technology, Supplement to *LC-GC North Am.* (Suppl.) **26**(54) (April 2008).

PHILIP MARRIOTT

Figure 5.17. Philip Marriott.

Professor Philip Marriott (Fig. 5.17) completed his B.Sc. (Hons.) and Ph.D. at LaTrobe University (Australia), where he studied packed-column and capillary gas chromatography of metal complexes, with application of flame photometric detection. In 1980 he accepted a position as a postdoctoral researcher at the University of Bristol, under Professor Geoffrey Eglinton (FRS), in the area of organic geochemistry, specializing in porphyrin gas chromatography and mass spectrometry. Some of this work included various bioporphyrin complexes. An investigation of the effect of complexation of different metals in the porphyrin macrocycle was undertaken. This work demonstrated the successful GC of some of the highest molar mass complexes to be chromatographed, with masses ranging up to ∼1500 amu.

From Bristol, a 5-year academic appointment at the National University of Singapore was then undertaken. During this period, the university was starting to establish the research program focus that would later see it become a leading Asian and International University. Research at this time was largely in the area of discovering and understanding the process of dynamic molecular interconversion in chromatography. From the initial observation of sterically hindered rotations in polyaromatic hydrocarbons, research was then extended to metal complexes and organometallics, and oximes.

After returning to Melbourne, and RMIT University in 1987, Professor Marriott is now Professor of Separation Science, and Deputy Director of the Australian Centre for Research on Separation Science (ACROSS). He currently has research interests in capillary electrophoresis, HPLC, and extraction methods, but most of his research is directed to understanding the principles and describing the processes of comprehensive two-dimensional gas chromatography, and undertaking a strong applications program in GC × GC (refer to the subsection on major research interests, below). Interest in the

GC × GC research area has lead to frequent invitations to present keynote and plenary lectures at many international conferences, and at companies and universities worldwide. In addition, workshops on GC × GC and MDGC technology are often presented.

Professor Marriott was the Honorary Treasurer of the Federation of Asian Chemical Societies, and Honorary Treasurer of the Royal Australian Chemical Institute (RACI). He has served as the Division Chair of the Analytical Division of the RACI.

Professor Marriott is a Fellow of the RACI, a Member of the American Chemical Society (ACS), and a Life Member of the Federation of Asian Chemical Societies (FACS).

Awards 2000–2007

Silver Jubilee Award, Chromatographic Society, London (2000)

Applied Research Award of the Royal Australian Chemistry Institute (2001)

Analytical Division Medal, Royal Australian Chemistry Institute (2004)

Karasek Award, Enviroanalysis (2007)

See Chapter 1, Table 1.3, d, h, k, n, p.

Major Research: Longitudinal Modulation Cryogenic Systems (LMCSs). The primary research topic of Professor Marriott, which has been the most innovative and productive over the past 10 years, is that of multidimensional gas chromatography (MDGC) and comprehensive two-dimensional gas chromatography (GC × GC). In both these cases we deal with coupled column methods. In particular, the technology of cryogenic modulation, developed and patented by Professor Marriott as the longitudinal modulation cryogenic system (LMCS; Fig. 5.18), has been the enabling technology that has allowed us to achieve these ends. It is conceptually simple, and has proved to be a fascinating but powerful tool. The functionality that can be brought to gas chromatography through use of the LMCS, and the diverse modes of operation, allows a range of novel methods to be developed.

PRINCIPLE OF OPERATION. The LMCS comprises a cold enveloping chamber with an open central channel located along its axis, which simply moves back and forth along a column, by about 3–4 cm. With applied cryogen, the column segment residing within the chamber is cooled (even in a high-temperature oven), but when the cold chamber is moved, the column segment immediately heats up so as to return the column to the prevailing oven temperature (Fig. 5.19). In this way, any inflowing solute (volatile at the prevailing oven temperature) will be trapped inside the phase or at the walls of the cold column. Moving the LMCS counter to the carrier flow allows the trapped analyte to then heat up and continue its travel toward the detector. This trap–elute (remobilize) process is the key to a range of chromatographic methods.

METHODS OF USE OF THE LMCS. The simplest method is one that involves whole chromatographic peak collection. If we trap a whole peak at the end of a first dimension column (^1D), then remobilize it into a second (^2D) short narrow column as a sharp band, we should get a much taller signal response. Consider a peak that is typically 10 s wide,

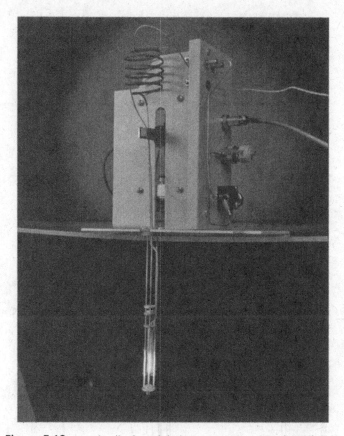

Figure 5.18. Longitudinal modulation cryogenic system (LMCS) unit.

but that becomes 0.25 s wide. This is an effective 40-fold narrowing, and so 40-fold increase in signal magnitude in the experiment.

Now allow the second column phase to differ from that of the first. If we co-collect a number of peaks into the one sampling event, there is every likelihood that all can now be resolved. The LMCS allows bands of the order of 10 ms to be injected into the second column, and this help immensely in subsequent resolution. In this mode, the ^2D column can be as long as one may wish, but for practical reasons we maintain a relatively short length of 5–10 m. Figure 5.20 shows a 1 min wide sample at the end of the ^1D column being focused and then eluted on a high-efficiency column in <45 s. Now, permit the LMCS unit to be used to "subsample" peaks. Thus we can slice up the ^1D peak into a series of packets based on the modulation or sampling period of the primary peak. This is according to the term we recently introduced, called the *modulation* ratio (M_R). This now accomplishes the requirements of comprehensive two-dimensional gas chromatography (GC × GC). We conducted the first studies in this area in 1996 (published in 1997). This was the first demonstration of a cryogenic approach being used for GC × GC, soon after this publication others were clearly encouraged to develop their own versions of this approach. GC × GC incorporates now a very short ^2D

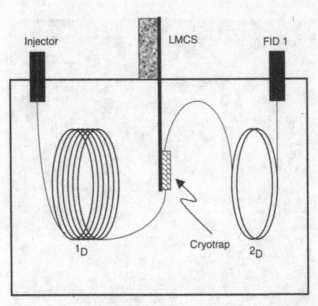

Figure 5.19. Schematic of the multidimensional gas chromatography (MDGC) GC × GC system.

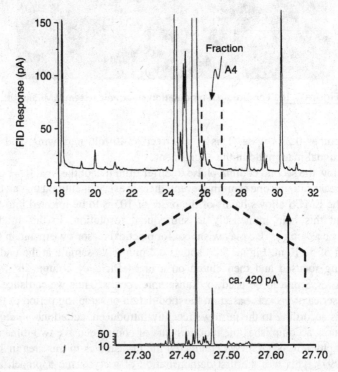

Figure 5.20. Example of targeted multidimensional gas chromatographic analysis.

column, usually 1–2 m long. This ideally allows each sampled peak group to be analyzed before the next is sampled. For complex mixtures, one might now propose that this should be a powerful approach to provide initial resolution in the first dimension column, and then an independent resolution in the second, to theoretically lead to a total peak capacity equal to the product of the individual capacities on each column; for example, allow $^1n = 500$; $^2n = 20$; $n_t \sim 10,000$. A further refinement of the use of LMCS is to simply loop the column through the cryogenic chamber twice—either in the same direction, or counter to the original direction. This is an effective way to isolate a specific target region within a sample chromatogram. It essentially allows one to store within the loop section the target region while the earlier solute is being eliminated from the column, but holds back any lower volatility solute by trapping these components at the entry strand of the loop. The target contents of the loop can then be applied to the ^2D column— which can be of any length desired. We have used this for olfactory analysis, and with a range of other techniques that are available only as "slow" spectroscopic detection.

Our research also has been directed to incorporating classical valve- or Deans' switch-type methods for MDGC. Thus we can isolate discrete heartcuts from a ^1D column and subject these to separate analysis on the ^2D column. Again, use of LMCS allows bands to be trapped and so maximize the efficiency of the ^2D column—there is no carryover of the band spread on the first column, and so this maximizes resolution. The most recent innovation in this approach is to use our loop modulation process with a Deans switch, which is essentially a dual-heartcut method. We predict that this should prove to be very useful for chemical discovery using various novel spectroscopic detection processes. This approach has just been reported by us, as a high-resolution MDGC method in conjunction with offline NMR, for absolute chemical structural characterization.

Supporting Applications—Demonstrating the Value of GC × GC. Clearly, a chromatographic method has to serve a useful, practical purpose if it is going to either have longevity, or be useful for routine application. We try to balance the development of new strategies of implementing our methods, with suitable demonstration of how these tools may be used. Selected examples follow.

DRUGS OF ABUSE (*DOPING CONTROL*). Drugs of abuse may be profiled by use of solid-phase microextraction (SPME) with GC × GC analysis. The volatile headspace products are sampled by SPME, and then the GC × GC approach allows their 2D presentation with considerable more clarity than that of a 1D analysis. It is apparent that small peaks in such a mixture can be better identified and recognized. Derivatised steroid drugs exhibit similar detection limits to those available in 1D GC-selective-ion monitoring SIM(MS) analysis, but now with a time-of-flight mass spectrometer (TOFMS) offering full mass spectral scanning. This therefore permits library matching, and aids complete identification.

DETECTION IN GC × GC. Our group was the first to demonstrate use of TOFMS for detection of resolved GC × GC analytes. Today, TOFMS is a mass spectral tool of choice for GC × GC, although we have also reported useful mass spectrometry detection and library matches through use of the slower scanning quadrupole MS method. This group also reported the first nitrogen–phosphorus detection of compounds with GC × GC. It was found that the detector geometry plays a role in permitting optimum performance, with extended jet and wide collector best to use. The group has experience with electron capture detection for a range of organo-halogen analytes. By use of a microfluidic valve, dual detection at the end of the second column was shown to be achievable, with not too much performance loss. The use of olfactory detection normally requires sufficiently long elution times to permit reliable sensory perception. Methods with the LMCS and loop system appear well-suited to this technology.

ALLERGENS BY GC × GC. The first demonstration of allergens by using GC × GC was by this group. Even though the European regulations report 26 allergens of interest in perfume and related products, the problem of even this small set of compounds is confounded by the complexity of the matrix in which they arise—in both number and abundance of potential interferences. We have also investigated the use of fast targeted MDGC for this analysis with success.

INCENSE SMOKE. Incense may be considered somewhat of a model for a range of samples that produce combustion products in the atmosphere. The analysis of both the underlying fragrance in crushed incense, and the smoke stream in combusted incense, demonstrated that the PAH load in the smoke to be considerable and many N-containing compounds were identified. ECD-active compounds were found, but low TOFMS response in the equivalent GC × GC–TOFMS study compared to the ECD made positive identification incomplete.

FUNGICIDES AND DUAL DETECTION. Fungicides were the ideal target for dual detection (see above). Mixtures representative of different solute classes were examined in the GC × GC system. It was found that the ratio of ECD-to-NPD response could be a useful metric for compound identification.

RETENTION INDICES. The basic equation that quantifies elution in gas chromatography—the retention index—was applied in a number of studies by our group to demonstrate how such a system could be applied to give a two-dimensional representation of the 2D chromatographic space—we call this the *chromatographic retention map*. The key to this method is to make successive injections during a chromatographic run, so that reference compounds (e.g., alkanes) are delivered to the ^2D column at different elution temperatures, so that an approximate exponential decay curve is generated. Interpolation in the appropriate manner allows both ^1D and ^2D indices to be acquired. By extending this approach, it has been proposed that modification to allow the alkanes to be directly introduced to the second column could permit ready generation of temperature-variable indices on any second column phase. This will be useful for rapid regeneration of such indices on any phase column.

Perspective of Chromatography for the Twenty-First Century. Chromatography is entering a period of rapid transformation—which is probably surprising, given its maturity. I envisage a future that will encompass considerably more hyphenation both of separation dimensions and expanded (spectroscopic) detection capabilities. The foundations for this have been established in the late twentieth century, and what we have seen is that multidimensional methods have undergone a significant resurgence in both popularity, and performance. This resurgence has been a consequence of much improved technical implementation of the basic instrumental components, development of new separation media, and the realization that discovery in the chemical sciences must rely on the generation of considerable more peak capacity than has been possible in the past. This is common to both gas and liquid chromatography formats. The improved performance must be matched by detection capabilities, suited to these multidimensional methods. As an example, NMR has been a powerful tool for chemical characterization in offline methods, but it is hoped that continued improvements in detection limits will make this technology more generally applicable. Routine "multimode" mass spectrometry approaches through hyphenation (e.g., IMS with TOFMS), in combination with multidimensional separations, will provide a degree of separation and specificity that will make many previously difficult identification problems tractable. Trends toward much greater separation will run parallel with increased miniaturization and field-deployable equipment.

Professor Marriott's most significant publications are listed below.

REFERENCES

1. P. J. Marriott and R. M. Kinghorn, Longitudinally modulated cryogenic system—a new generally applicable approach to solute trapping in gas chromatography, *Anal. Chem.* **69**, 2582–2588 (1997).

2. A. C. Lewis, N. Carslaw, P. J. Marriott, R. M. Kinghorn, P. Morrison, A. L. Lee, K. D. Bartle, and M. J. Pilling, A larger pool of ozone-forming carbon compounds in urban atmospheres, *Nature* **405** (6788), 778–781 (2000).

3. R. Shellie, P. Marriott, and P. Morrison, Concepts and preliminary observations on the triple dimensional analysis of complex volatile samples by using GC × GC-TOFMS, *Anal. Chem.* **73**(6), 1336–1344 (2001).

4. M. S. Dunn, P. J. Marriott, R. A. Shellie, and P. D. Morrison, Targeted multidimensional gas chromatography using a microswitching valve, rapid cryogenic modulation and fast gas chromatography, *Anal. Chem.* **75**, 5532–5538 (2003).

5. S. M. Song, P. Marriott, A. Kotsos, O. Drummer, and P. Wynne, Comprehensive two-dimensional gas chromatography with time-of-flight mass spectrometry (GC × GC-TOFMS) for drug screening and confirmation, *Forens. Sci. Int.* **143**, 87–101 (2004).

6. S. Bieri and P. J. Marriott, Generating multiple independent retention index data in dual-secondary column comprehensive two-dimensional gas chromatography, *Anal. Chem.* **78**, 8089–8097 (2006).

7. G. T. Eyres, S. Urban, P. D. Morrison, J. P. Dufour, and P. J. Marriott, Method for small-molecule discovery based on microscale preparative multidimensional gas chromatography isolation with nuclear magnetic resonance spectroscopy, *Anal. Chem.* **80**, 6293–6299 (2008).

DAVID C. A. NEVILLE AND TERRY D. BUTTERS

Figure 5.21. Terry Butters.

Dr. Terry D. Butters (Fig. 5.21) is Reader in Glycobiology at the Oxford Glycobiology Institute, Department of Biochemistry. The discovery that enzyme inhibitors could be applied to the treatment of glycolipid lysosomal storage disorders in his laboratory at the Oxford Glycobiology Institute, Oxford University, has led to an approved therapeutic for the treatment of Gaucher disease. Dr. Butters is the joint recipient of the Gaucher Association Alan Gordon Memorial Award and the Horst-Bickel Award (1999). The development of new therapies for disease utilizing novel, small-molecule inhibitors of protein and lipid glycosylation remains the major focus of Dr. Butters' current research.

Figure 5.22. David Neville.

Dr. David C. A. Neville (Fig. 5.22) is a research scientist at the Oxford Glycobiology Institute, Department of Biochemistry. Dr. Neville's research focuses on the development of new chromatography methods to facilitate the analysis of biologically derived oligosaccharides from protein and/or lipid sources. This has led to the development of a published method to analyze glycophingolipid-derived oligosaccharides. Previous research has also led to the publication of a method to specifically purify phosphopeptides from a mixture prior to mass spectrometric analysis. Dr. Neville's current research is focused on developing a single-column separation of oligosaccharides that combines both hydrophilic interaction and anion-exchange chromatographies.

See Chapter 1, Table 1.3, a, c, h, t, v.

Hydrophilic Interaction Chromatography (HILIC) and Hydrophilic Interaction Anion-Exchange Chromatography (HIAX) of Glycoprotein- and Glycolipid-Derived Oligosaccharides. The posttranslational modification of membrane and secreted proteins with N- and O-linked oligosaccharides is one of the most common and conserved processes in cellular systems. Additionally, the modification of lipid moieties, i.e., dolichol, to form the highly conserved N-linked oligosaccharide precursor glucose$_3$mannose$_9$$N$-acetylglucosamine$_2$–dolichol (Glc$_3Man_9$GlcNAc$_2$–dolichol), phosphatidyl inositol to form glycosylphosphatidyl inositol (GPI) anchors or ceramide to form complex glycosphingolipids (GSL), is also a common feature of cellular systems. The proteins, substrates and reactants of these processes encompass the cellular glycosylation machinery/pathways. The structural diversity of these oligosaccharides means that the analysis of cell-derived oligosaccharides is a complex and demanding process. (1) the oligosaccharide must be released structurally intact from the protein or lipid by the use of chemical or enzymatic means, and (2) the unambiguous structural analysis of the released oligosaccharide requires multiple methods, most commonly a mixture of HPLC, linkage-specific glycosidase digestions, and mass spectrometry.

Biologically important glycosylated molecules are available typically in limited quantities. Therefore, to increase the detection sensitivity prior to HPLC, the most common method of analysis employed, derivatization of the reducing terminus of the released oligosaccharide with a fluorescent moiety (for review, see Ref. 2) is performed. Many different fluorophores can be used but for ease of handling, sensitivity, reaction conditions, and ease of purification of fluorophore-labeled oligosaccharide, we, at the Glycobiology Institute, routinely use anthranilic acid (2-aminobenzoic acid, 2-AA) [4] or 2-aminobenzamide [3]. Depending on the fluorophore of choice, the subsequent HPLC analyses can be undertaken using (1) weak or strong anion exchange, (2) hydrophilic interaction, (3) reverse phase, or (4) porous graphitized carbon chromatographies [2]. Where possible, the retention times of the oligosaccharides are compared to that of an external standard and elution is expressed in terms of glucose unit (GU) values (see Fig. 5.23), and GU values obtained with differing column matrices may be used to generate oligosaccharide retention chromatography databases. Additionally, it is possible to combine differing chromatographies, anion exchange and hydrophilic interaction, using a single-column or a tandem-column arrangement. This has been termed *mixed-mode chromatography* [3,4]. However, it has not, up to now, been possible to obtain GU values for charged oligosaccharides when using mixed-mode methods. Most commonly, the method of choice for separating oligosaccharides is by use of HILIC (for review, see Ref. 1).

We have overcome this shortcoming by testing of a number of different HPLC matrices from a number of manufacturers. It was possible, by careful choice of eluant, to develop a method that can separate 2-AA-labeled oligosaccharides into differing pools according to the number of sialic acid residues present and to effect separation of the external standard across the gradient employed in the chromatography. The column employed was a Dionex AS11 column and gave superior resolution when compared to the more commonly employed TSKgel Amide-80 column. The Dionex column matrix is a crosslinked latex covered divinylbenzene/ethylvinylbenzene polymer-based column with alkanol quaternary ammonium functional groups and is normally used for trace anion analysis by strong anion exchange (see http://www.dionex.com for more information). This has allowed use of the column in a combined HILIC and strong anion-exchange mode, which we have termed *hydrophilic interaction anion-exchange* (HIAX) *chromatography*. It is effective for 2AA-labeled oligosaccharide separations [1].

Figure. 5.23 shows the separation of an oligosaccharide pool (neutral and charged GSL-derived, and N-linked neutral and charged oligosaccharides from ovalbumin from fetuin, respectively) using an AS11 or Amide-80 column. As can be seen in Figure 5.23a there was a significant increase in the number of peaks attributable to charged, i.e., sialylated, N-link-derived oligosaccharides following analysis using the AS11 column. A minimum of 24 species were present in the AS11-derived profile, whereas only 13 species were apparent in the Amide-80-derived profile (Fig. 5.23b). There was also significant

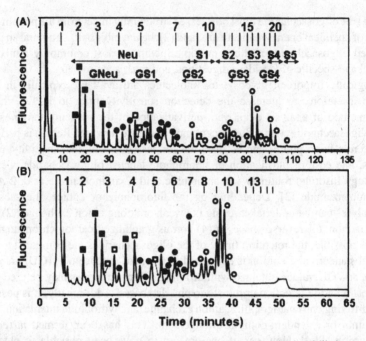

Figure 5.23. HPLC of 2AA-labeled oligosaccharides using either hydrophilic interaction with anion-exchange chromatography (HIAX) using a Dionex AS11 column (A) or hydrophilic interaction chromatography (HILIC) using a TSKgel Amide-80 column (B). The comparative separation of neutral (■) and charged (□) glycosphingolipid (GSL)-derived oligosaccharides, and neutral (●) and charged (○) glycoprotein-derived N-linked oligosaccharides are shown (only major species are highlighted). The retention times of the external standard, in glucose units (GU), are indicated for each profile. Neu, S1, S2, S3, S4, and S5 show the elution periods of neutral and mono-, di-, tri-, tetra-, and penta-sialylated 2-AA-labeled N-linked oligosaccharides, respectively. There is some overlap between the Neu and S1 oligosaccharides. GNeu, GS1, GS2, GS3, and GS4 show the elution periods of neutral and mono-, di-, tri-, tetrasialylated 2-AA-labeled GSL-derived oligosaccharides.

separation between the mono- to tetrasialylated (G1–G4) species using the AS11 column, whereas it was not immediately apparent which peaks were which in the Amide-80, separation. However, the retention on an Amide-80 column is related to oligosaccharide size, charge, and linkage with a general elution of mono- to tetrasialylated species. The GSL-derived also separated according to charge (GS1–GS4), though not in the same positions as the N-link-derived species. This was due to the increase in the number of contacts between the larger branched N-link-derived species when compared to the smaller linear GSL-derived species. When GSL-derived oligosaccharides were analyzed following Amide-80 there was overlap in the elution of oligosaccharide of similar monosaccharide numbers, irrespective of whether neutral or charged. Therefore, the introduction of the Dionex AS11 column for 2-AA-labeled oligosaccharide analysis greatly facilitates the analysis of cellular derived oligosaccharides.

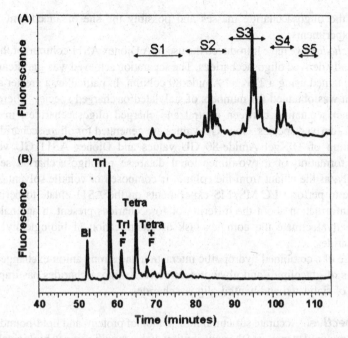

Figure 5.24. HIAX chromatography of α1 acid glycoprotein before (A) and after (B) removal of nonreducing terminus sialic acid residues. S1–S5 denote the elution periods of 2AA-labeled N-linked oligosaccharides containing one to five sialic acids residues, respectively. Bi, Tri, Tri + F, Tetra, and Tetra + F correspond to the elution positions of 2AA-labeled biantennary, triantennary, fucosylated triantennary, tetraantennary, and fucosylated tetraantennary oligosaccharides, respectively.

As stated earlier the complex nature of oligosaccharides can make for difficult analysis of cellular-derived oligosaccharides. This is clearly demonstrated in Figure 5.24a. N-linked oligosaccharides from α1 acid glycoprotein were analysed following separation over a Dionex AS11 column either before (a) or after (b) desialylation. Although the desialylation is partially incomplete, in that some monosialylated species remain, it is apparent that there was a minimum of 30 sialylated species present. Following desialylation five major neutral species were present: biantennary, triantennary, fucosylated triantennary, tetraantennary, and fucosylated tetraantennary oligosaccharides, respectively. This demonstrates that the addition of sialic acid generates a large number of species. This can be explained by the incomplete addition of sialic acid to the nonreducing termini, or hydrolysis of the sialic acid linkages, to give one to four sialic acid residues on the differing neutral structures present. The identification of which peaks correspond to which structure can be performed only following further analyses. This may require purification of the individual peaks and subsequent desialylation and chromatography. By use of linkage-specific enzyme digests, it may also be possible to determine the complete oligosaccharide structure and sequence. The column eluants may also be analyzed following mass spectrometric analysis to determine the oligosaccharide masses and possibly the site of attachment following MS/MS experiments.

In conclusion, we have introduced the use of A Dionex AS11 column for the analysis of biologically derived oligosaccharides. The separation achieved was significantly better than that obtained using a TSKgel Amide-80 column. In particular, a greater amount of information was obtained for numbers of sialylated or charged species. There is also a more distinct separation between neutral and charged oligosaccharides, in particular for the GSL-derived species. Once GU values are generated for oligosaccharide species, a combination of TSKgel Amide-80 GU values and Dionex AS11 GU values will allow the formation of a two-dimensional database of oligosaccharide separations. Additionally, as the eluant from the column is composed of volatile solvents, it should be possible to perform LC-MS/MS experiments on the AS11 eluate to derive further structural information about the differing oligosaccharides present in an analysis. This should greatly facilitate the complex task of identification of biologically important oligosaccharides.

The use of a combined hydrophilic interaction and strong anion-exchange mode for the analysis of glycolipid- and glycoprotein-derived oligosaccharides has improved the separation of sialylated and neutral oligosaccharides.

Perspective. Accurate structural information of protein- and lipid-bound oligosac-charides is required to evaluate the contribution of this modification to biological function.

Improved methods for the nondestructive release of the oligosaccharide, fluor-escence labeling, and separation using HPLC techniques has made a complex analytical problem easier to solve and provided tools that are accessible to most laboratories. Significant challenges remain and improvements in the chromatographic resolution of glycan species in a predictive and rapid fashion will be an important goal. The application of advanced techniques will aid in the identification of disease-related changes to protein and lipid glycosylation leading to an understanding of the pathological mechanisms and therapeutic opportunities.

REFERENCES

1. P. Hemstrom and K. Irgum, Hydrophilic interaction chromatography. *J. Separation Sci.* **29**, 1784–1821 (2006).
2. K. R. Anumula, Advances in fluorescence derivatization methods for high-performance liquid chromatographic analysis of glycoprotein carbohydrates, *Anal. Biochem.* **350**, 1–23 (2006).
3. G. R. Guile, P. M. Rudd, D. R. Wing, S. B. Prime, and R. A. Dwek, A rapid high-resolution high-performance liquid chromatographic method for separating glycan mixtures and analyzing oligosaccharide profiles, *Anal. Biochem.* **240**, 210–226 (1996).
4. D. C. A. Neville, V. Coquard, D. A. Priestman, D. J. M. te Vruchte, D. J. Sillence, R. A. Dwek, F. M. Platt, and T. D. Butters, Analysis of fluorescently labelled glycosphingolipid-derived oligosaccharides following ceramide glycanase digestion and anthranilic acid labelling, *Anal. Biochem.* **331**, 275–282 (2004).

YOSHIO OKAMOTO

Figure 5.25. Yoshio Okamoto.

Yoshio Okamoto (Fig. 5.25) received his bachelor (1964), master (1966), and doctorate (1969) degrees from Faculty of Science, Osaka University. He joined Osaka University, Faculty of Engineering Science, as Assistant Professor in 1969 and spent 2 years (1970–1972) at the University of Michigan as a postdoctoral fellow with Professor C. G. Overberger. In 1983, he was promoted to Associate Professor, and in 1990 he moved to Nagoya University as Professor. In 2004, he retired from Nagoya University, and thereafter he has been Guest Professor at EcoTopia Science Institute, Nagoya University. He has also been appointed as the Chair Professor of Harbin Engineering University in China since 2007. He received the Award of the Society of Polymer Science, Japan (1982); the Chemical Society of Japan Award for Technical Development for 1991; the Chemical Society of Japan Award for 1999; the Molecular Chirality Award (1999); the Chirality Medal (2001); the Medal with Purple Ribbon (Japanese Government) (2002); the Fujiwara Prize (2005); and the Thomson Scientific Research Front Award 2007 (2007), among others.

Development of Chiral Packing Materials Based on Structure Control of Optically Active Polymers. In 1979, the formation of stable one-handed helical poly(triphenylmethyl methacrylate) (PTrMA) was found through the helix-sense-selective polymerization of a prochiral methacrylate using chiral anionic initiators. The chiral polymer exhibited unexpected high chiral recognition of various racemic compounds when used as the chiral packing material (CPM) for HPLC, which was commercialized in 1982 as the first chiral column based on an optically active polymer. This success encouraged us to develop further useful CPMs based on polysaccharides, cellulose, and amylose. By using these polysaccharide-based CPMs, particularly phenyl-carbamate derivatives, nearly 90% of chiral compounds can be resolved not only analytically but also preparatively, and several chiral drugs have been produced using the CPMs.

INTRODUCTION. A pair of enantiomers (optical isomers) often exhibits different physiological and toxicological behavior in biological systems. Therefore, in the past 30 years, the development of chiral drugs with a single enantiomer has attracted great attention in drug industries, and the market for chiral drugs has grown tremendously. The separations of enantiomers by high-performance liquid chromatography (HPLC) have also remarkably advanced since the 1970s and this method has become one of the practically useful methods particularly for the determination of enantiomeric excess (ee). In this short article, I will recount mainly how the currently most popular CPMs based on optically active polymers have been developed in our group since the 1970s.

DISCOVERY OF ONE-HANDED HELICAL POLY(TRIPHENYLMETHYL METHACRYLATE). Since Natta found the helical structure of isotactic polypropylene in the solid state, many stereoregular polymers, such as polystyrene, poly(methyl methacrylate), and poly(vinyl ethers), have been known to have helical conformation in the solid state. However, these polymers cannot stably maintain their helical structure in solution because of high molecular motion of the polymer chains. This also means that the synthesis of one-handed helical polymers with a stable conformation from typical vinyl monomers cannot be readily realized, and therefore, no one predicted such a possibility of obtaining one-handed helical polymers via the homopolymerization of achiral vinyl monomers before we succeeded in the asymmetric synthesis of one-handed helical polymer from a bulky monomer, triphenylmethyl methacrylate (TrMA) (Fig. 5.26).

The first attempt in the helix-sense-selective polymerization of TrMA was carried out in early 1979, and the change in the optical rotation was followed during the polymerization in toluene at −78°C. The rotation did not change at the initial stage, but gradually increased after 1 h and reached a very large positive value, suggesting the formation of the (+)-PTrMA with a prevailing helicity. The system was then gradually warmed, and even at room temperature the optical activity showed almost no change, indicating that the helical structure can be stably maintained.

CHIRAL RECOGNITION. When we succeeded in synthesizing the PTrMA with a high one-handedness, we intended first to evaluate its chiral recognition ability by liquid chromatography with the insoluble polymer with a high molecular weight. The insoluble polymer particles were finely fractionated by sieving, and 20–44-mm particles were packed in an HPLC column. The column showed a rather good plate number as expected, but it was not stably operated for a long time because of the clogging of the terminal filter of the column by the polymer particles. The PTrMA is too brittle to use for this purpose. However, when the polymer was coated on a macroporus silica gel surface, we could

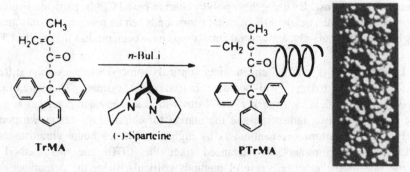

Figure 5.26. Asymmetric synthesis of one-handed helical poly(triphenylmethyl methacrylate).

obtain a useful CPM. The PTrMA-coated CPM was easily packed in an HPLC tube by a slurry method. The packed column showed a high plate number and could be stably used for the resolution of many racemic compounds. The column was commercialized in 1982 as the first polymer-based one.

DEVELOPMENT OF CPMs BASED ON POLYSACCHARIDE DERIVATIVES. Through the development of the CPMs for HPLC with the helical polymethacrylate, we considered that the separation by chiral HPLC will become a very attractive method for both the analysis and preparation of chiral compounds and desired to develop practically valuable CPMs with a high performance. Our attention was given to polysaccharides, particularly to cellulose.

In 1984, we synthesized cellulose trisphenylcarbamate (CTPC) through the reaction of microcrystalline cellulose (Avicel) with phenyl isocyanate and evaluated it as the CPM for HPLC by the above method devised for the PTrMA-based CMP. When we analyzed racemic compounds on the CTPC coated on macroporous silica gel, we were very surprised at its excellent ability. This success resulted in the preparation of many phenylcarbamate derivatives with various substituents on the phenyl group. Until now, we have synthesized more than 100 derivatives, including the derivatives with different substituents regioselectively. Among the many derivatives, cellulose tris(3,5-dimethylphenylcarbamate) (CDMPC, commercial name Chiralcel OD) affords the most powerful CSP with a high chiral recognition for a broad range of compounds.

Amylose was also examined in the same way as described above for cellulose, and once again tris(3,5-dimethylphenylcabamate) (ADMPC, commercial name Chiralpak AD) was found to be the most attractive. With these polysaccharide-based CPMs, nearly 90% of chiral compounds may be resolved. CPMs of new generation immobilized polysaccharide derivatives. In 1987, the first immobilization of the polysaccharide derivatives was examined in my group using diisocyanates, which can bind between aminopropyl silica gel and a cellulose derivative and also between cellulose derivatives. Since then, several immobilization methods have been reported both by us and by other groups, and some immobilized CPMs have been recently commercialized. We recently reported an attractive method using 3-(triethoxysilyl) propyl isocyanate as the binder. This method is applicable to most phenylcarbamate derivatives of cellulose and amylose, and the introduction of a few percentage of 3-(triethoxysilyl)propyl groups is sufficient to attain the high immobilization of the polysaccharide derivatives. The immobilized CPMs show chiral recognition similar to that of the coated CPMs because of the incorporation of a very low immobilized residues.

Future Perspective. As described above, the polysaccharide-based CPMs show very high chiral recognition and today, these are the most popular CPMs. Many chiral compounds can be analyzed with these CPMs. However, these are still not sufficient for the preparative resolution at an industrial scale, which usually asks a high productivity. To attain a high productivity, a CPM must show a high chiral recognition to a target compound. Until now, we have intended to develop the CPMs with a broad applicability. However, for a large-scale separation, we must find out the efficient methods for synthesizing the CSP with a specifically high chiral recognition for a target compound.

REFERENCES

1. Y. Okamoto and T. Nakano, Asymmetric polymerization, *Chem. Rev.* **94**, 349 (1994).
2. Y. Okamoto and E. Yashima, Polysaccharide derivatives for chromatographic separation of enantiomers, *Angew. Chem. Int. Ed.* **37**, 1020 (1998).
3. C. Yamamoto and Y. Okamoto, Optically active polymers for chiral separation, *Bull. Chem. Soc. Jpn.* **77**, 227 (2004).
4. T. Ikai, C. Yamamoto, M. Kamigaito, and Y. Okamoto, Immobilized polysaccharide-based chiral stationary phases for HPLC, *Polym. J.* **38**, 91 (2006).

JANUSZ PAWLISZYN

Janusz Pawliszyn (Fig. 5.27), native of Gdansk, Poland, obtained his engineering B.Sc. and M.Sc. degrees in bioorganic chemistry at the Technical University of Gdansk (1978). He moved to the United States in 1979, where he completed his Ph.D. degree in Analytical Chemistry at Southern Illinois University in 1982 with Professor John Phillips, introducing the concept of concentration modulation in multiplex gas chromatography. During 1982–1984 he worked at the University of Toronto as a postdoctoral fellow with Professor Michael Dignam, designing a photothermal deflection fourier transform infrared (FT-IR) instrument to study electrochemical interfaces.

Figure 5.27. Janusz Pawliszyn.

He is an author of over 400 scientific publications and a book on solid-phase micro-extraction. He is a Fellow of Chemical Institute of Canada, Editor of *Analytica Chimica Acta*, *Trends in Analytical Chemistry*, and a member of the Editorial Board of the *Journal of Separation Science*, *Analyst*, and *Chemia Analityczna* and the Comprehensive Analytical Chemistry book series. He received the 1995 McBryde Medal, the 1996 Tswett Medal, the 1996 Hyphenated Techniques in Chromatography Award, the 1996 Caledon Award, the 1998 Jubilee Medal from the Chromatographic Society, UK, the 2000 Maxxam Award, the 2000 Varian Lecture Award from Carleton University, the Alumni Achievement Award for 2000 from Southern Illinois University, the Humboldt Research Award for 2001, the 2002 COLACRO Medal, and the 2003 Canada Research Chair. In 2006 he was elected to the most frequently cited chemists by ICI. In 2008 he received the A. A. Benedetti-Pichler Award from the Eastern Analytical

Symposium, the Andrzej Waksmundzki Medal from the Polish Science Academy, and the Manning Principal Award. He presently holds the Canada Research Chair and NSERC Industrial Research Chair in New Analytical Methods and Technologies.

Professor Pawliszyn began his independent scientific career in 1984 as an Assistant Professor at Utah State University, developing new concepts of universal and absorption concentration gradient detection based on probe laser-beam deflection. He applied this scheme to investigate electrode surfaces, transport through membranes, and detection in high-efficiency separation techniques, including capillary electrophoresis. He also proposed the use of laser desorption to facilitate rapid gas chromatographic separation of poorly volatile polymeric compounds. In his research, he explored the application of new electrooptical devices such as optical fibers and light-emitting and laser diodes. Continuation of this work, a few years later, led to the invention and development of solid-phase microextraction. In 1988, Professor Pawliszyn moved to the University of Waterloo, where he progressed through the ranks and in 1997 was promoted to the position of Full Professor. He developed a strong analytical program focusing on the area of analytical separation and sample preparation. He is internationally recognized, particularly in the latter area, for his fundamental contributions to supercritical-fluid, solid-phase, and membrane extraction techniques. Several other sample preparation techniques, such as hot-water extraction and accelerated solvent extraction, were proposed after his and his coworkers' fundamental work on high-temperature supercritical-fluid extraction. In the area of analytical separation, he introduced the concept of whole-column detection, which was used recently as a basis for a commercial product by a Toronto-based company, Convergent Bioscience. He also demonstrated use of cross-correlation chromatography in combination with membrane extraction. This work led to the development of membrane extraction with a sorbent interface, a sample preparation technique for continuous monitoring of organic compounds in onsite environments.

During his work, Dr. Pawliszyn developed strong relationships with manufactures of analytical instrumentation, which allowed him to commercialize his new concepts. The most successful idea, solid-phase microextraction, was introduced to the market by Supelco and Varian in 1994. In 1995, Dr. Pawliszyn became an Industrial Research Chair in Analytical Chemistry, originally supported by NSERC, Supelco, and Varian for a 5-year term. Since then the program has been extended twice, involving a long list of companies including the Canadian based division of Eli Lilly, Merck-Frosst, Smiths Detection, NoAb Biodiscoveries, and Convergent Bioscience. The collaboration also involved several international companies: Firmenich (Switzerland), Leap Technologies (USA), Leco (USA), Gerstel (Germany), PAS Technologies (Germany), and Shinwa Chemical Industries (Japan).

Professor Janusz Pawliszyn's main scientific contributions include novel sample preparation technologies. He introduced and developed such techniques as solid-phase microextraction (SPME), membrane extraction with a sorbent interface (MESI), needle trap (NT), and thermal pumping for supercritical fluids. He also made significant contributions to development of optical detection system for capillary electrophoresis and capillary isoelectric focusing, such as concentration gradient and whole-column imaging detection (WCID).

The main goal of the sample preparation research in Dr. Pawliszyn's laboratory is to move the analysis, or at least sample preparation process, onsite where the investigated system is located, rather than transporting the samples to the laboratory. The proposed devices are small and very simple in principle. To make his technology available to the users, Dr. Pawliszyn is collaborating with a range of companies specializing in sample preparation technology and the design of portable analytical instrumentation. The novel techniques introduced by Professor Pawliszyn and his team are recognized to be very practical and are being developed into products or have already been made commercially available. The solvent-free solid-phase microextraction (SPME) technology has had the most significant impact on the analytical community by simplifying the sample preparation step. In this technique a small coated fiber mounted inside a syringe needle is used to replace typical extraction procedures that involve the use of copious amounts of organic solvents. The extraction can be fully automated for laboratory application or adopted for onsite or in vivo applications. Devices based on this technology are presently available from Supelco, and automated systems are available for gas chromatographic instruments from a majority of manufactures. The technique is used worldwide for a range of analytical applications from basic research in flavor and fragrance to continuous monitoring of drinking-water quality. The impact of the SPME concept on the scientific community has been substantial considering that 3 out of the top 20 articles published in *Analytical Chemistry* over the last decade have been SPME articles originating form the Dr. Pawliszyn's laboratory. Currently the primary focus of the SPME research is on in vivo sampling involving not only animals but also humans. The main reason for interest in this area is driven by the ability of SPME to extract only components of interests from the bloodstream or tissue without removing any matrix. SPME, using an externally coated extraction phase on a microfibre mounted in a syringe-like device seems to be a logical target for the development of such tools. As in any microextraction, compounds of interest are not exhaustively removed from the investigated system. On the contrary, conditions can be devised where only a small proportion of the total compound is removed, thus avoiding a disturbance of the normal balance of chemical components. Also, because it is a syringe-like device that can be physically removed from the laboratory environment for sampling, it is amenable to the monitoring of a living system in its natural environment, avoiding the need to move the living system to an unnatural laboratory environment.

In addition to commercialization of SPME, Supelco has also been involved in manufacturing the thermal pump for supercritical-fluid extraction (SFE) developed at the University of Waterloo (UW). MESI is another solvent-free sample preparation technique able to simplify continuous monitoring in the onsite environment. The technique is under development as a part of a microscale ion mobility spectrometer (Smiths Detection). Professor Pawliszyn's research focus is on a fundamental understanding of the extraction processes and then the application of this knowledge to realize advantages in developing new techniques and methodologies.

The idea of whole-column detection originates from the early work on universal and selective concentration gradient imaging. This concept has been extended to adsorption and fluorescence modes. Whole-column detection has been commercialized by a new startup company, Convergent Bioscience of Toronto, as the imaging capillary isoelectric focusing system. We also proposed an ampholyte-free approach to capillary isoelectric focusing, an approach that is very useful in preparative scale separation of proteins and preconcentration of proteins prior to mass spectrometric analysis, since the interfering ampholytes are eliminated.

Dr. Pawliszyn is actively disseminating the developed technology to users by organizing courses, symposia, and conferences. His SPME Theory and Practice course has been taught biannually at the University of Waterloo since 1996 and has became a permanent fixture at the PITTCON and HTC meetings. This course has also been taught at several universities and institutions at locations around the globe, including Australia, New Zealand, Germany, France, the UK, Italy, Poland, Czech Republic, Belgium, Finland, Sweden, Norway, and China. He initiated ExTech, an annual meeting dedicated to new developments in extraction technologies. He is an author of over 350 scientific publications and a book on solid-phase microextraction. The impact of his research and inventions on the practice of analytical chemistry has been substantial as demonstrated by the ISI Highly Cited status, and it is expected to grow as new sample preparation technologies developed in his laboratory are more widely accepted.

Research Activity and Biographical Sketch. The primary focus of Professor Pawliszyn's research program is the design of highly automated and integrated instrumentation for the isolation of analytes from complex matrices and the subsequent separation, identification and determination of these species. Currently his research focuses on elimination of organic solvents from the sample preparation step to facilitate onsite monitoring and analysis. Several alternative techniques to solvent extraction are investigated, including use of coated fibers, packed needles, membranes, and supercritical fluids. Dr. Pawliszyn is exploring application of the computational and modeling techniques to enhance performance of sample preparation, chromatographic separations, and detection. The major area of his interest involves the development and application of imaging detection techniques for microcolumn chromatography and capillary electrophoresis.

REFERENCES

1. J. Pawliszyn, Sample preparation— *Quo Vadis? Anal. Chem.* **75**, 2543–2558 (2003).
2. G. Ouyang and J. Pawliszyn, A critical review in calibration methods for solid-phase microextraction, *Anal. Chim. Acta* **627** 184–197 (2008).
3. F. Musteata and J. Pawliszyn, Bioanalytical applications of SPME, *Trends Anal. Chem.* **26**, 36–45 (2007).
4. G. Ouyang and J. Pawliszyn, SPME in environmental analysis, *Anal. Bioanal. Chem.* **386**, 1059–1073 (2006).

ROBERTO SAELZER AND MARIO VEGA

Figure 5.28. Roberto Saelzer.

Professor Roberto Saelzer (Fig. 5.28) is a Professor at the University of Concepcion, (Chile) working since 1974 at the Food Science and Nutrition Department belonging to the Faculty of Pharmacy. His undergraduate studies were undertaken at the University of Concepcion obtaining his Pharmaceutical-Chemist title from the University of Chile in 1971. Professor Saelzer has a Master's degree in Nutrition from the University of Barcelona (Spain). He is a teacher of pregraduate students of Chemistry and Pharmacy and Nutrition and Dietetics in the topics of food chemistry and food control.

Since 2000 he has been the Associate Director of Teaching of the University, where he coordinates all the undergraduate programs. Previously he occupied the Vice Dean's cargo of the Faculty of Pharmacy for two consecutive periods.

Figure 5.29. Mario Vega.

Professor Mario Vega (Fig. 5.29) is also a Professor at the University of Concepcion, belonging to the same Department since 1972. He studied Chemistry and Pharmacy at the University of Concepcion and got his Pharmaceutical Chemist Title from the University of Chile in 1972. He is Master of Science in Human Nutrition. He teaches food science and nutrition to undergraduate students of Pharmacy, Biochemistry and Nutrition. He is involved in technical assistance for the food industry in the area regarding quality control and contaminants in both cases for finished products and raw materials.

Since 2006 he has served as chair of the Food Science and Nutrition Department of Faculty of Pharmacy.

See Chapter 1, Table 1.3, b4, p, r.

Quantitative Thin Layer Chromatography—Food Analysis. Both academics have been involved in the last 25 years (since the early 1980s) in development and application of quantitative thin-layer chromatography for food analysis. Their current research work is concerned with mycotoxins and antibiotics analysis in human food, animal feed, and salmon. They have had an active contribution in the diffusion of the application of HPTLC for the quantitative analyses in Chile and Latin America in the framework of COLACRO activities (Latin American Congress on Chromatography), namely pre-Congress courses, round tables, and technical seminars, activities that were recognized with the COLACRO Medal award in Buenos Aires, Argentina in 2000.

Their research activities allow under- and postgraduate students the possibility to learn about modern thin-layer chromatography (HPTLC) and how to apply it to solve analytical problems. The most recent aspects of their research work include their participation in the MYCOTOX project (ref ICA4-CT-2002-10043) entitled *The Development of a Food Quality Management System for the Control of Mycotoxins in Cereal Production and Processing Chains in Latin America South Cone Countries*, which started at the beginning of 2003. It involved partners from France, the UK, Argentina, Brazil, Chile, and Uruguay. The overall objective of the project was to improve the competitiveness of domestically and internationally traded cereals by controlling the occurrence of mycotoxins in maize and wheat products used as human food and animal feed. The project was based on a multidisciplinary approach, including analytical, technological, and socioeconomic components in order to ensure, jointly with all stakeholders, both quality and safety throughout maize and whole-wheat chains. At the present time they are participating in a research project focused on ochratoxin A in Chilean wines.

Principal Research and Fields of Application. Our laboratory has long been involved in method development using HPTLC as the tool of choice for quantitative evaluations in the fields of food and feeds analyses. As we belong to a pharmaceutical faculty, we are also involved in research in the field of pharmaceutical applications, herbals, cosmetics, forensic and environmental applications, in supporting others with the devices for and the knowledge of the HPTLC technique, and in giving to students and young researchers the opportunity to engage in research projects using our laboratory facilities.

Our efforts to demonstrate the reliability of HPTLC have been fruitful, and many food industries have incorporated this technique to their quality control laboratories, on the grounds of its flexibility, speed, and cost-effectiveness. All of this occurred because we have been successful in changing the image of a conventional qualitative and old-fashioned analytical tool into a modern one in the state-of-the-art technique, complementary to GC and HPLC. Training courses to keep the users up-to-date with applications have been part of our responsibilities; these courses have been provided for Chilean and citizens others from Latin America.

Also we have published our results applying HPTLC in several journals and through conferences, seminars, and short courses using as a diffusion platform the COLACRO held in different countries of North and South America. Suitable instruments are required to obtain the best results; in this respect it is important to keep in mind the permanent support given by the manufacturers of adequate devices affordable both for the university and for the industry that need it. Automation and optimization of the software and instrumentation facilitates the TLC procedure, giving separations such as those shown in Figure 5.30.

Other applications of HPTLC in quality control include analysis of compounds such as antibiotics in fish feed and tissue, mycotoxins such as aflatoxins, deoxynivalenol, zearalenon, and ochratoxin A in wine and human plasma.

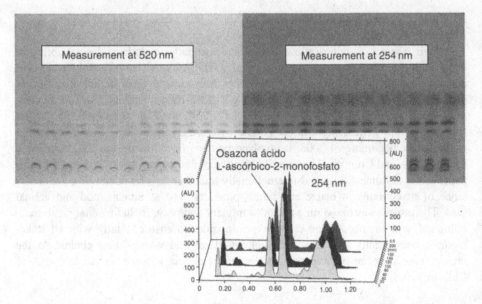

Figure 5.30. HPTLC analysis of vitamin C phosphate in a complex matrix such as salmon feed.

Perspective. Our perspective of HPTLC for the twenty-first century is to keep it as a quantitative tool with all the advantages that it possesses, such as side-by-side visual comparison of samples, standards, and unknowns, complemented by the recent advances in connecting HPTLC with mass spectrometry, which confirms the flexibility of planar chromatography.

Some Publications

1. M. Vega, R. Saelzer, C. Figueroa, and V. H. Jaramillo, Use of AMD-HPTLC for analysis of biogenic amines in fish meal, *J. Planar Chromatogr.* **1**, 72 –75 (1999).

2. R. Saelzer, G. Godoy, M. Vega, M. de Diego, R. Godoy, and G. Ríos, Instrumental planar chromatographic determination of benzopiazepines: Comparison with liquid chromatography and gas chromatography, *J. AOAC Int.* **84**(4), 1287–1295 (2001).

3. A. Pacin, S. Resnik, M. Vega, R. Saelzer, E. Ciancio Bovier, G. Ríos, and N. Martínez, Ocurrence of ochratoxin A in wines in the Argentinian and Chilean markets, *ARKIVOC* **xii**, 214–223 (2005).

4. D. von Baer, E. von Baer, P. Ibieta, M. Vega, and R. Saelzer, Isoflavones in Lupinus albus and Lupinus angustifolius: Quantitative determination by capillary zone electrophoresis, evolution of their concentration during plant development and effect on anthracnose causing fungus Colletotrichum lupine, *J. Chil. Chem. Soc.* **51**(4), 1025–1029 (2006).

5. M. B. Aranda, M. H. Vega, and R. F. Villegas, Routine method for quantification of starch by planar chromatography (HPTLC), *J. Planar Chromatogr.* **18**, 289–293 (2005).

6. M. H. Vega, E. T. Jara, and M. B. Aranda, Monitoring the dose of florfenicol in medicated salmon feed by planar chromatography (HPTLC), *J. Planar Chromatogr.* **19**, 204–207 (2006).

7. K. Muñoz, M. Vega, G. Rios, S. Muñoz, and R. Madariaga, Preliminary study of Ochratoxin A in human plasma in agricultural zones of Chile and its relation to food consumption, *Food Chem. Toxicol* **44**(11), 1884–1889 (2006).

8. G. E. Morlock and M. Vega-Herrera, Two new derivatization reagents for planar chromatographic quantification of sucralose in dietetic products, *J. Planar Chromatogr.* **20**, 411–417 (2007).

PATRICK J. SANDRA

Figure 5.31. Patrick J. Sandra

Pat J. Sandra (Fig. 5.31) (born October 20, 1946 in Kortrijk, Belgium) received his M.S. Degree in Chemistry in 1969 and his Ph.D. in Sciences in 1975 from Ghent University, Belgium. Since then, he has been on the Faculty of Sciences at the same university, where he is currently Professor in Separation Sciences at the Department of Organic Chemistry. In 1986 he founded the Research Institute for Chromatography (RIC) in Kortrijk, Belgium, a center of excellence for research and education in chromatography, mass spectrometry, and capillary electrophoresis. He is currently also an extraordinary professor at the Department of Chemistry at the University of Stellenbosch, South Africa; at the Department of Analytical Chemistry, Evora, Portugal; and Director of the Pfizer Analytical Research Center at Ghent University.

Pat Sandra is active in all fields of separation sciences (GC, LC, SFC, and CE), and major areas of his research are high-throughput, high-resolution gas chromatography (GC) miniaturization, hyphenation, and automation for study of chemicals, pharmaceuticals, natural products, food products, and pollution.

He is the author or co-author of more than 470 scientific papers with peer review and has written books on high-resolution GC, sample introduction in capillary GC, essential oil analysis, micellar electrokinetic chromatography and water analysis.

He was the recipient of the 1989 Tswett Award (Russia), the 1994 Dal Nogare Award (USA), the 1994 Martin Gold Medal (Chromatographic Society, UK), the 1995 Golay Award (USA), the 1996 Pharmaceutical Society Award 2005 (Slovak Republic), and the American Chemical Society (ACS) Chromatography Award 2005 (USA). He also received the Doctor Honoris Causa title in Pharmacy from the University of Turin, Italy (2004) and in Food Chemistry and Safety from the University of Messina, Italy (2007). He was appointed Honorary Professor at the

Dalian Institute for Chemical Physics, Chinese Academy of Sciences in June 2007. In November 2008 he received the American EAS award in New York and the CASSS Award in San Francisco for outstanding contributions to separation sciences.

See Chapter 1, Table 1.3, d, h, o, p, t, v, w.

My Science in Chromatography. During 2000–2005, several new developments were made in sample preparation. Stir-bar sorptive extraction (SBSE) [1] was fully automated, and the number of applications increased steadily [2]. Sorptive tape extraction (STE) was developed in collaboration with Chanel for the characterization of the impact of cosmetics on skin [3].

In 2003 in collaboration with Pfizer, the Pfizer Analytical Research Centre (PARC) was established at Ghent University, and I became more involved in fluid-based separation techniques. Keywords in our research were (and still are) high productivity, high productivity, and high resolution. Research included evaluation of ultra-high-pressure and elevated-temperature LC, the latter in collaboration with Selerity, a company based in Salt Lake City that is now SandraSelerity Technologies. A typical chromatogram of a high-efficiency LC analysis of a tryptic digest of serum after depletion of the most abundant proteins is shown in Figure 5.32 [4]. To increase peak capacity even further, we also started research in comprehensive LC × LC [5] and on long monolithic columns. The scientific output of PARC was quite successful, and more than 30 papers could be published. Our contributions to LC have been internationally recognized. We organized,

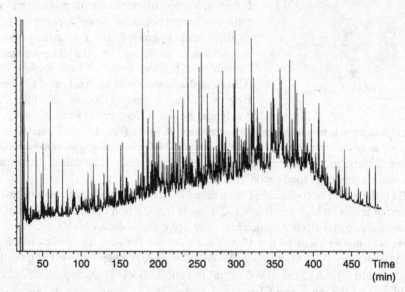

Figure 5.32. Tryptic digest of serum after removal of most abundant proteins: 12 × 25-cm × 2.1-mm-i.d. columns 300 Å Stablebond C18; temperature 60°C; flow rate 200 µL/min; gradient: A—0.1% TFA in 2% acetonitrile; B—0.1% TFA in 70% acetonitrile, 0—70% B in 500 min; 20 µL injection; DAD 214 nm; DP 500 bar; peak capacity 1230.

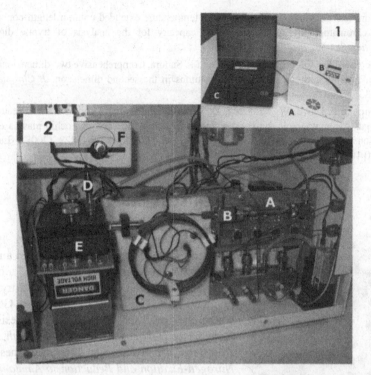

Figure 5.33. Micro-gas chromatograph. Panel 1: A—casing; B—keyboard; C—laptop for data acquisition; Panel 2: A—chemical trap; B—low dead volume connector; C—column with resistively heating jacket; D—optical fiber; E—plasma chip cartridge; F—spectrometer.

in collaboration with Prof. J. Crommen of the Université de Liège, Belgium, the HPLC Meeting 2007 in Ghent. More than 1200 participants attended the meeting.

At the Research Institute for Chromatography, we developed a portable micro-gas chromatograph equipped with a plasma emission detection microchip (Fig. 5.33) [6]. The system is presently evaluated for in-field monitoring. We also started a new division for metabolic studies (Metablys).

The following references are key publications.

REFERENCES

1. E. Baltussen, P. Sandra, F. David, and C. Cramers, Stir bar sorptive extraction (SBSE), a novel extraction technique for aqueous samples: Theory and principles, *J. Microcol. Separations* **11**, 737–747 (1999).

2. F. David, and P. Sandra, Stir bar sorptive extraction for trace analysis, *J. Chromatogr. A* **1152**, 54–69 (2007).

3. S. Sisalli, A. Adao, M. Lebel, and P. Sandra, Sorptive tape extraction—a novel sampling method for the in vivo study of skin, *LC GC Eur.* **19**(1) 33–39 (2006).

4. P. Sandra and G. Vanhoenacker, Elevated temperature-extended column length conventional liquid chromatography to increase peak capacity for the analysis of tryptic digests, *J. Separation Sci.* **30**, 241–244 (2007).

5. I. Francois, A. De Villiers, B. Tienpont, and P. Sandra, Comprehensive two-dimensional liquid chromatography applying two parallel columns in the second dimension, *J. Chromatogr. A* **1178**, 33–42 (2008).

6. B. Tienpont, F. David, W. Witdouck, D. Vermeersch, H. Stoeri, and P. Sandra, Features of a micro-gas chromatograph equipped with enrichment device and microchip plasma emission detection (μPED) for air monitoring, *Lab-on-a-Chip* **8**, 1819–1828 (2008) (Lab Chip DOI:10.1039/B811599K).

VOLKER SCHURIG

Figure 5.34. Volker Schurig.

Volker Schurig (Fig. 5.34) was born 1940 in Dresden (Saxony, Germany). He received a musical education at the "Dresdener Kreuzchor," a choir of 800 years' tradition. He studied chemistry at the Eberhard-Karls-University of Tübingen, Germany, where he received the Diploma (thesis title: *Reaction of Halomuconic Acid Esters with Sodium Iron Tetracarbonylate*), the Doctorate (thesis title: *Nitrogen-Fixation and Reduction to Ammonia*), and the Habilitation (thesis title: *Thermodynamics and Analytics of the Molecular Association of Donor-Molecules with Transition Metal Chelates by Complexation Chromatography*). He held postdoctoral positions at the Weizmann Institute of Science, Israel, with Professor Emanuel Gil-Av, and at The University of Houston, United States, with Professor Albert Zlatkis, working on selective separations by complexation gas chromatography with the main focus on attempts to separate enantiomers. He is Professor of Organic Chemistry (stereochemistry and separation methods) at the Eberhard-Karls-University of Tübingen, Germany. He has supervised more than 50 diploma and doctoral theses. He was Visiting Professor at the Université Paris-Sud in Orsay, invited by Professor Henri B. Kagan, and at the Weizmann Institute of Science, Israel, invited by Professor Emanuel Gil-Av. He retired in 2007. He is author and co-author of 390 general and specific scientific publications, including patents. He has lectured at 250 scientific institutions, industries and conferences, respectively. He organized the Third International Symposium on Chiral Discrimination in 1992.

Recent Awards

Honorary Membership of the Slovak Pharmaceutical Society (2001)

Fellowship Award of the Japanese Society for the Promotion of Science (2003)

The Swiss "Convention Intercantonale Romande pour L'enseignement du 3e Cycle en Chimie" Lecture (2003)

The M. J. E. Golay Award and Medal of Chromatography, Riva del Garda, Italy (2004)

Chirality Medal in Gold, New York, USA (2004)

Research Prize of the Transtec-Lantronix Foundation (Tübingen, Germany) on the topic miniaturized gas chromatography for the determination of extraterrestrial asymmetry (2005)

See Chapter 1, Table 1.3, a, e, o, q.

Major Research Achievements. An account on the major research achievements of Volker Schurig and his coworkers (Fig. 5.35)—as placed in historical context—has previously been published [1]. Recent research endeavors are summarized as follows. The separation of enantiomers on chiral transition metal chelates (Chirasil-Metal) has been continued. While the practical application of enantioselective complexation gas chromatography became less example important following the introduction of cyclodextrins as chiral selectors [2], the method offers interesting insights into the mechanisms of enantiorecognition [3]. Following the first of a series of enantiomerization phenomena in gas chromatography in 1984 [4] (Fig. 5.36), interconversion barriers of stereolabile drug substances have been determined [5].

The synthesis of three regioselective Chirasil-Dex stationary phases (permethylated-β-cyclodextrin bonded to poly(dimethylsiloxane) has been accomplished [6]. A

Figure 5.35. Ernst Bayer, Emanuel Gil-Av, and Volker Schurig (with Vadim Davankov in the background) at the 3rd International Symposium on Chiral Discrimination, Tübingen, 1992.

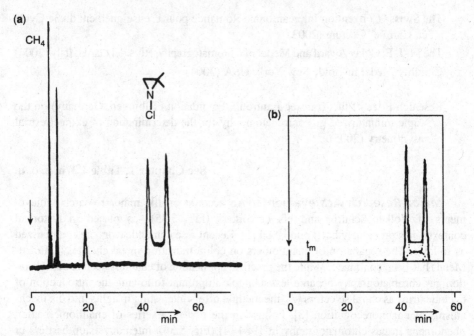

Figure 5.36. First observed enantiomerization phenomena in chromatography: 1-chloro-2,2-dimethylaziridine on nickel(II)-bis[3-(trifluoroacetyl)-(1R)-camphorate] in squalane. Comparison of the experimental (a) and simulated (b) interconversion profile [4].

commercially available capillary column coated with immobilized Chirasil-Dex (Varian, Inc.) is integrated in the COSAC experiment as part of the payload of the Rosetta mission of European Space Agency (ESA) heading toward the comet 67P/Churyumov-Gerasimenko in an effort to determine extraterrestrial homochirality in space. Miniaturization of enantioselective GC toward outer-space experiments has been investigated [7] (Fig. 5.37). Aiding research devoted to the amplification of chirality as a precondition of primordial life, the precise and accurate determination of minute deviations from the racemic composition of α-amino acids has been evaluated by enantioselective GC [8].

Chirasil-Dex-coated monoliths, prepared by different methods, including sol-gel chemistry, have been used for enantioseparation in capillary electrochromatography (CEC [9]). The online coupling of CEC with coordination ion spray–mass spectrometry (CIS-MS) has been developed for the separation of apolar enantiomers [10]. The first example of enantioseparations by open tubular CEC has been extended to a unified approach using a single column for enantioselective GC, SFC, LC and CEC as well as to the hyphenation of CEC-MS [11] (Fig. 5.38).

The present research is concerned with the investigations on large ring cyclodextrins, linear dextrins as chiral selectors, and mixed binary chiral stationary phases for enantioselective GC as well as the enantioseparation of unfunctionalized hydrocarbons.

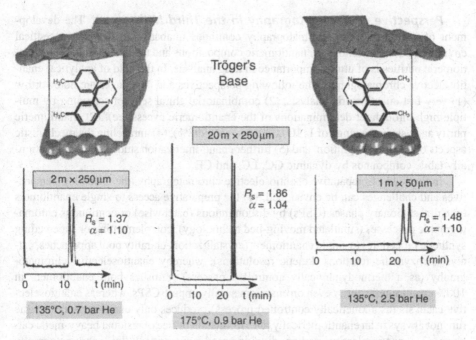

Figure 5.37. Miniaturization of enantioselective gas chromatography. Influence of column parameters on enantioselectivity α and enantioresolution R_s [7].

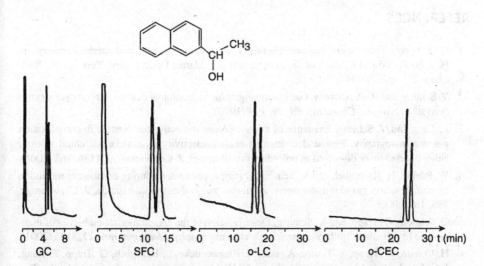

Figure 5.38. Unified enantioselective chromatography. Enantioseparation of 1-(2-naphthyl)ethanol on Chirasil-Dex (0.15 μm) by GC, SFC, LC, and CEC on the same 1 m (92 cm effective) $\times$ 0.05-mm-i.d. open-tubular fused-silica capillary column [11].

Perspective of Chromatography in the Third Millennium. The development of enantioselective chromatography continues unabated. In the pharmaceutical environment, the screening of enantiomeric compositions and the detection of high enantiomeric purities is of utmost importance in drug analysis. In the field of analytical enantioselective chromatography, the following perspectives and challenges are noteworthy: (1) very fast enantiomeric analysis, (2) combinatorial chiral selector screening by multiple high-throughput determinations of the enantiomeric excess (ee), (3) enantiomeric purity analysis in the range of 1 : 100,000 (ee = 99.998%), (4) unraveling the mechanistic aspects of enantiorecognition, and (5) further enantiomerization studies of configurationally labile compounds by dynamic GC, LC, and CE.

In the field of preparative enantioselective chromatography, the following perspectives and challenges can be envisioned: (1) the preparative access to single enantiomers on chiral stationary phases (CSPs) by discontinuous (batchwise) or continuous chromatographic processes (simulated moving-bed technology), complementing or superceding synthetic procedures to single enantiomers (crystallization, chirality pool approaches, catalysis, enzymatic reactions, kinetic resolutions), whereby enantioselective chromatography (as a thermodynamically controlled process) furnishes both enantiomers in 100% enantiomeric purity even on enantiomerically impure CSPs, whereas enantioselective catalysis (as a kinetically controlled process) produces only one enantiomer in one run, not always in an enantiomerically pure form and not free of residual heavy-metal catalyst; (2) enantioselective carbon dioxide-based supercritical fluid chromatography (SFC) will gain momentum as a "green" alternative to liquid chromatography; and (3) refinement of monoclonal antibodies and synthetic plastibodies (molecularly imprinted polymers, MIPs) as CSPs in enantioselective chromatography.

REFERENCES

1. V. Schurig, From metal-organic chemistry to chromatography and stereochemistry, in H. J. Issaq (Ed.), *A Century of Separation Science*, Marcel Dekker, New York Basel, 2001, Chap. 21, pp. 327–347.

2. V. Schurig and H.-P. Novotny, Gas chromatographic separation of enantiomers on cyclodextrin derivatives, *Angew. Chem. Int. Ed.* **29**, 939 (1990).

3. Z. Jiang and V. Schurig, Existence of a low isoenantioselective temperature in complexation gas chromatography. Profound change of enantioselectivity of a nickel(II) chiral selector either bonded to, or dissolved in poly(dimethylsiloxane), *J. Chromatogr. A* **1186**, 262 (2008).

4. W. Bürkle, H. Karfunkel, and V. Schurig, Dynamic phenomena during enantiomer resolution by complexation gas chromatography. A kinetic study of enantiomerization, *J. Chromatogr.* **288**, 1 (1984).

5. G. Schoetz, O. Trapp, and V. Schurig, Determination of the enantiomerization barrier of thalidomide by dynamic capillary electrokinetic chromatography, *Electrophoresis* **22**, 3185 (2001).

6. H. Cousin, O. Trapp, V. Peulon-Agasse, X. Pannecoucke, L. Banspach, G. Trapp, Z. Jiang, J. C. Combret, and V. Schurig, Synthesis, NMR spectroscopic charaterization and polysiloxane-based Immobilization of the three regioisomeric monooctenylpermethyl-β-cyclodextrins and their application in enantioselective GC, *Eur. J. Org. Chem.* 3273 (2003).

7. V. Schurig and H. Czesla, Miniaturization of enantioselective gas chromatography, *Enantiomer* **6**, 107 (2001).

8. C. Reiner, G. J. Nicholson, U. Nagel, and V. Schurig, Evaluation of enantioselective gas chromatography for the determination of minute deviations from racemic composition of α-amino acids with emphasis on tyrosine: Accuracy and precision of the method, *Chirality* **19**, 401 (2007).

9. J. Kang, D. Wistuba, and V. Schurig, A silica monolithic column prepared by the sol-gel process for enantiomeric separation by capillary electrochromatography, *Electrophoresis* **23**, 1116 (2002).

10. A. von Brocke, D. Wistuba, P. Gfrörer, V. Schurig, and E. Bayer, On-line coupling of capillary electrochromatography with coordination ion spray-mass spectrometry for the separation of enantiomers, *Electrophoresis* **23**, 2963 (2002).

11. V. Schurig and S. Mayer, Separation of enantiomers by open capillary electrochromatography on polysiloxane-bonded permethyl-β-cyclodextrin, *J. Biochem. Biophys. Meth.* **48**, 117 (2001).

FRANTISEK SVEC

Figure 5.39. Frantisek Svec.

Frantisek Svec (Fig. 5.39) was born on September 3, 1943 in Prague, Czechoslovakia. He studied polymer chemistry at the Institute of Chemical Technology (ICT), where he received his Ph.D. in 1969. He then remained in that school for another 7 years as an Assistant Professor. He moved to Institute of Macromolecular Chemistry of the Czechoslovak Academy of Sciences in Prague in 1976, where he went through the ranks to become Scientific Secretary of the Institute and Technology Transfer Manager. Despite the widespread skepticism and belief that no HPLC system could force a liquid to flow through a typical macroporous material, he started a search for novel materials with excellent flowthrough properties designed for separation. Thus, the first generation of rigid porous polymer monoliths in thin disk format was hatched.

In 1992 he accepted invitation to join faculty at Cornell University. Being at the new place, he started new projects. One led to extending the monoliths from disks to regular chromatographic column shape. Using his intimate knowledge of the fundamentals that control pore formation enabled him to design a very simple yet efficient approach to monolithic columns involving the in situ polymerization of a mixture of monomers and porogenic solvents directly within the housing of the separation column for rapid and highly selective HPLC separation of proteins, peptides, oligonucleotides, and synthetic polymers. Another project concerned the direct preparation of monodisperse beads for HPLC separations.

He then went west, following the old American tradition and anchored in 1996 at the Pacific coast at the University of California, Berkeley. This move enabled him to broaden the scope of his research. Here he conceived a great variety of porous beads with different surface chemistries, enabling new applications in HPLC. Following the success of the monolithic technologies, he extended their applications to numerous other areas. Although remaining affiliated with the university, in 2005, he accepted position of Facility Director at the newly constructed Molecular Foundry in the Lawrence Berkeley National Laboratory. Here he focuses his attention on capillary and microfluidic formats for micro- and nanoscale HPLC, for which the monolithic stationary phases are extremely well suited. In addition to chromatographic studies involving use of monoliths in nanoscale HPLC, GC, CEC, and TLC, he also carries out research related to hydrogen storage in nanoporous polymers, and superhydrophobic polymer surfaces.

He also accepted the position of Visiting Professor of Analytical Chemistry at the University of Innsbruck, Austria, where he has taught for 3 consecutive years since 2003. In the same year, he was elected President of the California Separation Science Society (now CASSS). He is Editor-in-Chief of the *Journal of Separation Science* and member of international advisory boards of many major chromatographic journals. His work resulted in over 360 publications, including 56 book chapters and review articles, three books, and 75 patents. He has received numerous honors, including the M. J. E. Golay Award and Medal in Chromatography and the Eastern Analytical Symposium (EAS) Award for Achievements in Separation Science (2005), the ACS Award in Chromatography (2008), and the Dal Nogare Award in Chromatography (2009).

See Chapter 1, Table 1.3, e, h, i, n.

New Advances in Chromatography Column Technology: Porous Polymer Monoliths

INTRODUCTION. *Monoliths* are defined as continuous porous materials used mostly as separation media and supports that completely fill the separation device. They compete with well-established technologies such as separation columns packed with particulate porous stationary phases. In packed columns, the liquid flows readily through the interstitial voids between the particles where resistance to flow is the lowest, while the liquid present in the pores does not move and remains stagnant. Thus, the components of the separated mixture must diffuse from the main stream into the pore and back to the flow again. This mass transport is sufficiently fast for small molecules, but it is considerably slower for large molecules such as proteins, nucleic acids, and synthetic polymers because their diffusion coefficients are several orders of magnitude smaller than those of low-molecular-weight compounds. Since there are difficulties in achieving the high-speed chromatographic separation of large molecules using columns packed with typical porous particles, new approaches have been sought to accelerate mass transfer in porous packings. These efforts have led to the development of both smaller porous and nonporous particles as well as particles consisting of a nonporous core covered with a porous shell. In contrast, a monolithic column is formed by a cocontinuous phase of a solid material and pores. Now, the mobile phase must flow through the pores and the mass transport occurs mostly via convection, which is much faster than diffusion. The configuration of the monoliths enables independent optimization of the size of the flowthrough channels as well as the "depth" of the pores accessible only by diffusion and the simultaneous maximization of both column permeability and mass transport. Theoretical modeling and practical results demonstrated that this type of stationary phase affords highly efficient separations with low resistance to hydraulic flow.

Although the first thoughts concerning structures similar to those that we now call monoliths were already mentioned in works by the Nobel Prize winner R. Synge in the early 1950s, soft materials and technologies available at that time were unsuitable for their preparation. Experiments with hydrogel columns and open-pore polyurethane foams that followed a decade later were not very successful, either, and the idea of monolithic separation media had almost been forgotten. Their rebirth occurred in the late 1980s in work carried out by my group in Prague focusing on porous polymer monoliths in a disk format punched from a monolithic sheet or cut from monolithic cylinders [1]; this work was continued at Cornell University in the early 1990s by developing column formats with the monoliths prepared in situ [2] and by S. Hjertén in Uppsala, who used highly compressed ionizable polyacrylamide gels [3]. N. Tanaka's contribution to the family of monoliths was based on silica and emerged slightly later [4]. These three quite different approaches represent the foundation of most of the current monolithic technologies. Interestingly, all three groups experienced serious difficulties while trying to publish the early results, since reviewers did not recommend accepting the manuscripts for publication due to the lack of merit.

MONOLITHIC DISKS. The disk format was one of the first produced useful monolithic stationary phases originally designed for the rapid separation of proteins. In the mid-1980s, B. G. Belenkii studied chromatography of proteins in gradient elution mode using stationary phases with a variety of chemistries and column geometries and found that only a certain, often very short, distance is required to achieve good separation. This finding resulted in the concept of short separation beds. Since it was very difficult to create such beds from particulate sorbents, due to irregularities in packing density and excessive channeling, he was investigating different approaches. After extensive discussions, T. Tennikova and I, working together at the Institute of Macromolecular Chemistry in Prague, Czech Republic, designed and developed a new concept of stationary phase in the format of a monolithic disk that then enabled the predicted very fast separations [1]. From the very beginning, monolithic disks exhibited an excellent performance in separations of proteins and other large biological molecules using gradient elution. Figure. 5.40 shows a copy of an authentic chromatogram from 1989 presenting a fast separation of four proteins achieved in less than 90 s.

The preparation of the disks is simple. Typically, a piece of monolith is prepared in a flat or tubular mold, the sheet or cylinder is removed from the mold, and the porous polymer is punched or sliced to obtain a few-millimeters-thin discs. Most of these generic disks are prepared from reactive monomers, such as glycidyl methacrylate, where the epoxide group is subsequently modified to afford the desired

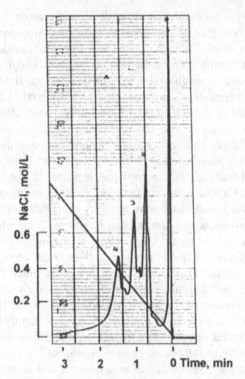

Figure 5.40. Authentic chart from 1989 demonstrating ion-exchange separation of myoglobin (1), conalbumin (2), ovalbumin (3), and soybean trypsin inhibitor (4) using monolithic disk. Conditions: poly(glycidyl methacrylate-co-ethylene dimethacrylate) disk 2 mm thick, 20 mm diameter modified with diethylamine, mobile-phase 0.01 mol/L Tris buffer pH = 7.6, 2-min gradient from 0 to 0.6 mol/L sodium chloride in the buffer.

interacting functionalities. This modification reaction is carried out with the final disks under controlled conditions in a flask. The modified disks are then placed in a specifically designed cartridge. In contrast to the cartridge, the disks are disposable. The variety of chemistries enables the application of the disks in widely diverse separation modes, such as reversed-phase, ion-exchange, hydrophobic interaction, or bioaffinity chromatography.

MONOLITHIC COLUMNS. I joined the faculty at Cornell University in the early 1990s and, with the help of J. Fréchet, initiated research on monolithic porous polymer columns formed by an in situ "molding" process in analytical size stainless-steel or glass columns [2]. The process was again very simple. A column tube sealed at the bottom was filled with polymerization mixture containing monomer, crosslinker, inert porogenic solvent, and a free-radical initiator, sealed, and heated in water bath to 60°C for 12–24 h. The seals were then replaced with chromatographic fittings and the column attached to a pump. After removing the porogenic solvents from the pores by pumping an organic

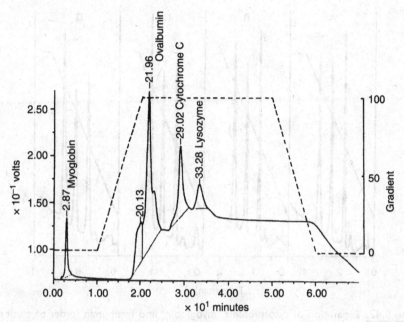

Figure 5.41. Authentic chart from 1992 demonstrating separation of four proteins using monolithic column. Conditions: poly(glycidyl methacrylate-co-ethylene dimethacrylate) monolithic column 30 × 8 mm i.d. modified with diethylamine to form 3-diethylamino-2-hydroxypropyl functionalities, mobile phase 0.01 mol/L Tris buffer pH = 7.6 for 10 min, followed by a 10-min gradient from 0 to 1 mol/L sodium chloride, UV detection at 218 nm.

solvent through the monolith, the column was ready for applications. Alternatively, the monolith prepared from reactive monomers could be functionalized in column by filling the pores with a reagent and enabling the reaction to run for a desired period of time. Figure 5.41 shows a set of chromatograms showing very fast separations that we obtained in 1992. The reversed-phase separations of proteins using constant gradient volume in 5-cm-long and 8-mm-i.d. poly(styrene-co-divinylbenzene) monolithic column was excellent. These experiments clearly demonstrated the speed and permeability that we could achieve albeit on the account of a high flow rate of up to 25 mL/min (Fig. 5.42).

In contrast to the disks introduced above, monolithic columns have a different aspect ratio (i.e., their length exceeds their diameter) and are polymerized *within* a conduit in which they remain all the time after the preparation is completed. They can be reproducibly fabricated in various formats, such as chromatographic columns and capillaries. The latter, that we currently prefer, is readily compatible with mass spectrometry and, therefore, becomes very popular. Obviously, it is much easier to polymerize liquid precursors and form the stationary phase in situ within a narrow-bore capillary than is packing it efficiently with microparticles.

Several methods such as direct copolymerization of functional monomers, chemical modification, and grafting of existing "generic" monolith are now available for the

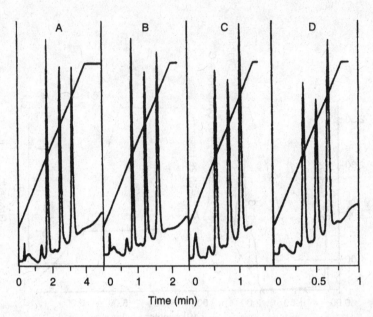

Figure 5.42. Separation of cytochrome c, myoglobin, and ovalbumin (order of elution) by reversed-phase chromatography on the continuous poly(styrene-co-divinylbenzene) rod column at different flow rates. Conditions: column, 50 × 8 mm i.d.; mobile phase, linear gradient from 20 to 60% acetonitrile in water; flow rate, 5 (A), 10 (B), 15 (C), and 25 mL/min (D).

preparation of monolithic stationary phases with desired surface chemistry. For example, directly polymerized poly(styrene–divinylbenzene) monoliths proved to be an excellent stationary phase for the very fast separation of proteins as shown above. Similarly, long alkyl methacrylate-based monolithic columns also prepared via direct copolymerization in a capillary are well suited for the separation of peptides in protein digests. Another option is the preparation of a monolith from monomer containing a reactive functionality such as glycidyl methacrylate and its subsequent modification to afford stationary phases for a variety of separation modes. In this technique, each single reactive site affords one new functionality. Recently, we added another method to the arsenal of known approaches that enables both preparation of monoliths and modulation of their surface chemistries while relying on UV-light-triggered processes. First, the "generic" monolith is prepared in a capillary using the UV-initiated polymerization process, which is faster than its thermally initiated counterpart. The desired surface chemistry of the monolith is then provided via grafting reaction of a functional monomer initiated again by UV light. One of the major advantages of grafting is that numerous functionalities emanate from each surface site, thus dramatically increasing the column loading capacity.

CONCLUSION. The era of monoliths started with the development of disks and larger chromatographic columns for liquid chromatography. After two decades from their inception, the monolithic columns are commercially available in a variety of formats that include disks and tubes from BIA Separations (Ljubljana, Slovenia), as well as standard columns and capillaries produced by Dionex (Sunnyvale, CA). However, the monoliths are also making their mark in other areas of chromatography. For example, electrochromatography, thin-layer chromatography, gas chromatography, and chromatography in microfluidic chips are separation techniques that benefit from the specifics of the monolithic materials [5–7]. Many of these applications were exemplified in the book summarizing the state-of-the-art the we published at the beginning of this century [8]. Nowadays, we also start seeing growing use of monolith in nonchromatographic applications such as reactive filtration, solid-phase extraction, and enzyme immobilization [9,10] and it is conceivable that they will find many more uses in the future.

Perspective. Since it is unlikely that interest in increasingly rapid separations of growing numbers of samples may decline, the war with diffusional limitations of mass transfer within stationary phases will continue. Development of various approaches, including decreasing size of particulate packings, formation of novel bead formats, and monoliths, appears to lead the way. We need to wait to see which of these techniques will prevail. All have their advantages and weaknesses. It is also possible that a completely new technology will emerge. It may be condemned at the beginning but accepted eventually just as we experienced with monolith. In addition to the use of "brutal force" of increasing inlet pressure and/or flow rate, throughput can also be enhanced via parallelism. An ideal format for this approach would be microfluidic chips that can include several parallel channels in which separations can run simultaneously. This technology may lead to realization of the lab-on-a-chip concept adding more functions to the separations.

Zooming in on monoliths, I trust they will be finding new applications with many completely different from chromatography. For example, the initial results with immobilized enzyme reactors containing several enzymes in tandem demonstrate that systems mimicking metabolic paths can be formed and used in environmentally friendly production of valuable compounds. Formation of monolithic thin layers appears to attract attention as well. In combination with photopatterning, they will enable two-dimensional separations of peptides, preparation of superhydrophobic materials, and diagnostic devices with flow channels confined just by a large difference in surface tension.

REFERENCES

1. T. B. Tennikova, F. Svec, and B. G. Belenkii, High-performance membrane chromatography: a novel method of protein separation, *J. Liquid Chromatogr.* **13**, 63–70 (1990).
2. F. Svec and J. M. J. Frechet, Continuous rods of macroporous polymer as high-performance liquid chromatography separation media, *Anal. Chem.* **64**, 820–822 (1992).
3. S. Hjerten, J. L. Liao, and R. Zhang, High-performance liquid chromatography on continuous polymer beds, *J. Chromatogr.* **473**, 273–275 (1989).
4. H. Minakuchi, K. Nakanishi, N. Soga, N. Ishizuka, and N. Tanaka, Octadecylsilylated porous silica rods as separation media for reversed-phase liquid chromatography, *Anal. Chem.* **68**, 3498–3501 (1996).

5. S. Eeltink and F. Svec, Recent advances in the control of morphology and surface chemistry of porous polymer-based monolithic stationary phases and their application in CEC, *Electrophoresis* **28**, 137–147 (2007).

6. R. Bakry, G. K. Bonn, D. Mair, and F. Svec, Monolithic porous polymer layer for the separation of peptides and proteins using thin-layer chromatography coupled with MALDI-TOF-MS, *Anal. Chem.* **79**, 486–493 (2007).

7. F. Svec and A. A. Kurganov, Less common applications of monoliths: III. Gas chromatography, *J. Chromatogr. A* **1184**, 281–295 (2008).

8. F. Svec, T. B. Tennikova, and Z. Deyl (Eds.), *Monolithic Materials: Preparation, Properties, and Applications*, Elsevier, Amsterdam, 2003.

9. F. Svec, Less common applications of monoliths: I. Microscale protein mapping with proteolytic enzymes immobilized on monolithic supports, *Electrophoresis* **27**, 947–961 (2006).

10. F. Svec, Less common applications of monoliths: Preconcentration and solid-phase extraction, *J. Chromatogr.* **841**, 52–64 (2006).

PETER C. UDEN

Figure 5.43. Peter C. Uden.

Peter Uden (Fig. 5.43) is a graduate of the University of Bristol, UK, from which he received the B.Sc. (1961) and Ph.D. (1964) degrees in Inorganic Chemistry. He was granted the degree of D.Sc in 2001 to mark his career research accomplishments. He was Research Associate and Lecturer at the University of Illinois, Champaign—Urbana (1964–1966) and Lecturer at the University of Birmingham, UK (1966–1970). He joined the faculty at the University of Massachusetts, Amherst in 1970, and was promoted to Professor in 1978. He became Emeritus Professor in 2004 and remains active in research and academic life. At Bristol he began his lifelong interest in the then-emerging technique of gas chromatography under the GC pioneer, Frederick Pollard.

His research has spanned a wide range, with a continued focus on inorganic and organometallic topics and elemental speciation in general. His interest in interfaced chromatography with atomic emission and atomic mass spectrometry has continued for more than 30 years. Topic areas have included trace metal complex analysis; petrochemical analysis, notably in the shale oil field; determination of halogenated species in water; pyrolysis and thermal extraction; and clinical trace analysis in the treatment of neonatal issues. In recent years a primary effort has been in organoselenium speciation pertaining to nutritional and clinical analysis.

He has been research advisor to over 110 doctoral and master's students with whomhe has published more than 160 research papers, books, and book chapters.

More than 20 of his graduates have entered the field of college teaching, and many others have attained high-level research and management positions in industry, notably in the pharmaceutical field.

Peter Uden received the Benedetti-Pichler Award from the American Microchemical Society, the Chernyaev Medal from the Russian Academy of Sciences, the UK Royal Society of Chemistry Award in Analytical Reactions and Reagents, and a Fellowship of Japan Society for the Promotion of Science. He has served as Chairman of the IUPAC Commission on Chromatography and Analytical Separations and Chair of the Separations and Chromatography Subdivision of the ACS division of Analytical Chemistry. He was a UNESCO consultant in analytical program development in Brazil and World Bank consultant for a number of years on graduate programs and instrumentation in the Peoples' Republic of China.

Awards after 2000 Peter C. Uden include the 2007 Eastern Analytical Symposium Award for "Achievements in Separation Science."

See Chapter 1, Table 1.3, d, j, p, t, w, v.

Research Achievements. It may be argued that chromatographic science has made a preeminent impact on analytical chemistry during the past 75 years with the increasingly complex challenges of chemical separations, identification, characterization, and quantification ensuring its centrality in the future. I was fortunate to be introduced as student to the then-emerging technique of gas chromatography in 1959 by the chromatographic pioneer, Frederick Pollard at the University of Bristol in England. Before that time most research and application in gas chromatography (GC) had been made in the challenging petroleum chemistry field. Inorganic GC was in its infancy, but we were enthusiastic at Bristol, and I have never lost my early affection for this field.

The twin foci of chromatography—separation and detection—have always developed in parallel. Detection has developed from early "universal" response modes to selective and sensitive techniques with quantitative analyte identification and characterization as clear goals. At Bristol our research in the new field of organometallic chemistry led the GC separation studies of both σ-bonded compounds of the silicon group (Si, Ge, Sn, and Pb) and of π-bonded compounds such as ferrocene. We reported preparation and GC application of a hexadecyl-bonded silica stationary phase, but did not have the foresight to try it in a reversed-phase HPLC experiment!

During time spent at the University of Illinois in Urbana at Champaign working with the eminent inorganic coordination chemistry, John C. Bailar, Jr., I learned of the pioneering work of a former Bailar student, Robert Sievers, in transition metal diketonate chelate GC. Subsequently, at the University of Birmingham in England and later at the University of Massachusetts, we prepared and examined GC characteristic of novel bidentate metal chelates of divalent cobalt, copper, nickel, palladium, and platinum with thioketones, iminoketones, dithiocarbamates, etc. and achieved viable packed-column separations. A range of tetradentate divalent metal chelates of diaminodiketone ligands had very high thermal and reaction stability, and GC was possible with no degradation seen even at picogram-level electron capture detection. In many ways, successful

GC of these compounds necessitated as taxing an example of optimization procedures as had yet been seen. Quantification of transition metal ions as their derived chelates also proved challenging and later these inorganic systems were also separated successfully by both normal and reversed-phase HPLC.

In 1974 we initiated research in two chromatographic application areas that proved to be major foci for a decade or more. The global petroleum energy crisis of the early 1970s called for new initiatives in the United States. We moved into high-resolution GC and HPLC characterization studies of "oil shale pyrolysis" and shale oil chemical class separations. At the same time environmental forces were engendering national concern on "drinking water" safety in the context of chlorination disinfection byproducts. High-resolution CG and HPLC were needed along with halogen-selective detection at the subnanogram level. It was from this need that we began to explore element specific GC detection. Two alternative concepts arose for possible atomic emission detection. A DC argon plasma (DCP) proved valuable for metal compounds, but a microwave-induced/sustained helium plasma (MIP) gave excellent sensitive and selectivity for nonmetals as well. We built prototype MIP detection systems and applied them effectively to trace volatile halogenated compounds in water sources and as derived from humic acid chlorination. Haloforms, haloorganic acids, and haloaldehydes were studied, and trichloracetic acid and trichloracetaldehyde were observed for the first time as disinfection byproducts.

The atomic emission detections (now denoted as AED) were applied to a wide range of analytical problems ranging from polymeric silicone pyrolysis to geochemical ocean sediment characterization as well as our central field of organometallic and metal chelate GC. Metal exchange and redistribution reactions were characterized and high temperature (to 400°C) GC-AED applied to metalloporphyrins. The AED was introduced as a commercial instrument (Hewlett-Packard), and we collaboratively extensively in its evaluation and promotion. Quantification specification of inorganic and organic mercurials in standard reference materials was carried out in collaboration with the US NIST (National Institute of Standards and Technology).

In the 1990s our attention turned toward speciation studies of selenium and to an extent of arsenic in seleniferous plants, yeast, and nutritional supplements, and relationships and characterization by GC-AED with Se- and S-specific detection. A range of compounds containing Se and S were observed in *alium* vegetables, such as garlic and onions, and also in exhaled human breath, including previously unobserved compounds. In a complementary fashion, we developed optimal reversed-phase HPLC methods utilizing ion-pairing perfluoroaliphatic acids for selenoamino acids and related analytes. Interfaced HPLC-ICPMS (inductively coupled plasma mass spectrometry) enabled effective speciation for non-volatile polar compounds (Fig. 5.44). Quantitative extraction procedures enabled measurement of nonvolatile compounds, while parallel chloroformate derivitization of selenoamino acids gave confirmation of resolved species. Among the most important discoveries was the presence of a previously unknown Se–S amino acid, S-(methylseleno) cysteine, present in selenized yeast from reaction of dimethyldiselenide with cysteine. This and similar compounds may have key roles in the utilization of selenium supplements in cancer chemoprevention and other medicinal functions. An essential part of these studies is the detailed study of elemental quantification by various

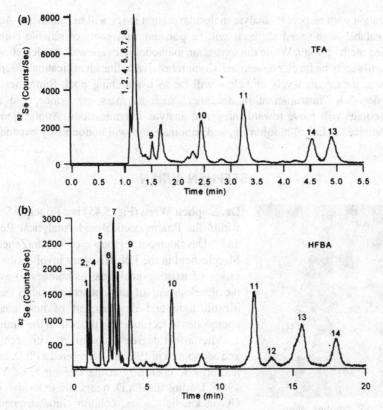

Figure 5.44. Selenium (Se)-specific HPLC-ICP-MS chromatograms using 0.1% trifluoroacetic acid (a) and 0.1% heptafluorobutanoic acid ion-pairing agent (b). 1—selenous acid (selenite); 2—selenic acid (selenate); 4—methylseleninic acid (methylseleninate); 5—selenolanthionine; 6—trimethyl selenonium cation; 7—selenocystine; 8—selenocystathionine; 9—Se-methylselenocysteine; 10—Se-2-propylselenocysteine; 11—selenomethionine; 12—unknown, 13—γ-glutamyl-Se-methylselenocysteine; 14—Se-allylselenocysteine. [Reprinted from *J. Chromatogr.* **866,** 51 (2000), with permission from Elsevier.]

atomic special methods. The goal of compound-independent calibration (CIC) is a critical one and requires exhaustive statistical analysis to determine the effects of functionality on elemental spectral response in analytical plasmas.

It goes without saying that our contribution to the development of chromatography would have been unable without the involvement of more than 100 graduate students and colleagues during 45 years. Perhaps they have enjoyed this still fertile field as much as I have over the years.

Perspectives of Professor Uden. The essential experimental features of gas and liquid (and supercritical-fluid) mobile-phase chromatographies are established and will continue. Optimization with respect to speed, resolution, and sample capacity will be a continued focus with a viable two-parameter combinations being paramount.

Optimization with respect to analyte molecular weight range will be enhanced. Accurate and repeatable high-speed methods will be pursued with goals of reliable computer-controlled methodology. While the separation methodologies are well set, detection techniques will surely be further developed. Comprehensive online identification of separated analytes at nanogram levels or below will be an overarching goal. Simplification of information-rich "instrumentation detectors" such as mass, resonance, and atomic spectroscopies will move toward integrated analyte determinations. Application fields in biochemicals, clinical monitoring, and nanomaterials will undoubtedly expand.

STEPHEN WREN

Figure 5.45. Stephen Wren.

Dr. Stephen Wren (Fig. 5.45) is a Principal Scientist within the Pharmaceutical and Analytical Research and Development group of AstraZeneca at Macclesfield in the UK. His work involves the application of existing analytical techniques to assist in the development of new pharmaceuticals, and the identification and development of new analytical approaches to facilitate this process in the future.

After a first degree in chemistry with technology and economics in 1982, he completed a Ph.D. in physical organic chemistry at the University of York in 1986. During the Ph.D. years the extensive use of GC in kinetic studies, column chromatography for reagent purification, and NMR, MS, and IR for product characterization led to his general interest in analytical chemistry.

Following work on environmental analysis at the Laboratory of the Government Chemist in London, he joined the former ICI Pharmaceuticals in Macclesfield, UK.

Dr. Wren was awarded the Silver Jubilee Medal of the Chromatographic Society in 2006.

See Chapter 1, Table 1.3, h, i, l, p, q, r, v.

Research Achievements. Dr. Wren has been active in research in analytical chemistry for a number of years, both via individual investigation and by collaboration with leading academics, with whom he has helped to supervise eight Ph.D. students. He has authored or co-authored 30 peer-reviewed publications, most of them covering separation science, especially in the fields of capillary electrophoresis and ultra performance liquid chromatography (UPLC). Other fields of work include the use of deuterium exchange MS to assist in the structural elucidation of unknowns by separation techniques hyphenated to MS, and investigations into the use of nuclear quadrupole resonance for pharmaceutical analysis.

One area of interest is in trying to improve understanding of the physical processes that underlie separations, and then exploiting that understanding to develop improved separation methods. An example of this is in using capillary electrophoresis for the separation of enantiomers by the use of buffer additives such as cyclodextrins. A series of papers and a book explored how a simple competition model could be used to explain complex behaviors and optimize separation conditions. The theory developed showed that there is an optimum concentration of the chiral selector, which is inversely related to the affinity of the analyte for the selector.

In the field of UPLC, a series of papers showed that the technique could be routinely applied to pharmaceutical analysis and offered significant advantages in speed and separating power. Comparative work showed that equivalent results could be obtained using either conventional HPLC or UPLC, and that hyphenation of UPLC to MS enabled the identification of low-level impurities.

SELECTED REFERENCES

1. S. A. C. Wren, Optimization of pH in the electrophoretic separation of 2-, 3-, and 4-methylpyridines, *J. Microcol. Separations* **3**, 147 (1991).
2. S. A. C. Wren and R. C. Rowe, Theoretical aspects of chiral separation in capillary electrophoresis I. Initial evaluation of a model, *J. Chromatogr.* **603**, 235 (1992).
3. M. A. Survay, D. M. Goodall, S. A. C. Wren, and R. C. Rowe, Self-consistent framework for standardising mobilities in free solution capillary electrophoresis: applications to oligoglycines and oligoalanines, *J. Chromatogr. A* **741**, 99 (1996).
4. S. A. C. Wren, Prediction of isotope patterns for partially deuterated analytes in HPLC/MS, *J. Pharm. Biomed. Anal.* **34**, 1131 (2004).
5. S. A. C. Wren, Peak capacity in gradient ultra performance liquid chromatography (UPLC), *J. Pharm. Biomed. Anal.* **38**, 337–343 (2005).
6. S. A. C. Wren and P. Tchelitcheff, Use of ultra-performance liquid chromatography in pharmaceutical development, *J. Chromatogr. A* **1119**, 140–146 (2006).

EDWARD YEUNG

Edward Yeung (Fig. 5.46) received his A.B. in Chemistry from Cornell University and his Ph.D. in Chemistry from the University of California at Berkeley. Since then, he has been on the chemistry faculty at Iowa State University, where he is currently Robert Allen Wright Professor and Distinguished Professor in Liberal Arts and Sciences. His research interests span both spectroscopy and chromatography. He has published in areas such as nonlinear spectroscopy, laser-based detectors for chromatography, capillary electrophoresis, trace gas monitoring, single-cell and single-molecule analysis, DNA sequencing, and data treatment procedures in

Figure 5.46. Edward Yeung.

chemical measurements. He is an Associate Editor of *Analytical Chemistry*. He served on the editorial advisory board of *Progress in Analytical Spectroscopy*, the *Journal of Capillary Electrophoresis*, *Mikrochimica Acta*, *Spectrochimica Acta Part A*, the *Journal of Microcolumn Separation*, *Electrophoresis*, the *Journal of High Resolution Chromatography*, *Chromatographia*, and the *Journal of Biochemical and Biophysical Methods*. He was awarded an Alfred P. Sloan Fellowship; was appointed Honorary Professors of Zhengzhou University, Zhongshan University, Xiamen University, and Hunan University; and was elected Fellow of the American Association for the Advancement of Science. He received the ACS Division of Analytical Chemistry Award in Chemical Instrumentation, four separate R&D 100 Awards, the Lester W. Strock Award, the Pittsburgh Analytical Chemistry Award, the L. S. Palmer Award, the ACS Fisher Award in Analytical Chemistry, the Frederick Conference on Capillary Electrophoresis Award, the ACS Award in Chromatography (2002), the International Prize of the Belgian Society of Pharmaceutical Sciences (2002), the Eastern Analytical Symposium Award in Separation Science (2003), the Iowa Inventor of the Year Award (2004), the Ralph N. Adams Award in Bioanalytical Chemistry (2005), the Golay Award (2006), and the Chicago Chromatography Discussion Group Merit Award (2006).

See Chapter 1, Table 1.3, a, e, l, p.

Research on Molecular Dynamics at Liquid–Solid Interface. Recent work in our group has been directed toward testing various chromatographic theories at the sub-micrometer-length scale and down to the single-molecule limit. Understanding molecular dynamics at liquid–solid interfaces is important in many areas of chromatography, such as retention and peak distortion. Chromatographic separation is usually attributed to the differential rate of migration induced by solute distribution between the stationary and mobile phases. From the fundamental perspective, the sorption isotherm, a plot of the solute concentration in the stationary phase against that in the mobile phase at a constant temperature, can provide important knowledge of the underlying thermodynamic functions. Methods for experimental determination of the bulk adsorption isotherms from chromatograms are well established. However, details at the molecular level are still missing. From the practical perspective, the effectiveness of a separation is a function of the component's capacity factor, as well as the column's efficiency and selectivity. Capacity factor of a component is of central importance for the description of chromatographic migration and is a measure of the degree to which it partitions into the stationary phase. A priori prediction of the capacity factor will greatly facilitate the optimization of chromatographic conditions.

Direct observation of single-molecule motion [1–4], imaging, and interactions at a liquid–solid interface is one technique for studying rate theory and band broadening in chromatography and capillary electrophoresis (CE). While it is well known that both electrostatic and hydrophobic interactions govern protein adsorption at a liquid–solid interface [5], relatively little is known about how these interactions influence the individual protein molecules in the interfacial layer. These fundamental properties are the foundation for many chromatographic [6] and electrophoretic protein separation theories

[7]. If both electrostatic and hydrophobic interactions govern protein adsorption at liquid–solid interfaces, which interaction is stronger? What factors influence the behavior of individual protein molecules on the fused-silica surface and on the capillary inner wall? Does statistics properly average out the stochastic behavior of a collection of individual molecules? These questions can be answered with the real-time dynamics and imaging of single molecules at fused-silica/liquid interfaces [3–8].

As a model protein, we selected R-phycoerythrin (RPE) [9], which is a 240-kDa autofluorescent protein consisting of seven subunits and ~30 chromophores. Single proteins of RPE were imaged by a total internal reflection–fluorescence microscopy technique [1–10]. Such images restrict observation to about 150 nm from the surface, thereby allowing us to interrogate the surface and the liquid layer immediately next to the surface. The distribution of individual protein molecules at the fused-silica–water interface was analyzed as a function of pH and buffer composition. The results of the real-time imaging experiments were also compared to the elution behavior of proteins in CE. Finally, the correlation of pI obtained with isoelectric focusing experiment and single-protein imaging has been investigated.

The plot of the total number of protein molecules recorded versus pH for 16.6 pM RPE in the pH range of 4.0–8.2 is shown in Figure 5.47. Generally, for pH < pI of protein, a substantial increase in the number of molecules is observed at the water/ fused-silica interface. However the total number of RPE molecules recorded at pH 4 (<pI of RPE) decreased compared to pH 5 and 5.5. Since we can detect motion of the proteins in a video sequence, we can also tell if the individual protein molecules were randomly located in the solution inside the evanescent-field layer (EFL) or were adsorbed at the solid surface. As expected, the residence time of individual protein molecules in the

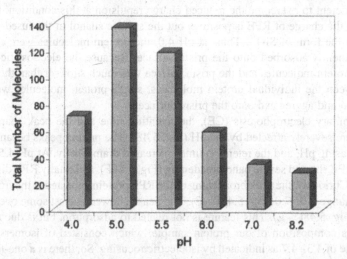

Figure 5.47. Histogram of numbers of single-protein molecules recorded versus pH for 16.6 pM RPE in pH 8.2–4.0 buffer solutions. Each data point represents the mean value of number of molecules calculated from three independent measurements.

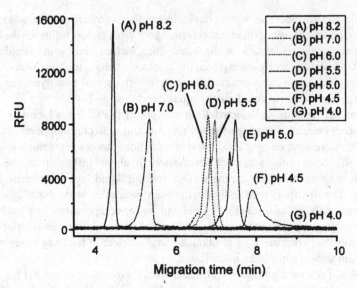

<u>Figure 5.48</u>. Electropherograms of RPE showing the pH effect. CE conditions: capillary, 58 cm × 75 µm i.d. (48 cm to the detector); applied voltage, +20 kV; RFU, relative fluorescence unit.

interfacial layer gradually increased below pH 7.0, due to the decreasing charge repulsion between the negatively charged surface and the negatively charged protein. All of the protein molecules were strongly adsorbed on the fused-silica surface at pH 5.0 even though the pIs of RPE isomers are less than the pH. This means that hydrophobic attraction is sufficient to overcome the reduced charge repulsion at this condition. Below pH 4.1 (<pI), the charge of RPE is positive, but the surface silanol of the fused-silica wall remains in the form of SiO⁻. Thus, at pH 4.0, the protein molecules were completely and permanently adsorbed onto the prism surface. Because the electrostatic attraction between protein molecules and the prism surface was much stronger than the repulsive force between the individual protein molecules, single-protein molecules were finally precipitated and aggregated onto the prism surface.

In capillary electrophoresis (CE), the retention time and the peak shape of RPE protein were severely affected by the pH (Fig. 5.48). The protein peaks became broader with decreasing pH, and the retention times increased dramatically. At pH 4.5, the peak shape of RPE showed severe band broadening (Fig. 5.48F), and finally RPE did not elute at pH 4.0. Curiously, the peak broadening of the RPE protein appeared at the front of peak in the pH range 5.0–7.0 (Fig. 5.48B–E). The peaks showed at least some overlap at pH 5.0–6.0 (Fig. 5.48C–E). This feature is not related to adsorption, but is due to the heterogeneous composition of our protein sample, which consisted of isomers spanning the pI range of 4.3–4.7 as indicated by isoelectric focusing. So, there is a one-to-one correlation between the single-molecule behavior and the electrophoretic peaks.

In a subsequent study [11], we employed simulation to further elucidate the adsorption properties of RPE on the inner wall of cylindrical fused-silica capillary columns under the influence of an electric field. Using CE, the capacity factor of RPE at pH 5.0 was determined by the mobility-based adsorption isotherm [12]. Using total internal reflection fluorescence microscopy, motions of single RPE molecules were imaged at the water/fused-silica interface, and a reliable counting method was developed for the adsorbed RPE molecules. We demonstrate the potential application of single-molecule imaging as a means to measure the capacity factor and desorption rate of a molecule on a chromatographic surface.

Over 30 sets of images taken at different surface positions and/or with different samples were processed for each concentration. Using the counting results and the estimated total number of RPE molecules in the observation region, the capacity factor was calculated to be $k'_0 = 0.25$, with $a \sim 20\%$ relative standard deviation. This capacity factor can be directly compared with those determined in the CE experiments [11]. A combined plot of k'_0 versus $1/r_c$ (where r_c is the radius of the capillary in mm) from the two kinds of experiments gives a straight line with a slope of 1.22, an intercept of 0.014, and a correlation coefficient of 0.998. These values are thus in good agreement with one another.

The desorption rate can also be estimated from the changes in counts over time. The differences in count between video frames were then multiplied by the ratio of the total adsorbed molecule number to the number of adsorbed molecules remaining since the first frame. In this way, we determined that molecules have nearly constant desorption rates ($\sim$40 per frame or every 200 ms). This behavior was expected because equilibrium had been established on the surface.

REFERENCES

1. X. H. Xu and E. S. Yeung, Direct measurement of single-molecule diffusion and photodecomposition in free solution, *Science* **275**, 1106–1109 (1997).

2. Y. Ma, M. R. Shortreed, and E. S. Yeung, High-throughput single-molecule spectroscopy in free solution, *Anal. Chem.* **72**, 4640–4645 (2000).

3. X.-H. Xu and E. S. Yeung, Long-range electrostatic trapping of single-protein molecules at a liquid–solid interface, *Science*, **281**, 1650–1653 (1998).

4. M. R. Shortreed, H. Li, W.-H. Huang, and E. S. Yeung, High-throughput single-molecule DNA screening based on electrophoresis, *Anal. Chem.* **72**, 2879–2885 (2000).

5. V. Chan, S. E. McKenzie, S. Surrey, P. Fortina, and D. J. Graves, Effect of hydrophobicity and electrostatics on adsorption and surface diffusion of DNA oligonucleotides at liquid/solid interfaces, *J. Colloid Interface Sci.* **203**, 197–207 (1998).

6. T. Kues, R. Peters, and U. Kubitscheck, Visualization and tracking of single protein molecules in the cell nucleus, *Biophys. J.* **80**, 2954–2967 (2001).

7. S. P. Radko, A. Chrambach, and G. H. Weiss, Asymmetry of protein peaks in capillary zone electrophoresis: effect of starting zone length and presence of polymer, *J. Chromatogr. A* **817**, 253–262 (1998).

8. S. H. Kang, M. R. Shortreed, and E. S. Yeung, Real-time dynamics of single-DNA molecules undergoing adsorption and desorption at liquid–solid interfaces, *Anal. Chem.* **73**, 1091–1099 (2001).

9. S. H. Kang and E. S. Yeung, Dynamics of single-protein molecules at a liquid/solid interface: Implications in capillary electrophoresis and chromatography, *Anal. Chem.* **74**, 6334–6339 (2002).

10. T. Funatsu, Y. Harada, M. Tokunaga, K. Saito, and T. Yanagida, Imaging of single fluorescent molecules and individual ATP turnovers by single myosin molecules in aqueous solution, *Nature*, **374**, 555–559 (1995).

11. N. Fang, J. Li, and E. S. Yeung, Mobility-based wall adsorption isotherms for comparing capillary electrophoresis with single-molecule observations, *Anal. Chem.* **79**, 6047–6054 (2007).

12. N. Fang, J. Li, and E. S. Yeung, Quantitative analysis of systematic errors originated from wall adsorption and sample plug lengths in affinity capillary electrophoresis using two-dimensional simulation, *Anal. Chem.* **79**, 5343–5350 (2007).

KEY PUBLICATIONS

N. Fang, J. Li, and E. S. Yeung, Mobility-based wall adsorption isotherms for comparing capillary electrophoresis with single-molecule observations, *Anal. Chem.* (accelerated article) **79**, 6047 (2007).

S. H. Kang, M. R. Shortreed, and E. S. Yeung, Real-time dynamics of single-DNA molecules undergoing adsorption and desorption at liquid-solid interfaces, *Anal. Chem.* (accelerated article) **73**, 1091 (2001).

X. Gong and E. S. Yeung, An absorption detection approach for multiplexed capillary electrophoresis using a linear photodiode array, *Anal. Chem.* **71**, 4989 (1999).

X.-H. Xu and E. S. Yeung, Direct measurement of single-molecule diffusion and photodecomposition in free solution, *Science* **275**, 1106 (1997).

J. A. Taylor and E. S. Yeung, Multiplexed fluorescence detector for capillary electrophoresis using axial optical fiber illumination, *Anal. Chem.* **65**, 956 (1993).

(*Note*: Tables 4.3, 4.5, and 5.1–5.8 identify almost all chromatography awardees. In addition to those just mentioned, several awardees and other chromatography contributors are located in subsequent chapters because of pertinent subject relevance.)

6

HISTORY AND DEVELOPMENTS IN CHROMATOGRAPHIC COLUMN TECHNOLOGY AND VALIDATION TO 2001

Ernst Bayer

Research Center for Nucleic Acid and Peptide Chemistry, Institute for Organic Chemistry, University of Tübingen, Germany

Walter G. Jennings

University of California, Davis, J&W Scientific, Inc., Folsom, California

Ronald E. Majors

Agilent Technologies, Wilmington, Delaware

J. Jack Kirkland

Little Falls Analytical Division, Hewlett-Packard Co., Newport, Delaware

Klaus K. Unger

Institute for Inorganic and Analytical Chemistry, Johannes Gutenberg University, Mainz, Germany

Heinz Engelhardt

Department of Analytical and Environmental Instrumentation, Universität Des Saarlandes Im Stadtwald, Saarbrücken, Germany

Gerhard Schomburg

Department of Chromatography and Electrophoresis, Max-Planck-Institut für Kohlenforschung, Mülheim an der Ruhr, Germany

William H. Pirkle

Department of Chemistry, University of Illinois, Urbana

Christopher J. Welch

Process Research Laboratory on Automation & Robotics, Merck & Co., Inc., Rahway, New Jersey

Daniel W. Armstrong

Department of Chemistry and Biochemistry, University of Texas at Arlington, Texas

Jerker O. Porath

Center for Surface Biotechnology, Uppsala Biomedical Center, Uppsala University, Uppsala, Sweden

Jan B. Sjövall

Department of Medical Biochemistry and Biophysics, Karolinska Institute, Stockholm, Sweden

Charles W. Gehrke

Department of Biochemistry and the Experiment Station Chemistry Laboratory, College of Agriculture, University of Missouri, Columbia

CHAPTER OUTLINE

6.A. INTRODUCTION (Ernst Bayer)

In this chapter and S11 in the Supplement (available online at http://www.chemweb.com/preprint), we have collected the summaries of some chromatographers, who participated in the development of chromatographic column technology with emphasis mainly on gas and liquid chromatography. The authors discuss the important characteristics of retention and selectivity of the analyte and the factors involved. Of course, this does not cover the whole range of this field, and we did not require a unified style or format for the contribution. The intent was to collect the individual views, and trace them back to the initiatives and beginnings of important developments.

An excellent introduction into this field is the contribution of W. G. Jennings, who starts with the early days of gas–liquid chromatography with packed columns. At that time, the development of stationary phases was mainly for the separation of flavor components and petroleum products, relatively nonpolar compounds. Therefore, the liquids and support materials were not very well defined. Crushed and sieved firebricks (Sterchamol) to diatomaceous earth supports were used, which were good for hydrocarbons. However, the desire to extend GC to more polar compounds, such as many natural products, required defined and purified silica, as well as less volatile liquids, such as poly(siloxanes) or poly(ethylene glycol). The research on more defined silica and liquids began and set the basis for the use of silica later in HPLC.

However, when M. J. E. Golay published his paper on capillary gas chromatography, the first step in miniaturization of chromatographic separation methods was achieved, and together with the continuously ongoing automatization, changed analytical chemistry. I still remember the first capillary chromatograms in our laboratory. We were fascinated, sitting in front of the recorder, while observing how an extract of biological samples showed more than 200 peaks coming one after another in less then one hour. Of course, many details had to be solved to install capillary gas chromatography. The contributions of W. G. Jennings and G. Schomburg give a good summary. However, beside their contributions and the authors cited in their chapters, we have to acknowledge also the important early work of L. S. Ettre, C. Cramers, and P. Sandra.

The research on stationary phases became very critical when HPLC was introduced as a routine method. Retrospectively, it is astonishing that it took so long a time to apply the wide knowledge accumulated in the theory of gas chromatography to conventional liquid chromatography. To move from conventional low-efficiency LC to high performance, HPLC theory suggested the reduction of the particle diameter of the stationary phase and choosing the optimum flow rate of the mobile phase. Beside the necessary instrumental development using higher pressure and appropriate detectors, the availability of small-diameter particles was also mandatory.

Diameters between 5 and 20 μm, especially on silica supports, were developed; their preparation was and still is a great issue. As pioneers in the field of silica supports, J. J. Kirkland (Section 6.C.1) and K. K. Unger (Section 6.C.2) summarized their basic work in this field. G. Schomburg (Section 6.C.4) and H. Engelhardt (Section 6.C.3) report on their work on polymer-encapsulated stationary phases. The wide variety of stationary phases for HPLC requires excellent characterization and testing procedures, which is discussed in Engelhardt's (Section 6.C.3) contribution. HPLC started initially as normal-phase chromatography with predominantly silica and aluminum oxide as adsorbents and organic solvents as mobile phases. The year 1971 was decisive, because J. J. Kirkland (Section 6.C.1) at DuPont introduced the covalently bonded phases, which became the standard in HPLC for reversed-phase (RP) chromatography. More than 80% of the HPLC separations since then have been done with such bonded phases, which can be prepared by reacting the silanol groups of silica with chloro- or alkoxysilanes.

These bonded phases were much more stable against hydrolysis as compared to the covalently bounded brush supports with ester bonds, which were first introduced as covalently bonded RP phases in gas chromatography by I. Halász. Of course, the bonded RP phases are also much more robust than the phases with nonpolar liquids adsorbed on the surface of the supports, which already had been used in conventional liquid chromatography. Subsequently, after the pioneering work of J. J. Kirkland, many companies entered this field, and advanced the technology. In an excellent review by R. E. Majors, the present status of stationary phases for HPLC to 2001 is summarized (Section 6.B.3). He also reports on the newest development, the monolithic columns, for which a bright future is predicted in which he updates column technology to 2008 (see Section 6.F).

Characterization of the stationary phases is a topic in H. Engelhardt's (Section 6.C.3) research. The chemical nature of the bonded phases was for a long period not very well understood, and the recipes of their preparation were like black magic. With the development of solid-state *cross-polarization–magic-angle spinning–nuclear magnetic resonance* (CP-MAS-NMR) spectroscopy, the structure of bonded phases could be investigated. In a basic paper [1], we assigned the structure of bonded phases and subsequently could also use NMR spectroscopy to investigate the interaction of analytes and solvents with the alkyl groups of the RP phases [2]. The application of NMR spectroscopy was very successful for investigation of the chromatographic process. In collaboration with my gifted Ph.D. student, J. Tallarek, and G. Guiochon, University of Tennessee and the NMR group of Prof. H. van Ass in Wageningen, we were able to analyze the flow profiles in packed beds of chromatographic columns with different stationary phases [3]. The mechanism of perfusion packings [4] could also be studied.

Besides the dominance of silica-bonded alkyl phases, many specialized stationary phases with high selectivities are on the market. As pointed out by R. Majors at PITTCON 2000, seven new chiral columns were introduced. The direct separation of enantiomers has attracted chemists since the dawn of stereochemistry. However, E. Gil-Av achieved a real break-through since 1965 [5] (see Chapter S9 on the Internet and Chapters 3 and 4 in this book) [5]. He synthesized derivatives of α-amino acids with $>$CO— and —NH functions as stationary phases for conventional column and capillary gas chromatography. Many chiral analytes could be separated. However, a serious restriction was temperature stability. Even not all enantiomers of the common amino acids of proteins could be separated. Therefore we introduced the polymer in 1977, Chirasil-Val [6], where the chiral selector tertiary butylvaline amide was covalently immobilized to a polysiloxane. The immobilization of chiral selectors to polysiloxanes was subsequently used for many chiral selectors in GC, and in HPLC. An advantage of Chirasil-Val is that the D- and L-enantiomeric phases are available, and that all derivatives of all α-L-amino acids show stronger retention and appear after the D-enantiomer on Chirasil-L-Val. Therefore, GC can be used to assign the stereochemistry of α-amino acids. Another approach to separate enantiomers was the development of selectors, which interact by coordination with metals as achieved by V. Schurig [7]. The stability of such selectors on the basis of complexation as well as cyclodextrins later [8] has been improved by bonding to siloxanes.

Separation by coordination has also been used in HPLC by adding the metal complex as an additive to the eluent [9]. V. A. Davankov (see Chapter 5) has the chiral complexation reagents covalently linked to polymeric backbones. Polymeric phases with chiral groups have been prepared by Blaschke, but they suffer from insufficient stabilities on packed beds, due to the compressibility with increasing pressure. Therefore, they did not find broad application. The same is true for imprinted phases, described first by Wulff. These imprinted or structured phases are fascinating. Polymerization is done with the analyte. Under the assumption that the functional groups of the polymer arrange themselves in an optimum position around the analyte, high selectivities can be expected. It has been demonstrated that the concept is sound; however, up to now no phases with a decent separation efficiency have been obtained.

The article by W. H. Pirkle and C. J. Welch starts with an overview of the early enantiomer separations with classical liquid chromatography until 1974, and then continues his extremely valuable chiral HPLC phases, well known as Pirkle-type phases (see Section 6.C.5). Even if his chiral selectors still contain amide groups, additional functions for $\pi-\pi$ attractions and steric repulsions are incorporated, and allow separation of many groups of enantiomers.

The contribution of D. Armstrong includes a review on his very fruitful development of cyclodextrin- and glycopeptide-derived chiral phases for HPLC (see Section 6.C.6). The first experiments added the chiral selector to the mobile phase. This method was expensive, and therefore Armstrong immobilized derivatized cyclodextrins, which immediately found wide acceptance for HPLC separation of enantiomers. The rapid progress was possible because there was already considerable knowledge accumulated from chiral gas chromatography where derivatized cyclodextrins were bonded and used as chiral selectors [10–12].

Despite the fact that stationary phases based on silica are dominant in HPLC, there are some areas where phases based on polymers are of interest. The disadvantage of polymer materials is their compressibility, which makes their use at higher pressures impossible. Therefore, no small particle diameters are possible, and consequently the high efficiency of the silica-based stationary phases cannot be attained. This is also the reason why the classical polymers used for ion-exchange separation did not enter the field of HPLC. The higher stability of polymers over a wide range of pH always tempted one to develop polymer-coated silica. The contributions of G. Schomburg and H. Engelhardt address this field of research (see Sections 6.C.3 and 6.C.4).

Domains where polymers are still used extensively today are in size-exclusion chromatography and affinity chromatography. The inventors of Sephadex, J. Porath and Per Flodin, present a view on the history of these polymers and their various applications (see Section 6.C.7). J. Sjövall complements these views by reporting on his preparation of stationary phases by functionalizing and crosslinking Sephadex (see Section 6.C.8).

This chapter does not claim to be a complete review of all stationary phases prior to 2001. It was rather the intention to let scientists show how they actively participated in the development, summarize their research, and predict future developments. In Chapter 5 and subsequent chapters, today's chromatographers bring the story to 2009. Let history teach especially our young scientists how to achieve excellent research.

REFERENCES (for Ernst Bayer)

1. E. Bayer, K. Albert, J. Reiners, and M. Nieder, Characterization of chemically modified silica gels by ^{29}Si and ^{13}C cross-polarization and magic angle spinning nuclear magnetic resonance, *J. Chromatogr.* **264**, 197–213 (1983).

2. B. Pfleiderer, K. Albert, K. D. Lork, K. K. Unger, H. Brückner, and E. Bayer, Correlation of the dynamic behavior of n-alkyl ligands of the stationary phase with the retention times of paracelsin peptides in reversed phase HPLC, *Angew. Chem. Int. Ed. Engl.* **28**, 327–329 (1989).

3. U. Tallarek, D. van Dasschoten, H. van Ass, G. Guiochon, and E. Bayer, Direct observation of fluid mass transfer resistance in porous media by NMR spectroscopy, *Angew. Chem. Int. Ed.* **37**, 1882–1885 (1998).

4. F. Regnier, Perfusion chromatography, *Nature* **350**, 634 (1991).

5. E. Gil-Av., B. Feibush, and R. Charles-Sigler, Separation of enantiomers by gas-liquid chromatography with an optically active stationary phase, *Tetrahedron Lett.* **1966**, 1009 (1966).

6. H. Frank, G. J. Nicholson, and E. Bayer, Rapid gas chromatographic separation of amino acid enantiomers with a novel chiral stationary phase, *J. Chromatogr. Sci.* **15**, 174 (1977).
7. V. Schurig, Enantiomerentrennung eines Chiralen Olefins durch Komplexierungs Chromatographie an einem Optisch Aktiven Rhodium (1) Komplex, *Angew. Chem.* **89**, 113 (1977).
8. V. Schurig, D. Schmalzing, U. Mühleck, M. Jung, M. Schleimer, P. Mussche, C. Duvekot, and J. C. Buyten, Gas chromatographic enantiomer separation on polysiloxane-anchored permethyl-β-cyclodextrin (chirasil-dex), *J. High Resol. Chromatogr.* **13**, 713 (1990).
9. E. Gil-Av., Selectors for chiral recognition in chromatography, *J. Chromatogr. Library* **32** 126 (1985).
10. Z. Juvancz, G. Alexander, and G. Szejtli, Permethylated β-cyclodextrin as stationary phase in capillary gas chromatography, *High Resol. Chromatogr. Commun.* **11**, 105 (1987).
11. W. A. König, S. Lutz, and G. Wenz, Modified cyclodextrin—a new highly enantioselective stationary phase for gas chromatography, *Angew. Chem. Int. Ed. Engl.* **27**, 979 (1987).
12. V. Schurig and H.-P. Novotny, Separation of enantiomers on diluted permethylated β-cyclodextrin by high-resolution gas chromatography, *J. Chromatogr.* **441**, 155 (1988).

6.B. SUPPORTS, STATIONARY AND BONDED PHASES

6.B.1. Interview of Joseph Pesek, Editor of *Journal of Separation Science* (JSS) (2008), with Professor Walter G. Jennings

With this article, the *Journal of Separation Science* presents a new feature interviews with scientists who have made substantial contributions to our field over the last several decades. We begin this series with Walt Jennings, a major innovator in the development of capillary gas chromatography.

Professor Dr. Walter Jennings has pursued successful careers in both academia and industry. He is an Emeritus Professor of the University of California, where he completed a 35-year career on the Davis Campus. He constructed his first gas chromatograph in 1954, and authored a number of books on chromatography. His most recent book, *Analytical Gas Chromatography*, 2nd ed., W. Jennings, E. Mittlefehldt, and P. Stremple, Academic Press (1997) has been hailed as "a classic."

Figure 6.B.1. Walter Jennings

He has also served as editor for several multiauthored books, and published some 300 scientific papers. Many of the scientists who worked with Professor Jennings during this period are now well-known academic professors in their own right; some occupy eminent positions in instrument companies, and others direct the investigations of research departments in areas as diverse as flavor, forensic, petrochemical, pharmaceutical, and environmental analysis.

In collaboration with one of his doctoral students, he founded J&W Scientific, Inc. in 1974. J&W grew to become the world's largest supplier of fused-silica columns, and was purchased in 1986 by Fisons plc. In 1996, a management group led by Professor Jennings repurchased the company. The company is now a part of Agilent Technologies.

Professor Jennings has spent sabbatical leaves in Austria, Germany, and Switzerland. Under the auspices of the International Atomic Energy Agency, he has served as Adviser in Gas Chromatography to the governments of, and spent considerable time working in, Bulgaria and Poland. He is recognized as a "Senior American Scientist" by the German government, and in 1973 received the prestigious "Humboldt-Preis" award from the Alexander von Humboldt Foundation, consisting of a substantial endowment, and a one-year appointment as a Senior American Scientist in residence.

His other awards include the Award of Merit from the Chicago Chromatography Discussion Group, the National Chromatography Award of the Northeast Regional Chromatography Discussion Group, the L. S. Palmer Award of the Minnesota Chromatography Forum, the Founders Award in Gas Chromatography administered by the Beckman Corporation, and others from the French Association of Analytical Chemists, the University of Bologna, the Taiwanese Food Chemists Society, and the Society of Flavor Chemists. In 1996, Professor Jennings received the M. E. Golay Award at the 18th International Symposium on Capillary Chromatography; in 1997, the Keene Dimick Award at the Pittsburgh Conference, and the, A. J .P. Martin Gold Medal from the Chromatographic Society (UK) for his "contributions to the field of chromatography"; in 1998, he received the Silver Jubilee Award of the 19th International Symposium on Capillary Chromatography at Riva del Garde, Italy. In 1999, Professor Jennings was honored by his alma mater, the University of California, Davis, by the bestowal of their highest award to individuals, the Award of Distinction. At the 2002 Pittsburgh Conference he received the Dal Nogare Award for his contributions to gas chromatography, and in 2004, the Anachem Award at the meeting of the Federation of Analytical Chemistry and Spectroscopy Societies (FACSS). In 2006 he received the CASSS Award from the California Separation Science Society and in 2008 became the receipient of the LC-GC Lifetime Acheivement Award.

In each year since the early 1970s, he has instructed more than 30 extracurricular courses in gas chromatography, at points all over the world. It has been estimated that over 30,000 chromatographers from all walks of life have attended these courses. Walt is a member of the honorary board of JSS, having been an editor previously for the journal as well as for the *Journal of High Resolution Chromatography*.

Recently I had an opportunity to sit down with my distinguished friend in his California home and ask him a few questions about his career. Here are the highlights of that conversation:

JP : Why did you study chemistry?

WJ: In high school, I never took a book home; I could make Cs in most subjects and Bs in chemistry and math, with no effort. I had two excellent teachers, one in chemistry and one in math, that recognized I was just coasting with no effort, tried to motivate me, but to no avail. Then I was drafted, and the army straightened me out. By this time I realized I had some potential and after $3\frac{1}{2}$ years in the military, started college on the GI bill. I was much more mature than my classmates, I was eager for knowledge, I studied.

JP : Who influenced you the most as an undergraduate student?

WJ: Four professors in lower division, organic, physical, and inorganic chemistry who I could talk to.

JP : How did you become interested in chromatography?

WJ: While I was still working on my PhD thesis in 1953, I was offered a position at the lowest step on the academic ladder. I knew that unless I established myself in a new field that was clearly differentiated, I would always be a grad student. In surveying the field, I met Keene Dimick at the USDA lab in Albany. He had just begun construction of his first GC and offered to be my mentor.

JP : What was your first professional job as a chemist?

WJ: When I received my degree in 1954, I was appointed Instructor in Food Science and Assistant Chemist in the Experimental Station at the University of California, Davis.

JP : What do you consider to be your major achievements?

WJ: Difficult question. I'm gratified that I will leave this world some 30+ scientists, former graduate students, most of whom have made their own sizable contributions. I am proud of establishing what has become, and still is, the prime producer of capillary GC columns. When Ray Dandenau invented the fused-silica column, it was welcomed by industrial analysts, but caused consternation among those who had based their careers on glass capillary columns. Ray credited me as the one person most responsible for converting the field in general into fused silica.

JP : What do you believe are the major breakthroughs in your field during the past few decades?

WJ: Bonded stationary phases, the fused-silica column, new developments in column deactivation, the vast improvements in reducing column bleed, developments in MS detection, including the ability to run and store SIM and full-scan modes simultaneously, and the new two-dimensional GC.

JP : What new developments do you expect to come in the near future in GC?

WJ: The weak point in two-dimensional GC is quantitation. It should be possible to devise programs that will scan these scattered blobs and summing together those near neighbors possessing structural similarities.

JP : What new developments do you expect to come in the near future in chromatography in general?

WJ: There is a need, particularly in the two-dimensional field as practiced in the petroleum industry, for high-temperature polar columns. Column manufacturers have not yet achieved this goal.

JP : How would your friends or family describe your personality?

WJ: I would hope as outgoing, friendly, easily approachable, and cooperative.

JP : What motivates you?

WJ: An insatiable curiosity. I want to know more.

JP : What are your interests with respect to art and culture?

WJ: My taste in music runs from early (pre-1930) jazz through classical and grand opera. I detest rock, country, and rap. In art, I am partial to etchings, oil paintings, and sculpture. I have accumulated a very modest collection.

JP : Do you like to travel?

WJ: I enjoy working with people once I get there, but I detest travel per se, particularly with all the recent restrictions. I just turned down a request to go to Riva del Garda this year.

JP : What is the most interesting place you have visited?

WJ: I have traveled a great deal. I'm most comfortable in Germany or Alsace, but I was probably most impressed by Istanbul and China.

JP : What keeps you busy when you are not engaged in science?

WJ: I listen to music almost constantly. I am still active building furniture and wood- and metalworking. I maintain a well-equipped workshop with a thread-cutting machine lathe, vertical mill, and a full line of woodworking equipment.

JP : What is your favorite book?

WJ: Dickens' Pickwick Papers.

JP : What other career would you have chosen if you could not be a scientist?

WJ: Probably teaching. An older professor once told me that as an academician, my job was *to explain phenomena*. I do enjoy devising simpler explanations.

JP : What advice would you give to young scientists just starting their careers?

WJ: As you approach a subject first, be thoroughly conversant with the literature. Learn what people have already done. Spending time on the Web and in the library can pay big dividends.

I should also report that Walt is a great lover of good wine. After our conversation we enjoyed a glass or two of one of his favorite wines.

6.B.2. Column Development—an Abbreviated History (Walter G. Jennings)

The possibility of using a vapor as the mobile phase in chromatography was suggested by A. J. P. Martin and R. L. M. Synge in 1941; A. T. James and Martin demonstrated the validity of that suggestion in 1952 [1]. Within the next 2 years, several petroleum analysts were building instruments and testing the limits of this new technique. Shortly thereafter, the first commercial gas chromatographs opened the technique to a much broader audience and applications multiplied rapidly.

The 1950s was the era of packed columns, diatomaceous earths, crushed firebrick, glass and plastic beads, which were all explored as support materials. As stationary phases, we tried materials as diverse as high-boiling-point esters, ionic and nonionic detergents, the Apiezon greases, and a number of ill-defined high-boiling-point fluids that had been abandoned in the chemical cupboard by our retiring predecessors. While a few knowledgeable investigators tried pacifying or deactivating the support materials prior to coating, others either ignored that complication or depended on the stationary phase per se to satisfy sorptive sites.

Support materials were usually coated in a slurry containing a measured amount of stationary phase in a low-boiling-point solvent. The solvent was commonly removed by rotating a round-bottomed flask under vacuum; fluidized-bed drying and the use of a transverse airflow in density sorting of uncoated and coated particles are concepts that emerged later. Some workers prepared columns possessing reasonable numbers of theoretical plates per unit length, but overall column lengths—and the total number of theoretical plates—were limited by the high-pressure drop imposed by the packing.

The number of theoretical plates required to separate any two solutes under a given set of conditions can be very closely estimated by the resolution equation, and is the product of three multipliers: the degree of resolution desired (R_s), the retention factor (k) of the more retained solute, and the separation factor (α) of the two solutes:

$$N_{\text{req}} = 16 R_s^2 \left(\frac{k+1}{k}\right)^2 \left(\frac{\alpha}{\alpha-1}\right)^2 \tag{6.1}$$

When rearranged to the form

$$R_s = \frac{1}{4}\left(\frac{k}{k+1}\right)\left(\frac{\alpha-1}{\alpha}\right)\sqrt{N} \tag{6.2}$$

it is apparent that component resolution is controlled by the retention factor (k), the separation factor (α), and the number of theoretical plates (N). Most packed columns generated only a few thousand theoretical plates. The restricted values of N were partially offset through manipulation of α. Contrary to the behavior of siliceous glasses, packings will accept a diversity of stationary-phase materials. It was the necessity to exploit separation factors (α) that led to the adoption of some 300–400 materials as "liquid phases"— stopcock greases, detergents, sugars, emulsifiers, and plasticizers—in packed columns. By the 1970s, some packed-column manufacturers were approaching "1000 theoretical plates per foot of column length," similar to what can be achieved with a 320-μm i.d. capillary. But the pressure drop problem remained; packings offered high resistance to gas flow, and this imposed a practical limitation on the length of the column.

Golay's open tubular column [2] offered the advantage of lower resistance to gas flow, and columns of much greater length became practical—but open tubular columns also presented new challenges. Their higher efficiencies and the lower flow volumes of carrier gas made the effects of some extracolumn influences much more critical. Attention was soon directed to the reduction or elimination of unswept or poorly swept areas in the gas flow path, injectors and detectors were redesigned, injection techniques were refined, and improved methods of interfacing the column to the chromatograph were developed. At this time, columns were usually made of metal capillary tubing, dynamically coated with a dilute solution of stationary phase in a low-boiling-point solvent, and dried with a gentle flow of inert gas (e.g., see Ref. 3). The inner surface of metal capillary tubing usually exhibits a high rugosity or "surface roughness" that is not conducive to the deposition of a uniformly thin even film of stationary phase. Those working with a broader range of stationary phases often experienced another problem with the open tubular column: as the column was heated beyond a very modest temperature (100–125°C), separation efficiency plummeted and the column died. To those experimentalists restricted to an academician's limited resources, such failures were a severe blow, which only exacerbated the original high cost of metal capillary tubing.

In 1960, Desty et al. [4] published the details of an elegantly simple machine for producing coiled glass capillary tubing. This offered a source of a more inert capillary tubing at a much lower cost, and with a smoother interior surface. Even more important was the fact that columns made of glass capillary tubing were transparent, permitting visual assessment of the stationary phase as it existed on the wall of the tube. We could now see that when these columns died, what had been a continuous thin and uniform film of stationary phase had changed into discrete globules randomly scattered over the surface. We had been coating open tubular columns with the same low-viscosity fluids that had been used for packed columns. The higher temperatures made them even less viscous and also weakened the attractive forces between the thin film of the stationary phase and the glass surface. As the cohesive forces of the phase became dominant, the film changed into scattered globules. Several workers simultaneously and independently realized that higher-viscosity stationary phases that maintained their high viscosities even at higher temperatures would encourage longevity in open tubular columns.

Polysiloxanes range from low-viscosity fluids—OV 101, SF 96, SF 550—to high-viscosity gums—OV 1, SE 30, and on through to the more highly crosslinked silicone rubbers, and ultimately the siliceous glasses. The length of the polymer chain and the degree of crosslinking between chains largely govern the physical properties. Most workers had employed the low-viscosity fluids, believing that diffusivity would be higher in those fluids than in the viscous gums. In the interest of more stable columns, we began using high-viscosity gums, and column longevity showed a significant improvement. S. J. Hawkes later demonstrated that solute diffusivities are actually higher in the gums than in the fluids [5].

The efficiency of the column-coating process was usually measured by comparing the value of H_{actual}, the height equivalent to a theoretical plate that the column actually generated, to $H_{\text{theor min}}$, the theoretical minimum value of H for a column of that radius and a solute of that k:[1]

$$\%\text{UTE} = 100\frac{H_{\text{theor min}}}{H_{\text{actual}}} \tag{6.3}$$

This reflects the degree to which the theoretical maximum efficiency of the total system has been achieved, namely, the "utilization of theoretical efficiency." At that time, the factors detracting from ideality in column performance were dominated by heterogeneities in the stationary phase film, leading to the use of the term "coating efficiency" (CE values were usually less than 50% in those days). These and similar terms can be justifiably criticized because they usually measure efficiency of the total system; that is, H_{actual} is affected by both column and extracolumn contributions (e.g., injection efficiency).

---

[1]The theoretical minimum value for H is calculated from the equation $H_{\text{theor min}} = r\{[(11k^2 + 6k + 1]/[3(1 + k)^2]\}^{1/2}$. This is an inexact estimate, in that its derivation ignores the C term of the van Deemter equation.

These developments introduced the era of "glass capillary gas chromatography." A variety of techniques were proposed for coating glass capillaries, and a number of workers demonstrated good columns prepared by several modifications of both dynamic and static methods. Studies on column deactivation, coating, and bonded stationary phases led to steady improvement in glass capillary columns. Many of us in the field gloried in the fact that users of these fragile columns formed an elite group, while their fragility worried most packed-column practitioners. In the universities and government laboratories where most of this work was being conducted, column breakage was rarely disastrous. But downtime can be much more serious in industry, where most managers preferred the stability and dependability of the more rugged packed column. Industry recognized the advantages offered by the capillary column—superior separations in shorter times in more inert systems—but most were understandably reluctant to convert their packed-column analyses to glass capillary chromatography.

In 1979, R. Dandenau and E. Zerenner announced their findings on fused silica [6]. Their chromatograms stimulated comments and discussion, but few in the audience at that third biennial Hindelang meeting realized the dramatic effect that fused silica would exercise on industrial acceptance of capillary gas chromatography. When Dandenau folded a fused-silica column into the shape of a bow tie on the speaker's podium, it signaled the end of the era of glass capillary columns. Only a few of us who had been struggling to convert a reluctant industry to capillary columns immediately realized the impact that this would exert on industrial chemists. Fused-silica columns were obviously much more robust, and as column manufacturers adopted the new material, many segments of industry began the conversion to capillary analyses. Another factor that slowed the conversion was the greatly reduced flow volume of carrier gas used for capillary columns. This affects everything from injection to detection, and requires the elimination of large and/or dead volumes in the gas flow path that were much less significant to the packed column because of its higher volumes of gas flow. This difference between packed and capillary columns required a learning curve for the uninitiated, and was actually the reason for the introduction of the 0.53-mm "Megabore" column, whose flow volumes were much higher.

Even so, regulatory agencies that had promulgated their methods on packed columns, waited another decade before they joined this conversion.

Columns now often approached, and not infrequently achieved, 100% of their theoretical efficiencies [Eq. (6.3), above], and some of us became a bit smug. Partly to distinguish ourselves from our packed-column inferiors, we emphasized N, theoretical plate numbers (or "efficiencies"), and ignored α—the separation factor [Eq. (6.2), above]. Eventually, we were humbled by demands for the detection of ever-decreasing amounts of more and more active analytes, and today we deal with high alpha phases in high-efficiency columns, some of which are meticulously "tailored" to accomplish optimal separations of specific mixtures. Again, we experienced a learning curve.

In polysiloxane-type phases, alphas are adjusted by substituting polar groups for some of the methyl substituents, but these substitutions can change some of our chromatographic parameters. For example, D_s, solute diffusivity in the stationary phase, decreases with the increasing substitution of increasingly polar groups:

$$H = \frac{2D_M}{u} + u\left[\frac{2k}{3(k+1)^2}\frac{d_f^2}{D_s} + \frac{[1+6k+11k^2]}{96(k+1)^2}\frac{d_c^2}{D_M}\right] \tag{6.4}$$

Referring to Eq. (6.4), the importance of C, the resistance to mass transport from stationary to mobile phase in the M. J. E. Golay equation, varies directly with the magnitude of d_f, the thickness of the stationary phase film, and indirectly with D_s, diffusivity of the phase. With polydimethylsiloxane, columns whose film thicknesses (d_f) range from 0.1 to 0.4 μm exhibit essentially identical efficiencies as indicated by H_{min}. As d_f exceeds 0.4 μm, H increases—and, of course, N decreases—but this change is fairly minor until d_f approaches 1.0 μm. In more polar phases, D_s decreases, and this break occurs not at 0.4 μm, but somewhere below 0.20 μm. While this can be countered by decreasing the thickness of the stationary-phase film, thinner-film columns require smaller samples and reduce the dynamic range of the system. In addition, the propensity of the stationary phase to both undergo thermal degradation and generate higher bleed usually varies directly with the degree of polar substitution. The selectivity of a polysiloxane-based phase can be quite sensitive to substitution, and it is not unusual to find that phase with the lower substitution not only possesses greater thermal stability and generates lower bleed but can also yield better separation. Figure 6.1 illustrates the separation of a series of pesticides on polysiloxane columns with 50% and 35% trifluoropropyl substitution.

With the development of the benchtop mass spectrometer, more and more analysts began using mass selective detectors, and more recently, ion trap detectors. The high sensitivity of these detectors, and in the case of the ion trap, the limited ion capacity of the trap, require the lowest possible background signal from column bleed for maximum sensitivity to the analyte. Although never as widely used, atomic emission (AED) and chemiluminescence detectors have the same requirement for low background from column bleed. Those using these bleed-sensitive detectors required still lower-bleed columns, in answer to which a whole new stationary phase chemistry emerged.

As the temperature of a gas chromatographic column is increased to a higher plateau, the chemical bonds of the stationary phase are subjected to increased thermal stress, degradation fragments are produced at an increased but constant rate, and the baseline signal rises and remains steady as long as that temperature and flow are maintained. True column bleed does not generate peaks or humps; a signal that rises and then falls as the temperature and gas flow remain constant must have a "point of origin"—as opposed to being simultaneously generated throughout the entire heated length of the column. Annoying to any bleed-sensitive detector, this increased background signal can be the limiting factor in GC/MS, and is especially troublesome with the newer ion trap mass spectrometers.

Bleed signal from polysiloxane columns is generally attributed to cyclic siloxanes that usually arise from thermal and/or oxidative degradation of the phase, but contaminants in the detector or in the gas lines supplying the detector, plasticizers, and other materials outgassing from septa and ferrules, as well as fingerprint oils from column installation, also contribute to what we commonly perceive as column bleed. The higher-bleed levels from columns coated with conventional phases often masks the signal from these extracolumn sources, but their significance increases and may become limiting in the case of low-bleed columns. At temperatures exceeding 320°C, even a pristine detector from which the column has been removed, and the fitting capped, invariably generates 1–2 pA signal, which is mistakenly perceived as column bleed. In column-to-column comparisons of bleed rates and upper temperature limits,

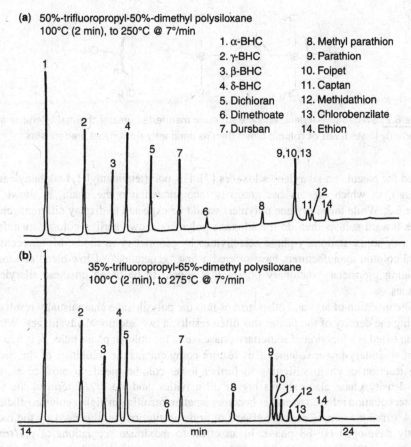

Figure 6.1. Chromatograms of pesticide standards on (a) 50% trifluoropropyl dimethyl siloxane and (b) 35% trifluoropropyl dimethyl siloxane columns. Note that the bleed level of the less substituted column (35%) is markedly less at 275°C than that generated by the 50% substituted column at 250°C, while the less substituted column also delivers improved separation.

it is important to recognize that for a given quality of stationary phase at a given temperature and carrier gas flow, the level of column bleed is a function of the mass of stationary phase in the gas flow path. Under similar conditions, shorter, smaller-diameter, thinner-film columns should always exhibit lower bleed levels than do longer, larger-diameter, thicker-film columns.

Whether derived via thermal stress, the back biting reaction, or oxidation, the degradation fragments from polysiloxanes consist mainly of cyclic siloxanes, dominated by trimers and tetramers of (—Si—O—) to which methyl (or other) substituents occupying the remaining two bonds of the tetravalent silica atom may cleave, or remain attached. In a conventional 95% dimethyl–5% diphenyl–polysiloxane, the phenyl groups are pendant to the siloxane chain. In 1946, a DuPont chemist, M. Sveda, working on bulk polymers,

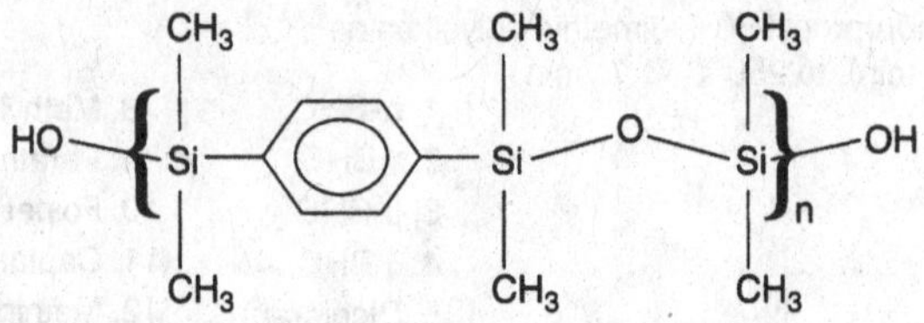

Figure 6.2. Sveda's silyl-arylene siloxane structure manifested greater thermal tolerance and significantly lower levels of column bleed than its pendantly-substituted predecessors.

applied for patents on silarylene siloxanes [7]. His poly(tetramethyl-1,4-silphenylene–siloxane), in which the arylene group is incorporated into the chain, is shown in Figure 6.2. While the silarylene polymers would be expected to display different selectivities toward solutes than do the siloxane polymers shown earlier, column manufacturers can adjust stationary-phase selectivities by proprietary methods. More recently, several column manufacturers have offered a first "generation" of low-bleed columns, containing proprietary stationary phases that appear to be based on these silarylene siloxanes.

The insertion of aryl and other groups into the polysiloxane chain usually results in lowering the density of the phase; this often results in two additional advantages. While column bleed is a function of stationary-phase *mass*, the column phase ratio, β, is a function of stationary-phase *volume*. This feature complements the inhibition of the back-biting reaction of chain stiffening to further lower column bleed. In most cases, the lower-density phase also exhibits higher diffusivities, and the d_f^2/D_s term of the van Deemter equation referred to earlier becomes smaller, resulting in higher column efficiencies. A striking example of these effects of aryl substitutions is illustrated by the more recently developed HP-88 phase. In an effort to maximize separations of *cis/trans* fatty acids, it becomes apparent that because the *cis* isomer is less "balanced" than the *trans* isomer, the *cis* isomer must possess a higher dipole moment. Hence, several high-cyanopropyl phases have been developed for *cis–trans* separations. Unfortunately, while the efficacy of *cis/trans* separation varies directly with the degree of cyanopropyl substitution, so does the propensity for column bleed. In addition, stationary-phase diffusivities (D_s) vary indirectly with the degree of cyanopropyl substitution, and this has an adverse effect on column efficiency, which sometimes necessitates longer columns.

Some of the pathways that have been postulated for siloxane degradations involve reaction between two normally separated groups in the siloxane chain (e.g., see Ref. 8). This implies that the chain must fold to bring the reacting groups into closer proximity. Dvornic and Lenz [8] suggested that such reactions might be inhibited by insert on into the chain groups that would render the chain more rigid and restrict its flexibility (Fig. 6.2). The incorporation of this suggestion into the synthesis of silarylene and silcarborane polymers has led to a second generation of low-bleed stationary phases; these are characterized by still greater thermal and oxidative resistance. Reese et al. [9] suggested a number of suitable groups (Fig. 6.3). In retrospect, the increased thermal resistance that Sveda observed in his poly(tetramethyl-1,4-silphenylene–siloxane) is very probably attributable to chain-stiffening.

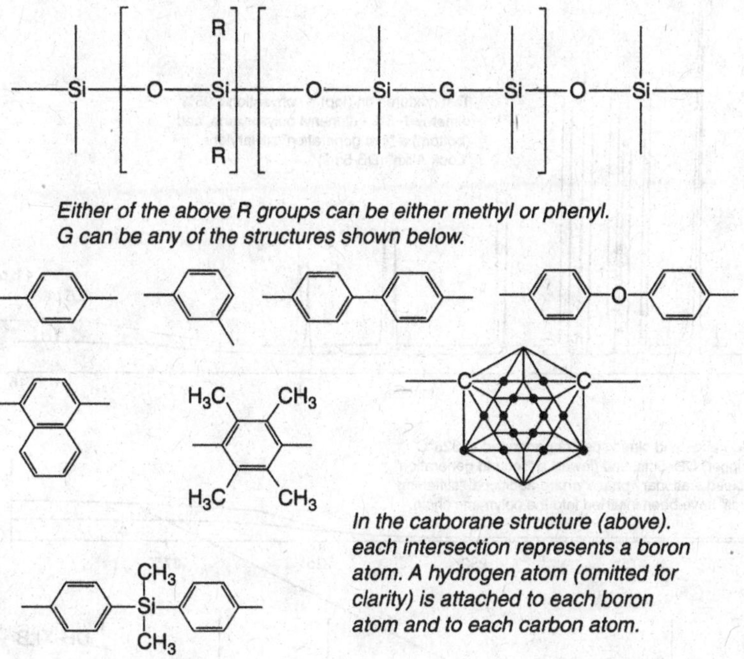

Either of the above R groups can be either methyl or phenyl.
G can be any of the structures shown below.

In the carborane structure (above).
each intersection represents a boron
atom. A hydrogen atom (omitted for
clarity) is attached to each boron
atom and to each carbon atom.

Figure 6.3. Inclusions suggested as polysiloxane "chain stiffening" groups. Adapted from "Analytical Gas Chromatography," W. Jennings, E. Mittlefehldt, and P. Stremple, 2nd Edition, Academic Press, 1997, New York, NY, p. 100.

Figure 6.4(a) contrasts chromatograms generated on a conventional 95%-dimethyl-5%-diphenyl polysiloxane column (upper trace) and on a silphenylene–siloxane "first generation" low-bleed column (lower trace). Figure 6.4(b) compares this "first generation" low-bleed column (DB-5 ms) with a "second generation" low-bleed column (DB-XLB), where selected chain-stiffening inclusions have been incorporated into the silarylene–siloxane chain. The bleed spectra in the lower portion of panel (b) were obtained on a Finnigan MAT ITD at 325°C. The reduced bleed level is impressive, but more importantly, the bleed spectrum is simpler; the stiffening groups inhibit folding of the chain, necessary to the formation of higher-mass fragments. These qualities of the second-generation phases make them especially valuable for those utilizing bleed-sensitive detectors such as the ion trap mass spectrometer. Some of these phases also exhibit unique selectivities that have excited interest among those interested in a variety of problematic separations, including congener specific PCB analysis [10].

Even if there are no further advances in column technology, the future will almost surely see a continued push toward shorter, smaller-bore columns, capable of much faster analysis. Figure 6.5 shows an example where a 48-min analysis is cut to 4.8 min, with essentially no loss in the resolution of major components.

Certainly we will see further technological developments. Future developments in the tubing category may lead to surface modifications of the inner periphery of the

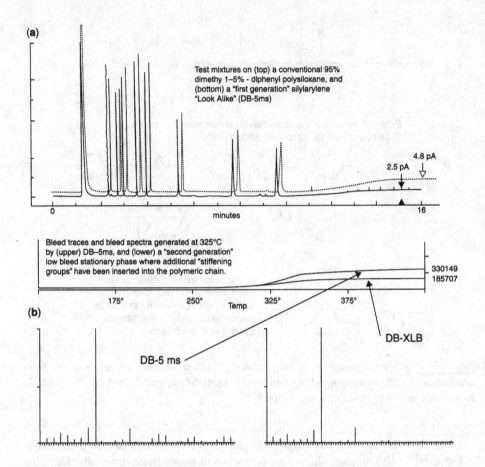

Figure 6.4. Three generations of high-efficiency fused silica WCOT columns. (a) This figure contrasts chromatograms of column test mixtures on a conventional 95%-dimethyl-5%-diphenyl polysiloxane column (dotted line), and a silphenylene siloxane "Look Alike" column (solid line, DB-5 ms). (b) This figure shows the bleed traces (top) and bleed spectra at 325°C (bottom) generated by the silarylene polymer above (DB-5 ms) and the second generation silarylene siloxane (DB-XLB) into whose polymeric chain additional stiffening groups have been inserted. Note the dearth of higher mass ions in this later trace.

silica tubing, yielding material possessing different (and perhaps a variety of) wetting characteristics, lowered activity toward solutes, and greatly increased tensile strength. In stationary phases, we have proceeded from the "silicone oils" and polyesters used earlier in packed columns through bonded polymeric phases, to low-bleed polymeric phases.

Figure 6.6 illustrates an experimental column whose manufacture employs proprietary techniques that permit deposition of a stable thick layer of an amorphous silica inside fused-silica tubing. Such columns differ from conventional PLOT or SCOT columns in

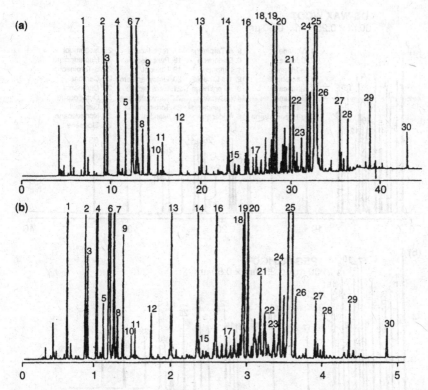

Figure 6.5. Chromatograms of western spearmint oil: (a) 44 minute analysis on a 50 m × 0.25 mm column, coated with a 0.25 μm film DB-Wax, programmed from 75°C (8 min) to 200°C at 4°/min, helium carrier at 25 cm/sec; (b) 4.8 min. analysis on a 10 m × 0.10 mm column, 0.20 μm film of DB-wax, programmed from 70° (0.5 min) to 120°C at 30°/min, to 200°C at 20°/min, hydrogen carrier at 55 cm/sec. *From Amer. Lab. (News Edition)* **30**(25), 1998, p. 38, and reproduced with permission from International Scientific Communications, Inc.

that they are more nearly "open tubular packed" columns, in which the deposited support is actually a continuous monolithic layer of porous silica. Certain low-viscosity, high-selectivity phases that have long resisted use in conventional WCOT columns [e.g., the immensely polar 1,2,3-tris-(2-cyanoethoxy)propane] can be affixed to and through this deposited layer.

The exploitation of these various factors previously discussed permit modern column manufacturers to occasionally offer columns that address a specific analytical need. Over the years a great deal of attention has been directed to cardiovascular disease and fatty acid analysis, and some of the dietary recommendations even seem contradictory. At present, there seems to be general agreement that *trans*–fatty acids are bad for heart disease, that some unsaturated fatty acids are essential, and that certain members of the latter group—notably the omega-3 and all *cis*–omega-6 fatty acids—may even possess some healing capabilities. All of this created a keen interest in *cis–trans*

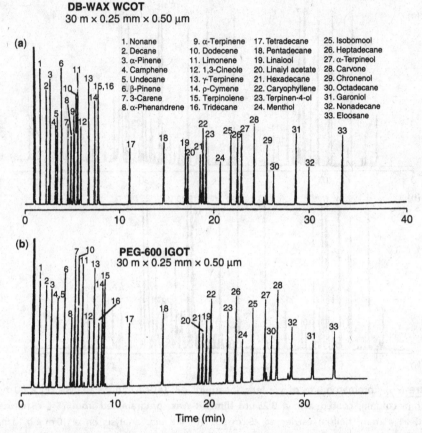

Figure 6.6. Series of standards (a) on a conventional WCOT column coated with DB-Wax, and (b) on a similarly dimensioned impregnated-gel open tubular (IGOT) column coated with PEG 600 (Aldrich Chemical Co., Milwaukee, WI). When employed as a packed column phase, PEG 600 has a recommended maximum temperature of 120°C; as an IGOT column, it exhibits a higher maximum temperature (>175°C). In packed columns, the latter phase displayed a unique selectivity for the constituents of many essential oils (e.g., monoterpenes, sesquiterpenes, oxygenates) that made it popular with natural product chemists. It is possible to impregnate the gel of IGOT column with phases such as this, that rarely persist as stable coatings on conventional WCOT columns. *From Amer. Lab. (New Edition)* **30**(25), 2998, p. 28 and reproduced with permission from International Scientific Communications, Inc.

separations. Because the less "balanced" *cis* configuration should exhibit a stronger dipole moment than its more balanced trans counterpart, highly polar stationary phases should yield better *cis–trans* separations, and several high-cyanopropyl phases have been developed for such separations. But as discussed earlier, both temperature tolerance and stationary-phase diffusivity (which affects column efficiency) usually vary indirectly, while column bleed varies directly with the polarity of the stationary phase. This dilemma

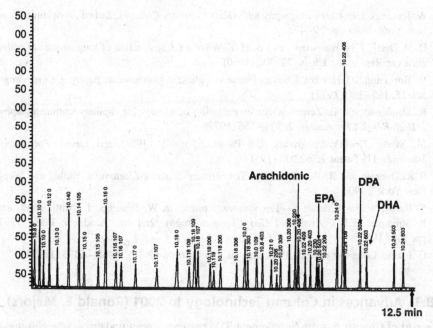

<u>Figure 6.7.</u> Chromatogram of a standard FAME test mixture compiled by Lipomics, Inc. on a 30 m × 0.25 mm HP-88 column. Further analytical details are considered proprietary by Lipomics. Analyst, Ryan Davis.

led one manufacturer to offer a stationary phase with the highest possible cyanopropyl substitution that also contained chain-stiffening substitutions (to enhance thermal tolerance, lower bleed intensity, and increase solute diffusivities). Lipomics (www. Lipomics.com) has developed sophisticated programs concerned with lipid metabolism based on analyses of a range of body tissues, and served as a beta test site in the development of this column. Figure 6.7 illustrates a typical analytical run. While both the precise composition of the stationary phase per se and some of the chromatographic details are considered proprietary, this serves as an outstanding example of the current state of the art in FAME analysis.

REFERENCES (for Walter G. Jennings)

1. A. T. James and A. J. P. Martin, Gas–liquid partition chromatography: The separation and microestimation of volatile fatty acids from formic acid to dodecanoic acid, *Biochem. J.* **50**, 697 (1952).

2. M. J. E. Golay, in V. J. Coates, H. J. Noebels, and I. S. Fagerson (Eds.), *Gas Chromatography Symp. 1957*, (East Lansing, MI), Academic Press, New York, 1958, pp. 1–13; D. H. Desty (Ed.), *Gas Chromatography Symp. 1958* (Amsterdam), Butterworth, London, 1958, pp. 139–143.

3. W. Jennings, *Gas Chromatography with Glass Capillary Columns*, 2nd ed., Academic Press, New York, 1980, pp. 39–43.

4. D. H. Desty, J. N. Haresnape, and B. H. F. Whyman, Construction of long lengths of coiled glass capillary, *Anal. Chem.* **32**, 302 (1960).

5. W. Burns and S. J. Hawkes, Choice of stationary phase in gas chromatography, *J. Chromatogr. Sci.* **15**, 185–190 (1977).

6. R. Dandenau and E. Zerenner, An investigation of glasses for capillary chromatography, *J. High Resol. Chromatogr.* **2**, 351–356 (1979).

7. M. Sveda, *Disilahydrocarbons*, US Patent 2,561,429 (1951) and *Linear Polyarylene Siloxanes*, US Patent 2,562,000 (1951).

8. P. R. Dvornic and R. W. Lenz, *High Temperature Siloxane Elastomers*, Hüthig and Wepf, New York, 1990.

9. S. Reese, within Chapter 4—The stationary phase, in W. Jennings, E. Mittlefehldt, and P. Stremple, (Eds.), *Analytical Gas Chromatography*, 2nd ed., Academic Press, 1997, pp. 100–105.

10. G. Frame, Congener-specific PCB analysis, *Anal. Chem.* **69**(15), A468–A475 (1997).

6.B.3. Advances in Column Technology to 2001 (Ronald E. Majors)

This part of Chapter 6, from Supplement S11, presents a brief update on new chromatography columns introduced at the 2000 Pittsburgh Conference [1] and some advances in the design of HPLC packings [2]; an update to 2008 is provided later in this chapter. In general, the products introduced have shown definite correlations to current research, development, and application activity in separation science. New introductions were made in the areas of high-performance liquid, reversed-phase, normal- and bonded-phase, ion-exchange, size-exclusion, large- and preparative-scale, and specialty preparative columns.

HPLC Columns. Reversed-phase HPLC maintains its dominance. A number of columns and families of columns are introduced each year at PITTCON. Significant numbers of reversed-phase columns with embedded polar functional groups, silica-bonded phases with high pH capability, fluorinated phases with new selectivities, and specialty phases for unique applications have been displayed.

Even though high-performance liquid chromatography (HPLC) column technology is somewhat mature, new developments continue. In addition to the annual coverage of new PITTCON column introductions, *LC-GC* updated the current status of HPLC column technology in 1994 with a "Column watch" that reviewed 25 years of commercial developments [3] and in 1997 with the supplemental issue "Current issues in HPLC technology" [4]. Improvements have occurred in packing material design, bonded-phase chemistry, column construction, and formats since 1997. In addition, users have a better understanding of the advantages and limitations of silica-based materials, and this understanding allows them to extend column life, even under what might be considered harsh chemical conditions for silica-based packings.

Improvements in Porous Packings. Porous packings have been in favor throughout HPLC's history. The transition from large porous particles and pellicular materials to small porous particles occurred in the early 1970s, when microparticulate silica gel (<10-μm d_p) came on the scene and appropriate packing methods were developed. Irregularly shaped microparticulate packings were in vogue throughout the 1970s until spherical materials were developed and perfected. The spherical packings could be packed more homogeneously than their irregular predecessors. In addition, these particles provided slightly better efficiencies. They could be manufactured in high purity. Indeed, the so-called type B silica that was low in trace metal content became the standard in the late 1980s, and now most silica-based materials are of this higher level of purity. Trace metals in silica gel cause interactions with certain compounds and can affect the acidity of residual silanols [4]. One early goal for the optimum use of HPLC packings was to achieve the best efficiency possible, thereby obtaining better overall chromatographic resolution [5]. It was long known that reducing the particle size improved packing performance. To better understand the various approaches to improve column efficiency, I'll briefly discuss the morphology of a porous packing material such as silica gel or alumina.

Diffusive pores dominate a typical porous packing, and the major surface area of the particle is within these pores. As depicted in the simplified picture of a porous particle in Figure 6.8a, solutes transfer from the moving mobile phase outside the particles to the stagnant mobile phase within the pores to interact with the stationary phase. After this interaction, the solute molecule must diffuse out of the particle and continue its journey down the column. This type of mass transfer occurs many thousands, perhaps many millions, of times as the differential separation process proceeds and the solute is eluted from the column. While the solute spends time in the diffusive pores, the mobile phase in which it was located originally traverses down the column ahead of the solute. This slow rate of mass transfer into and out of the porous particle is a major source of band broadening in HPLC. Using smaller particles shortens the pathlength of this diffusion process, improves mass transfer, and provides better efficiency.

However, congruent with an improvement in efficiency was a decrease in column permeability, that is, an increase in column backpressure. Although pumps can provide the necessary increase in pressure output at normal $1-2$ mL/min flow rates with commonly used solvents such as water, methanol, acetonitrile, and hexane, users found that columns were unstable when operated near pump pressure limits as high as 6000 psi (lb/in.^{2}). Although porous particles smaller than 3 μm have been introduced [6], they have not yet become mainstream media. The expected improvement in performance has not been realized, perhaps because of instrumental limitations. Users also have found that limited column life, high backpressures, and plugging problems—all practical considerations—have curtailed the use of columns with particles smaller than 3 μm.

The newest type of column introduced is the so-called "monolithic column." Monoliths are chromatographic columns that are cast as continuous homogeneous phases rather than packed as individual particles. Japanese manufacturers have been particularly active in this area, and they were among the first to investigate silica-based monoliths [1]. Monolithic phases generally possess flowthrough pores with macroporosity and mesopores, which are diffusive pores whose average pore diameter can be controlled. These columns show porosity higher than that of packed microparticulate columns; therefore, they have a lower pressure drop than do comparably sized columns under similar conditions.

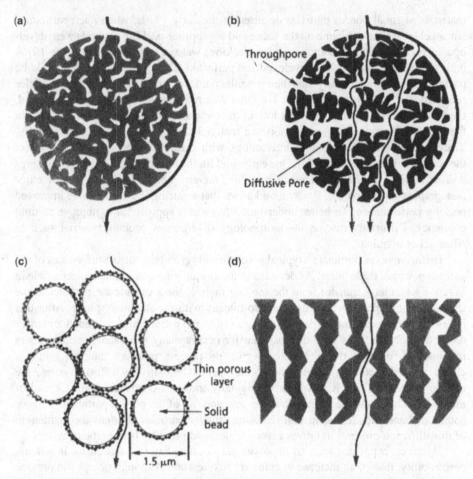

Figure 6.8. Schematics showing the flow characteristics and design of (a) totally porous particles, (b) perfusion packing, (c) nonporous silica or resin, and (d) monoliths and activated membranes.

Perfusion Packings.   Others have tried different approaches to improve efficiency. The perfusion packings developed by N. B. Afeyan and coworkers [7–9] and commercialized by PerSeptive Biosystems (now Applied Biosystems Group of Applera Corp.) in the 1980s provided improved chromatographic performance, particularly for larger molecules. Figure 6.8b shows a simplified presentation of a perfusion packing. Compared with the porous packing depicted in Figure 6.8a, the perfusion packing consists of two types of pores: diffusive and through pores. The diffusive pores are the same type present in the porous particles (Fig. 6.8a), and they provide the sorption capacity. The through pores allow mobile phase to pass through the packing itself, thereby increasing the rate of mass transfer in the mobile phase. Instead of predominantly flowing around the particle, a portion of the mobile phase flows through the particle, which allows the solute to spend less time undergoing the mass transfer process and thereby provides narrower peaks. The process actually is a combination of diffusion and convection. Commercial perfusion packings are polymeric particles of a larger particle size than the HPLC packings typically used; the smallest average particle size is approximately 12 μm. Compared with a porous packing of the same particle and pore size, the perfusion packings give better efficiency for large molecules [9]. The perfusion packings can be used at high flow rates compared with the older soft, organic porous packings such as fast-flow agarose or polydextrans, which tend to collapse at higher linear velocities. At these higher flow rates, the perfusion packings maintain their sample capacity, which makes them useful for preparative separations and purifications.

Nonporous Packings. The use of nonporous packings represents another approach to improve the rates of mass transfer. Nonporous packings are available in two types: nonporous silica and nonporous resin. As depicted in Figure 6.8c, the nonporous packings are very reminiscent of the older pellicular or porous-layer beads used in the early days of HPLC [10], but these materials have much smaller particle sizes, typically in the 1.5–2.5 μm d_p range. The older porous-layer beads had an average diameter of 40 μm, but they did provide better performance than did the porous packings with the same average particle diameter.

The thin porous layer allows much faster rates of mass transfer. Separations of only a few minutes can be achieved for both large and small molecules. Unfortunately, the thin layer of stationary phase also limits the capacity of the packing, which makes nonporous silicas and resins unsuitable for preparative separations. In addition, the backpressure from nonporous silica columns generally is much greater than that in columns with 3- or 5-μm d_p microparticulate HPLC porous packings. Backpressure is inversely proportional to the particle diameter squared.

Monoliths. Monolithic columns are still another approach to provide higher rates of mass transfer and lower pressure drops. These columns were reviewed in 1999 in *LC–GC* [11]. Several types of monolithic columns are available, including agglomerates of polyacrylamide particles, polymethyacrylate block, agglomeration of micrometer-sized silica beads, silica rod columns, and membranes of various types made by many manufacturers.

Monolithic columns (Fig. 6.8d) offer great potential in stable, easily replaced columns for analytical and preparative separations. Both silica- and polymer-based monoliths have been studied extensively. The silica-based materials were developed by N. Tanaka and coworkers [12] in Japan, introduced by Cabrera et al. [13] at HPLC 1998 in St. Louis, Missouri, and commercialized by E. Merck (Darmstadt, Germany) as the company's SilicaRod columns more recently renamed Chromolith. These columns are solid rods of silica monolith. Similar to the perfusion packings, they have both flow through pores with 1–2 μm macroporosity and diffusive pores called *mesopores*. The silica rods can be modified using the same derivatization chemistries that are used for regular HPLC packings for example, C18 bonded phase.

Sol-Gel Silica. Because silica gel is the most widely used base material for bonded phases in HPLC, expanding the pH operating range on both the acid and basic sides of the pH scale could benefit chromatographers. Two types of silica particles are used in commercial HPLC columns: sil-gel and sol-gel. *Sil-gel* particles, usually made by gelling soluble silicates or coalescing fumed silica, are characterized by higher porosities and irregular pore shapes with variable wall thicknesses. *Sol-gel* particles, which are made by aggregating silica-sol particles, have lower porosities and more regular pores with thicker walls defined by the surrounding solid silica-sol particles. Sol-gels generally are more mechanically stable than sil-gels. Both silica-gel types can withstand typical and mobile-phase buffers that are used on the acidic end of the pH scale. At low pH values, the unprotected siloxane bonded phases are more vulnerable and may be subject to catalyzed hydrolysis by the hydronium ion [14]. Because of their thinner pore walls, sil-gel particles appear to dissolve more quickly than sol-gel particles [15,16]. Accordingly, researchers anticipate developing more long-term stable methods at intermediate to high pH using columns made with sol-gel supports such as Hypersil (ThermoQuest, San Jose, CA), Spherisorb (Waters), and Zorbax (Agilent Technologies, Wilmington, DE) columns.

Future Directions in Packing Development. Silica-gel bonded phases undoubtedly will be used for a long time. Even though many new materials have surfaced, silica-based packings retain their dominant popularity in most laboratories. Their excellent efficiency, rigidity, lower cost, and ability to fit almost any HPLC mode will ensure their continued success. The trend toward using smaller, 3-μm particles packed into short columns will continue, and more than likely, 7.5–15 cm × 0.46-cm, 3-μm d_p packed columns will replace today's standard 25 × 0.46-cm, 5-μm d_p columns. These shorter columns with smaller particles can provide the same resolution as a longer column with larger particles. The driving forces will be high-throughput requirements—quality assurance, quality control, liquid chromatography-mass spectrometry (LC-MS) and LC-MS-MS, and combinatorial chemistry needs—with solvent savings and increased sensitivity as secondary benefits.

The monolithic columns should become more commercialized and affordable. If a 5- or 10-cm bonded silica monolith or silica rod column could be placed into a holder or housing and provide the efficiency and lifetime that research columns have shown, they would be ideal and easily replaced columns. A dead column could be removed easily and a new rod could be slid in place without using any special configurations or tools. The polymeric monoliths already come in housings, but the column lengths are rather short and the plate counts are insufficient to perform isocratic separations on complex mixtures.

New types of particles constructed from new materials will debut as chromatographers and manufacturers search for an ideal packing that will provide high recovery, excellent efficiency, low cost, and extraordinary stability. For example, titania is being studied as a base material for bonded phases and as a material for ion-exchange separations.

Researchers will continue to develop new copolymers that provide unique surface properties and better efficiency. One example is the hydrophilic polymers provided as

SEC packings, which can be used with either organic or aqueous mobile phases and converted from one solvent system to another. This type of column could provide a lower-cost alternative for chromatographers who analyze both types of samples. Bimodal SEC phases in which multiple pore sizes are available in one bead may give polymer chemists a packed bed that can cover a wide molecular weight range in one column. Hybrid inorganic–organic materials may offer a better compromise to solve stability problems for observed silica-based materials at higher pH.

Poroshell-type packings [17] may combine the advantages of rapid mass transfer (i.e., improved efficiency), a decent sample capacity, and good recovery for larger biomolecules.

Ion-Exchange and Ion Columns. Table III of Ref. 1 shows the 10 new ion-exchange and ion chromatography columns that were introduced at PITTCON 2000. In all, companies will show three cation-exchange columns, six anion columns, and a mixed-mode column based on titania. Continuing with the trends of the late 1990s, in 2000, polymer-based columns dominated the introductions. One company displayed a flash chromatography cation-exchange column based on a silica gel derivatized with a propyl sulfonic acid phase. Flash chromatography is a low-pressure preparative separation technique that uses large particles packed into glass columns of considerably wider bore than conventional HPLC columns. Polyetheretherketone (PEEK) or epoxy plastic column hardware and porous polymer frits that provide a metal-free environment are favored by the companies introducing ion-exchange columns, presumably for less demanding applications.

The analytical and preparative separation of proteins has always been a popular area for ion-exchange chromatography. Dionex uses phases of different functionality grafted onto polymer beads to provide specialized columns for the separation of closely related proteins. The ProPac WAX-10 column has a pellicular structure with a neutral, thin hydrophilic coating that is functionalized with a tertiary amine (WAX) or quaternary amine (SAX), and these phases provide anion-exchange separations of proteins with small differences in charge without nonspecific interactions with the packing surface. Figure 6.4 in Ref. 1 illustrates the separation capabilities of the ProPac SAX column for a sample of ovalbumin before and after treatment with alkaline phosphatase. A salt gradient was used to effect this separation.

Size-Exclusion Columns. Seven companies introduced 10 new products to product families for size-exclusion chromatography (SEC) [1, Table IV]. Columns for nonaqueous and aqueous SEC are equally divided. Most were polymer-based, but two silica-based packings made debuts. Most often, aqueous SEC columns are devoted to either biomolecules or water-soluble organic polymers. Several of the manufacturers now have columns that can provide both applications with a single column set. Polymeric size-exclusion columns that can be used with different solvents have also become important. Formerly, polymer-based columns were packed in a single solvent, and manufacturers recommended that the solvent never be changed for fear that the packing would swell or shrink and become unusable. The TSKgel Alpha Column series can be used with solvents ranging from water to nonpolar solvents; therefore, the same column can be used for both aqueous and nonaqueous size separations.

Large- and Preparative-Scale Columns. The definition of *preparative chromatography* has always been fuzzy; in essence, it is in the eyes of the beholder, because the mass of the injected or collected sample is dependent on the amount available. For some a few micrograms of material are sufficient for further characterization or use, but for others 10–100 g represents preparative amounts. This year at PITTCON 2000 manufacturers exhibited many columns that they call "semipreparative" or "preparative." Generally, these columns have inner diameters exceeding 4.6 mm. In fact, some columns have inner diameters as large as 100 mm and lengths of as much as 600 mm. Monolithic columns are excellently suited for protein chromatography, because the efficiency is relatively independent of velocity and the capacity is in a moderate range but still sufficient for preparative purposes. According to the nature of the material, stacked membranes have higher capacity, but efficiency is lost because of the additional grafting of ion-exchange groups and the design of the cartridge.

Specialty HPLC Columns. Specialty columns are HPLC columns developed for specific separations that are difficult to achieve on standard columns. Sometimes manufacturers will use a standard column, but test it specifically for a certain class of compounds and provide a recommended set of chromatographic conditions. In some cases, the specialty column comes as part of a total solution kit with reagents, standards, and method. Most specialty columns will come with test chromatograms from analyses performed in the factories before shipment, and some are guaranteed. Table V in Ref. 1 lists the 25 specialty columns that were exhibited at PITTCON 2000. Many of the specialty columns are application-specific—used primarily in the areas of bioseparations, environmental applications, pharmaceuticals, combinatorial chemistry, and natural products. The introduction of seven new chiral columns, representing 25% of the total, showing that this area was quite popular, mainly with pharmaceutical laboratories.

Chiral columns have begun to cover a wider gamut of modes and mobile phases. Originally, many chiral columns could be used only in organic solvents, but nowadays columns are available for aqueous solvents and organic–aqueous mixtures. Pirkle-type packings have been the most popular for separations based on π-donor or π-acceptor interactions between the chiral analyte and the functional groups around the asymmetric carbon. One company has expanded its line of columns based on Pirkle-type bonded phases. Another has expanded its columns based on teicoplanin stereochemistry and has introduced a column that shows very high selectivity for racemic amino acids and peptides using simple mobile phases.

PITTCON also attracts manufacturers of specialty columns for the pharmaceutical and biopharmaceutical industries. Columns for analyzing proteins, peptides, amino acids, nucleic acids and constituents, and carbohydrates always are popular introductions at PITTCON.

The high-throughput requirements of combinatorial chemistry and LC-MS in pharmaceutical chemistry have generated introductions of many specialty columns covered in Table V of Ref. 1. Often the requirements do not demand high-resolution separations, and short columns packed with small particles provide sufficient resolution. Several companies showed single or families of columns [1, Table I] that could be classi-fied as rapid resolution or fast LC columns for investigations of combinatorial libraries. Sometimes chromatographers use rapid gradients with these short columns to provide rapid separation of mixtures that contain compounds of markedly different polarity in a total cycle time of a few minutes.

Several columns for environmental analyses debuted in 2000. The ProntoSIL Enviro PHE column (Bischoff Chromatography) is specific for phenols separated by the method spelled out in US Environmental Protection Agency (EPA) Method 604/625. The Quat-Sep column (Analytical Sales and Services) will separate difficult quaternary amines such as paraquat and diquat. EPA Methods T0-5 and T0-11 specify 16 aldehydes and ketones that are unfavorable to the atmosphere. After derivatization of the carbonyl group with dinitrophenylhydrazine (DNPH), these derivatized colored compounds can all be separated on the Wakosil DNPH column (Wako Chemical). Figure 6.6a in Ref. 1 shows the separations of all 16 aldehydes and ketones using the Wakosil DNPH column; Figure 6.6b in Ref. 1 depicts an actual ambient air sample in which several of the carbonyl-containing compounds were detected.

An update to 2008 LC column technology is presented by R. E. Majors in Section 6.F below.

REFERENCES (for Ronald E. Majors)

1. R. E. Majors, New chromatography columns and accessories at the 2000 Pittsburgh confer-ence, Part I. *LC-GC North America* **18**(3), 262–280 (2000).

2. R. E. Majors, Advances in the design of HPLC Packings, *LC-GC North America* **18**(6), 586–598 (2000).

3. R. E. Majors, Twenty-five years of HPLC column development—a commercial perspective, *LC-GC North America* **12**(7), 508–518 (1994).

4. Current issues in HPLC column technology, *LC-GC North America* **5**, S1–S68 (1997).

5. K. K. Unger, J. N. Kinkel, B. Anspach, and H. Giesche, Evaluation of advanced silica packings for the separation of biopolymers by HPLC. I. Design and properties of parent silicas, *J. Chromatogr.* **296**, 3 (1984).

6. R. E. Majors, New chromatography columns and accessories at the 1995 Pittsburgh confer-ence, USA, *LC-GC North America* **13**(3), 202–219 (1995).

7. N. B. Afeyan, N. F. Gordon, I. Mazsaroff, L. Varady, S. P. Fulton, Y. B. Yang, and F. E. Regnier, Flow-through particles for the high-performance liquid chromatographic separations of biomolecules—perfusion chromatography, *J. Chromatogr.* **519**, 1–29 (1990).

8. F. Regnier, Perfusion chomatography, *Nature* **350**, 634–635 (1991).

9. N. B. Afeyan, S. P. Fulton, and F. E. Regnier, High-throughput chromatography using perfus-ive supports, *LC-GC North America* **9**(12), 824–832 (1991).

10. T. J. Barder, P. J. Wohlman, C. Thrall, and P. D. Dubois, Fast chromatography and nonporous silica, *LC-GC Mag. Separation Sci.* **15**(10), 918 (1997).

11. G. Iberer, R. Hahn, and A. Jungbauer, Monoliths as stationary phases for separating biopoly-mers, *LC-GC North America* **17**(11), 998–1005 (1999).

12. H. Minakuchi, K. Nakanishi, N. Soga, N. Ishizuka, and N. Tanaka, Octadecylsilylated porous silica rods as separation media for reversed-phase liquid chromatography, *Anal. Chem.* **68**, 3498–3501 (1996).

13. K. Cabrera, G. Wieland, D. Lubda, K. Nakanishi, N. Soga, H. Minakuchi, and K. Unger, *SilicaROD:* A new challenge in fast high-performance liquid chromatography separations, *Trends Anal. Chem.* **17**(1), 50–53 (1998).

14. J. J. Kirkland and J. W. Henderson, Reversed-phase HPLC selectivity and retention characteristics of conformationally different bonded alkyl stationary phases, *J. Chromatogr. Sci.* **32**, 473–480 (1994).

15. H. A. Claessens, M. A. vanStraten, and J. J. Kirkland, Effect of buffers on silica based column stability in reversed-phase high-performance liquid chromatography, *Chromatogr. A* **728**(1–2), 259–270 (1996).

16. J. J. Kirkland, Practical method development strategy for reversed-phase HPLC of ionizable compounds, *LC-GC North America* **14**(6), 486–500 (1996).

17. J. J. Kirkland, F. A. Truszkowski, C. H. Dilka, Jr., and G. S. Engel, Superficially porous silica microspheres for fast HPLC of macromolecules, Paper L072 presented at 23rd Int. Symp. High Performance Liquid Chromatography and Related Techniques (HPLC '99), Granada, Spain, May 30–June 4, 1999.

6.C. CONTRIBUTIONS BY OTHER CHROMATOGRAPHERS

The contributions are also available from Supplement S11 online at http://www.chemweb.com/preprint/.

6.C.1. Column Packings and Separations Technology (J. Jack Kirkland)

Earlier, separations were done with diatomaceous earth supports of the type used in GC, but with particle sizes in the 30–50 μm range. Solving the need for better column-packing materials was more difficult. Fortunately, expertise within DuPont again came to the rescue. In the adjacent laboratory was a close friend and a world-renowned expert in silica chemistry, Dr. Ralph K. Iler. Ralph had already consulted with me in developing PLOT columns for GC. The basic technology involved in this work was then adapted to create superficially porous particles, which consisted of a thin, porous shell coated on a solid silica core of about 25 μm. These particles showed excellent mass transport characteristics, resulting in the first packing materials that produced fast, high-quality HPLC separations in keeping with the theory proposed years before by Professor Cal Giddings and others. On the basis of this and other new technology, DuPont's Instrument Products Division in the late 1960s began to market HPLC instrumentation and columns based on the superficially porous particles, called *Zipax*® *controlled-porosity chromatographic support* [1].

In the late 1960s HPLC separations (other than ion exchange) generally were done by liquid–liquid chromatography. Here, the column support contained mechanically held liquid stationary phases for a partitioning process needed for separations. This technique was effective but awkward to use routinely, since the stationary liquid was difficult to maintain unchanged in the presence of a moving mobile phase. Realizing that a chemically-bonded stationary phase would offer a significant practical advantage, I undertook work with Dr. Paul C. Yates that ultimately resulted in Zipax particles coated with a covalently bonded, three-dimensional porous silicone polymer [2]. This new column-packing material allowed stable separations that were easily repeated routinely by operators. These new packings eliminated precolumns and presaturation of the mobile phase with the stationary phase, techniques required when using mechanically held liquid stationary phases. Columns of these materials soon were used routinely within DuPont laboratories, and versions of these materials were subsequently marketed by DuPont's Instrument Products Division under the Permaphase® trademark.

Theoretical studies by others in the early 1960s had predicted that much faster, higher-resolution separations would result from use of particle sizes much smaller than the traditional 25-μm materials that were used then. Practice with much smaller particles proved difficult. However, Prof. J. F. K. Huber again led the way and showed that rapid separations could be obtained with columns of 10-μm diatomaceous earth particles. Unfortunately, the techniques for preparing such columns were highly technique-dependent, and this approach did not result in a general following. Nevertheless, the success of this approach encouraged attempts to use small particles. Again with the assistance of Dr. Ralph Iler, I developed a method for preparing narrow particle size, using porous silica microspheres of <10 μm. However, for practical use, an efficient method for preparing packed columns of these materials was needed. Fortunately, a modification of the slurry packing method used previously in preparing columns of Permaphase packings [2] proved satisfactory. This development then led to the first successful columns with high-performance porous silica microspheres, and such columns were quickly used within DuPont in 1970 to produce separations [see J. J. Kirkland, Chap. 5 in C. W. Gehrke, R. W. Wixom, and E. Bayer (Eds.), *Chromatography: A Century of Discovery*, Elsevier, Amsterdam, 2001]. A paper on these new porous silica microspheres was given as my ACS Chromatography Award address at Boston in April 1972 [3], and DuPont subsequently made columns of these materials commercially available as Zorbax® porous silica microsphere chromatographic packing. Initial studies with these new particles used liquid–liquid partition chromatography, but liquid–solid (adsorption) technology quickly followed [4].

Interest in high-speed polymer separations then led to a series of fruitful collaborations with Dr. W. W. Yau, who was primarily concerned then with the fundamentals of polymer characterization. After I developed wide-pore silica microspheres suitable for high-speed size-exclusion polymer separations [5], Wallace Yau, Charles R. Ginnard, and I showed for the first time that only two properly designed pore sizes are required to produce columns with wide-range linear molecular weight calibrations [6]. This technology now is widely used to produce linear molecular weight calibration columns employed for most polymer characterizations by size-exclusion chromatography.

During the mid-1980s, continued interest in silica-based columns for HPLC led to important basic studies with another visiting young scientist, Dr. Jürgen Köhler. Out of this work came a much improved understanding of the surface chemistry of chromatographic silica supports [7]. This insight ultimately resulted in the development and patenting of a new form of silica column support, Zorbax Rx-SIL. This is an ultrapure, less acidic, chromatographically "friendly" material for separating a wide range of highly polar and ionizable materials, including peptides, proteins, etc. [8]. Starting in 1988, DuPont's Instrument Products Division again made new products commercially available based on this new ("type B") silica support.

During this time, cooperative research with another DuPont colleague, Dr. Joseph L. Glajch, resulted in approaches that were significant in the emergence of a systematic HPLC method development technology. Using Dr. Lloyd R. Snyder's selectivity triangle method for describing solvent selectivity, we devised systematic reversed-phase (and adsorption) chromatographic approaches for predicting separation resolution for a sample throughout the range of solvents selected for the separation [9]. This technology was used by DuPont in a new "Sentinel" HPLC instrument specifically designed for automatic HPLC method development.

Dr. Glajch and I also collaborated in another discovery that resulted in important chromatographic and commercial implications: sterically protected, covalently bonded silane stationary phases. Studies showed that column packings made with these materials were impressively more stable against stationary-phase hydrolysis and loss at low pH than conventional monofunctional silane stationary phases [10]. Materials with such stable phases were patented and commercialized by DuPont as StableBond™ chromatographic columns, which are widely used today.

Strong efforts have been made by researchers to understand the fundamental properties of column packing materials for HPLC, so that separation materials could be optimized for a wide range of needs. Because of its overall compromise of desirable qualities, porous silica remains as the preferred support material for HPLC columns. Users now mostly choose columns with the newer less acidic, highly purified ("type B") porous silica microspheres that have been specially prepared with low-activity surfaces. Columns of these silica particles elute basic compounds and other highly interactive structures, such as peptides and proteins efficiently and with good peak shapes. Covalently bonded silane stationary phases on these silicas are now prepared with excellent reproducibility and column efficiency; monofunctional coatings are generally preferred for these advantages.

Most reversed-phase HPLC separations of ionizable compounds are done at low pH, where ionizable solutes and residual silanol groups on the silica support are fully protonated. Under these conditions, most compounds exhibit excellent peak shapes and high separation efficiency. Excellent retention time reproducibility also can be expected, since the least change in solute interaction with the stationary phase can occur with small changes in operating parameters, such as percent organic phase, pH, and temperature. Because the siloxane bond that binds these groups to silica can be hydrolyzed at low pH, additional column stability is obtained by using bonded silanes with bulky side chains. These bulky groups sterically protect the silane against degradation by hydrolysis even at temperatures as high as 90°C. Some polymerized and bidentate-attached silanes also show good stability at low pH.

Small changes in operating conditions (pH, percent organic modifier, temperature) also can seriously affect separation resolution and reproducibility, so these parameters must be carefully controlled for good results. Since the solubility of the silica support in aqueous mobile phases begins to be appreciable above pH ~8, column degradation with continued use is caused by slow silica support dissolution with ultimate collapse of the packed bed. To reduce silica support dissolution and subsequent column failure, densely reacted and highly endcapped alkyl stationary phases have been developed. The resulting low-surface-energy coatings apparently present an effective high surface tension barrier that retards the rate of silica support dissolution in aqueous mobile phases, significantly increasing column lifetime and generally improving column performance. Column lifetimes are favored when longer alkyl ligands are used for the stationary phase. Studies have shown that column lifetime at intermediate and high pH also is greatly increased by using organic buffers, avoiding phosphate and carbonate buffers where possible. The rate of silica support dissolution also is reduced by using lower temperatures, so that separations no higher than 40°C are recommended.

REFERENCES (for J. Jack Kirkland)

1. J. J. Kirkland, Controlled surface porosity supports for high speed gas and liquid chromatography, *Anal. Chem.* **41**, 218 (1969).
2. J. J. Kirkland, High speed liquid–partition chromatography with chemically bonded organic stationary phases, *J. Chromatogr. Sci.* **9**, 206 (1971).
3. J. J. Kirkland, High-performance liquid chromatography with porous silica microspheres, *J. Chromatogr. Sci.* **10**, 593 (1972).
4. J. J. Kirkland, Porous silica microsphere column packings for high-speed liquid–solid chromatography, *J. Chromatogr.* **83**, 149 (1973).
5. J. J. Kirkland, Porous silica microspheres for high-performance size-exclusion chromatography. *J. Chromatogr.* **125**, 231 (1976).
6. W. W. Yau, C. R. Ginnard, and J. J. Kirkland, Broad-range linear calibration in high-performance size-exclusion chromatography using column packings with bimodal pores, *J. Chromatogr.* **149**, 465 (1979).
7. J. Köhler, D. B. Chase, R. D. Farlee, A. J. Vega, and J. J. Kirkland, Comprehensive characterization of some silica-based stationary phases for high-performance liquid chromatography, *J. Chromatogr.* **352**, 275 (1986).
8. J. Köhler and J. J. Kirkland, Porous silica microspheres having a silanol enriched surface, US Patent, 4,874,518, October 17, 1989.
9. J. L. Glajch, J. C. Gluckman, J. G. Charikofsky, J. M. Minor, and J. J. Kirkland, Simultaneous selectivity optimization of mobile and stationary phases in reversed-phase liquid chromatography for isocratic separations of phenylthiohydantoin amino acid derivatives, *J. Chromatogr.* **318**, 23 (1985).
10. J. J. Kirkland, J. L. Glajch, and R. D. Farlee, Synthesis and characterization of highly stable bonded phases for high-performance liquid chromatography column packings, *Anal. Chem.* **61**, 2 (1989).

6.C.2. Impact of Silica Chemistry on Separation Science/Technology (Klaus K. Unger)

During my graduate studies in chemistry at the Technische Hochschule, Darmstadt, Germany, I became fascinated by the research on finely divided and porous materials as well as on surface chemistry at the famous Zintl-Institut für Anorganische Chemie und Physikalische Chemie. During my Ph.D. thesis from 1962 to 1965, I developed procedures to adjust and to tailor the pore size of silicas made by the classical sol-gel process. These materials were applied as packings for size-exclusion chromatography (SEC) of synthetic polymers and colloidal sols. The aim was to search for rigid SEC packings as an alternative to soft polysaccharides and to semirigid crosslinked polymer gels.

In my habilitation thesis from 1965 to 1969, the focus of my research centered on chemical surface modification of silica with reactive silanes and the characterization of the adsorptive properties of surface modified silicas.

Having experienced classical column liquid chromatography with coarse silica in glass columns, I learned of the modern high-performance variant by contact with the late István Halász, who was at this time docent at the Johann-Wolfgang-Goethe-Universität Frankfurt/Main, Germany. István advised me in building HPLC equipment, and I transferred my knowledge in chemical surface modification to his laboratory.

Although the theoretical concepts in HPLC were highly advanced at the beginning of 1970, and clear directives could be given to make a separation more efficient and faster, the introduction of the method was still hampered by the lack of technical achievements. One of the primary requirements was to employ microparticles of rigid silica of $5-10$ μm particle size obtained by milling and size fractionation of larger particles. An appropriate sizing technology was, therefore, developed that enabled a cut to narrow-particle-size fractions by means of air elutriation by Alpine AG, Augsburg, Germany. A technique was also needed to pack these particles densely into stainless-steel columns with porous frits as endfittings. For this purpose the balanced density technique was applied, which employed a suspension of the silica particles in a high-density liquid. This suspension was pumped into the column at high pressure and high flow rate using an autoclave as a slurry reservoir. With these two achievements, highly efficient HPLC columns were packed with silicas in their native form and operated under straight phase conditions.

At the beginning of 1970, several research groups around the world attempted to manufacture spherical silica particles, such as J. J. Kirkland at DuPont de Nemours (USA), John Knox, University of Edinburgh, Scotland (UK), and P. Myers, Phase Separations Ltd. (UK). Spherical particles were assumed to generate a more stable and homogeneous column bed than did irregular-shaped ones and also to give a more favorable pressure drop–flow rate dependence than did irregular ones. This hypothesis was widely evidenced experimentally. Nowadays the majority of silica packings in HPLC are spherical in shape.

In 1973, we invented a process of making spherical silicas by a two-step process [1]. In the first step, tetraethoxysilane was partially hydrolyzed and condensed in an acidic ethanolic solution to polyethoxysiloxane (PES). In the second step, ethanol was removed and the PES was then completely hydrolyzed and condensed in a two-phase system containing ammonia solution with a pH of ~ 10 and an ethanolic PES solution. This process was performed under vigorous stirring. The PES was emulsified in the aqueous phase to liquid microdroplets that quickly solidified to silica hydrogel beads. The size of the beads was determined mainly by the kinematic viscosity of the PES and the speed of stirring using a specially designed stirrer. In this way silica beads were obtained with an average particle diameter between 5 and 20 μm. The pore diameter of the beads was varied between 5 and 15 nm with a specific surface area between 600 and 200 m^2/g. Still the batches had to be size classified into narrow cuts. Immediately, the question arose as to the ultimate minimum particle diameter in HPLC with respect to column efficiency, pressure drop, and analysis time. Optimization studies by Knox [2] revealed that the optimum particle size should be $\sim 3-4$ μm. To verify these predictions, we performed efficiency studies on columns packed with particles in the $1-10$ μm size range [3].

Parallel to the manufacture of silicas, the practical limitation in using native silicas in straight-phase chromatography became apparent: slow column equilibration and control of water content in n-hexane eluents. To circumvent these problems, so-called reversed-phase packings were developed. These materials were based on mesoporous silicas that were subjected to a reaction with n-octadecylchloro- or alkoxysilanes. In the period between 1975 and 1985, we studied in depth the silanization of the silica surface with respect to the kinetics, the surface stoichiometry, ligand density, and surface coverage [4,5]. In the following period reversed-phase chromatography with bonded n-octadecyl and n-octyl groups became the most widely applied packings in HPLC.

At the beginning of 1980, fundamental studies were undertaken to utilize mesoporous and macroporous silicas with an appropriate bonding chemistry for interactive biopolymer separations. In 1983, an international conference series was established in Washington, DC, named ISPPP (International Symposium on the Separation of Proteins, Polypeptides and Polynucleotides). In 2000 the 20th conference of this series took place in Ljubljana, Slovenia. The proceedings of these symposia were published as special issues in the *Journal of Chromatography* demonstrated the major achievements in this field (see Appendix 6A and, Supplement S-11 of online reference).

On the basis of the rich experience in bonding chemistry, our group focused on the synthesis of size-exclusion packings, ion exchangers, hydrophobic interaction, and affinity packings applied to protein and peptide separations [6]. From knowledge of the earlier work of W. Stoeber et al. [7], we succeeded in the synthesis of nonporous silica beads in the $1-2$ μm size range. By conducting surface reactions, we converted these beads into reversed-phase, hydrophobic-interaction, ion exchanger, and affinity packings. With these packings applied in short columns of 30 mm length and 4 mm i.d., we were able to perform extremely fast separation of biopolymers. For example, as shown in Reference [8], fragments from a tryptic digest of hemoglobin A could be separated on a 30 mm × 4.6 mm column packed with 1.5 μm MICRA nonporous silica packings surface modified with polystyrene under reversed phase gradient conditions within 150 seconds. A third area of active research was the coating of oxidic adsorbents with polymers as an alternative to silanization. This branch of activity was mainly initiated by my co-researcher Dr. A. A. Kurganov. He extended these studies in particular on packings based on titania and zirconia [9].

Around 1985, the synthesis of silica-rich zeolites such as the MFI type (ZSM-5, Silicalite I) attracted our interest. These zeolites were formed in the presence of structure-directing agents of so-called templates (e.g., tetraalkylammonium salts). Our objective was to understand the role of the template in the formation mechanism and to find the reaction conditions for template-free synthesis. Later we successfully developed a process to manufacture template-free ZSM-5 up to pilot scale.

In 1992, Mobil Oil Co. researchers presented the synthesis of ordered mesoporous silicas of the M41S family, which was a breakthrough in large-pore-size zeolite chemistry. One of the most prominent members of this mesoporous silica families, MCM-41, was formed by the self-assembled, cooperative interactions of silicates and organic detergents such as long-chain n-alkylammonium salts to form periodic mesostructured phases.

REFERENCES (for Klaus K. Unger)

1. K. K. Unger, J. Schick-Kalb, and K. F. Krebs, Preparation of porous silica spheres for column liquid chromatography, *J. Chromatogr.* **83**, 5–9 (1973).
2. J. H. Knox, Practical aspects of LC theory, *J. Chromatogr. Sci.* **15**, 352–364 (1977).
3. K. K. Unger and W. Messer, Comparative study on the performance of spherical and irregularly shaped silica and alumina supports having diameters in the 1 to 10 μm size range, *J. Chromatogr.* **149**, 1–12 (1978).
4. P. Roumeliotis and K. K. Unger, Structure and properties of n-alkyldimethylsilyl bonded silica reversed phase packings, *J. Chromatogr.* **149**, 211–224 (1978).
5. J. N. Kinkel and K. K. Unger, Role of solvent and base in the silanization reaction of silicas for reversed-phase high-performance liquid chromatography, *J. Chromatogr.* **316**, 193–200 (1984).
6. K. K. Unger, K. D. Lork, and J. Wirth, Development of advanced silica-based packing materials, in M. T. W. Hearn (Ed.), *The HPLC of Peptides, Proteins and Polynucleotides*, VCH Publishers, New York, (1991), pp. 59–117.
7. W. Stoeber, A. Fink, and F. Bohn, Controlled growth of mono-disperse silica spheres in the micron size range, *J. Colloid Interface Sci.* **26**, 62–75 (1968).
8. T. Issaeva, A. A. Kurganov, and K. K. Unger, Super high speed liquid chromatography of proteins and peptides on nonporous MICRA NPS RP packings, *J. Chromatogr. A*, **46**, 13–23 (1999).
9. A. Kurganov, U. Trüdinger, T. Issaeva, and K. K. Unger, Native and modified alumna, titania and zirconia in normal- and reversed-phase high performance liquid chromatography, *Chromatographia* **42**, 217–222 (1996).

6.C.3. Preparation and Characterization of Polyencapsulated Stationary Phases (Heinz Engelhardt)

A significant part of our research has been the preparation and characterization of bonded stationary phases for HPLC. Starting with studies on retention mechanisms [1], and the properties of polar phases in protein analysis [2], a new technique for the preparation of polymer encapsulated stationary phases was developed [3,4]. By this technique not only RP columns with excellent chromatographic properties can be prepared for the separation of basic analytes and better stability (up to pH 8.5) in aqueous solutions; it is also possible to prepare by this method ion exchangers [5] and stationary phases with a high selectivity and efficiency for enantiomeric separations. This is demonstrated in Figure 6.9, where the

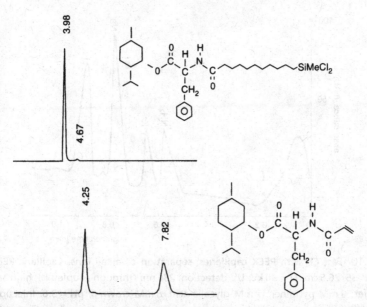

Figure 6.9. Comparison of enantioselectivity between monomeric and polymeric chiral stationary phases. Upper chiral selector: L-phenylalanine-D-methylesteramide. Sample: D,L-oxazolidinon. Upper chromatogram: Chiral silane bonded to silica; surface concentration, $2.37 \, \mu mol/m^2$; mobile phase, n-heptane-isopropanol (95 : 5); flow, 1 ml/min; $k = 0.87$; $\alpha = 1.00$. Lower chromatogram: Chiral acrylamide bonded on silica, modified with vinylsilane; surface concentration, $2.44 \, \mu mol/m^2$; mobile phase, n-heptane-tetrahydrofuran (90 : 10); flow, 1 ml/min; $k_i = 0.76$; $\alpha = 2.93$. This figure was reprinted with permission from the *Journal of Chromatography Library* series, vol. 64, *Chromatography: A Century of Discovery: 1900–2000*, Gehrke, C. W., Wixom, R. L., and Bayer, E., (Eds.), Chapter 5, page 174.

same selector has been bonded to the silica surface by a bristle-type reaction and by poly-encapsulation. Only in the latter case could enantioselectivity be achieved.

The characterization of packed columns for their efficiency and separation performance paralleled the work in stationary-phase synthesis. A commonly applicable test procedure has been developed [6] that allows one not only to determine the hydrophobic retention and selectivity of packed columns but also to select columns for their potential to separate basic solutes with symmetric peaks and high efficiency [7]. With the retention behavior of at least six solutes, the columns can be characterized unequivocally [8]. Meanwhile, a test also has been developed that permits the measurement of the surface contamination of stationary phases with metal ions. It could be demonstrated that the columns are collecting metal ions on their surface originating from the chromatographic equipment, column hardware, and eluents.

These test procedures can also be used to validate chromatographic columns. The validation of chromatographic instruments; for example, flow accuracy measurements of pumps, or the linearity of detectors [9], is a useful extension of the work on stationary-phase characterization.

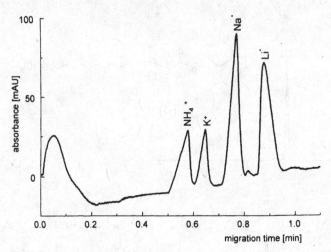

Figure 6.10. Fast CE with PEEK capillaries: separation of alkali ions. Capillary: PEEK i.d. = 75 μm; $L = 35/26.5$ cm; $U = 30$ kV; UV detection: 230 nm (through coupled HP high sensitivity cell). Buffer: 9 mM pyridine, 12 mM glycolic acid, 5 mM crown-6; pH = 3.6. Injection: 5 s at 35 mbar. (This figure was reprinted with permission from the *Journal of Chromatography Library* series, vol. 64, *Chromatography: A Century of Discovery: 1900–2000*, Gehrke, C. W., Wixom, R. L., and Bayer, E., (Eds.), Chap. 5, page 176.)

While working in separation science, capillary electrophoresis and capillary electrochromatography are challenging new areas in research. Coating of capillaries to diminish wall adsorption and to modify the electroosmotic flow has been a consequence from our work in modifying surfaces of silica to the transfer to fused-silica capillary surfaces [10,11]. To circumvent some of the intrinsic problems of fused-silica capillaries, the potential of organic polymeric capillaries in CE has been studied [12]. The potential of PEEK (polyetheretherketone) for fast analysis is demonstrated in Figure 6.10. Our main effort in CE was in the separation of small molecules, the separation of enantiomers, but lately also in the analysis of synthetic polyelectrolytes [13].

REFERENCES (for Heinz Engelhardt)

1. K. Karch, I. Sebestian, I. Halász, and H. Engelhardt, Optimization of reversed-phase-separation, *J. Chromatogr.* **122**, 171–184 (1976).

2. H. Engelhardt and D. Mathes, High-performance exclusion chromatography of water-soluble polymers with chemically bonded stationary phases, *J. Chromatogr.* **185**, 305–319 (1979).

3. H. Engelhardt, H. Löw, W. Eberhardt, and M. Maub, Polymer encapsulated stationary phases: Advantages, properties and selectivities, *Chromatographia* **27**(11/12), 535–543 (1989).

4. H. Engelhardt, M. Krämer, and H. Waldhoff, Enhancement of protein detection and flow injection analysis, *J. Chromatogr.* **535**, 41–53 (1990).

5. F. Steiner, C. Niederländer, and H. Engelhardt, Optimization of alkali-alkaline earth cation separation on weak cation-exchangers, *Chromatographia* **43**(3/4), 117–123 (1996).

6. H. Engelhardt and M. Jungheim, Comparison and characterization of reversed phases, *Chromatographia* **29**(1/2), 59–68 (1990).

7. H. Engelhardt, H. Löw, and W. Götzinger, Chromatographic characterization of silica based reversed phases, *J. Chromatogr.* **544**, 371–379 (1991).

8. H. Engelhardt, M. Arangio, and T. Lobert, A chromatographic test procedure for reversed-phase HPLC column evaluation, *LC-GC* **15**(9), 856–866 (1997).

9. H. Engelhardt and Ch. Siffrin, Means of validation of HPLC systems and components, *Chromatographia* **45**, 35–43 (1997).

10. J. Kohr and H. Engelhardt, Capillary electrophoresis with surface coated capillaries, *J. Microcolumn Sep.* **3**, 491–495 (1991).

11. H. Engelhardt and M. A. Cunat-Walter, Preparation and stability tests for polyacrylamide-coated capillary electrophoresis, *J. Chromatogr. A* **716**(1–2), 27–33 (1995).

12. H. Bayer and H. Engelhardt, Capillary electrophoresis in organic polymer capillaries, *J. Microcolumn. Sep.* **8**(7), 479–484 (1996).

13. H. N. Clos and H. Engelhardt, Separations of anionic and cationic synthetic polyelectrolytes by capillary gel electrophoresis, *J. Chromatogr. A* **802**(1), 149–157 (1998).

6.C.4. Chemistry of Coated and Modified Support Surfaces— Contributions to Progress in Highly Selective and Efficient Separations Technology (Gerhard Schomburg)

This report presents contributions developed at the Max-Planck Institut to the highly selective and efficient separations by the development of miniaturized, multicolumn, upscaled, and MS-coupled instrumentation and to online data processing of analytical data, and our research on the chemistry of coated and modified support surfaces in analytical, preparative, chromatographic, and electrophoretic systems.

Introduction. The major topics of about 40 years of work in different groups of the author in the analytical laboratories of the Max-Planck-Institut für Kohlenforschung (coal research) from 1956 to 1996, and at the Max-Planck-Institut für Strahlenchemie, were instrumental and methodological developments for the miniaturized version of analytical gas chromatography (GC), liquid chromatography (LC), capillary electrophoresis (CE), supercritical-fluid chromatography (SFC) in open tubular columns, and automated cyclic large-scale GC and LC. Also studied was the necessity of miniaturization in GC, LC and CE for highly selective and efficient analytical separations of complex mixtures of small and large, nonpolar and polar, as well as charged molecules at high resolution and efficiency. This, in general, required the development of suitable instrumentation, which had to contain detectors for the monitoring of the separations at adequate sensitivity and analytical performance. In this context, highly selective (e.g., chiral) stationary-phase coatings on support surfaces of porous small particle materials and in open tubular columns had to be generated. Gradients of temperature in GC and of mobile-phase composition in LC make fast separations of samples with wide retention range at suitable resolution possible. At the Max-Planck Institut, analysis of higher performance with regard to resolution, analysis time, and detection of traces at high signal-to-noise ratios (SNRs) for the analysis of such samples were achieved in systems of coupled capillary columns for GC [1], of packed or micropacked columns in LC, and multidimensional analytical (also large-scale) GC and LC separations.

The coupling of mass spectrometric equipment to the columns mainly in GC instruments (together with my colleague, D. Henneberg) [2], and the computer handling of huge amounts of digitized analytical data arising from the different separation methods were other successful steps in instrumental development, together with my colleague, E. Ziegler [3]. Contributions to instrumental and methodological developments in chromatography and capillary electrophoresis, which were research topics in many analytical laboratories in the scientific world, were achieved by the different groups of G. Schomburg during the 40 years of his activities at the Max-Planck Institut. In addition, the research programs in GC, LC, SFC, and CE performed in the author's laboratories were strongly directed to miniaturization.

Miniaturization was adopted in systems for analytical separation, first in gas chromatography after the 1958 introduction by M. J. E. Golay of the open tubular columns, in which coating layers of thermostable GC-suited polymers on the inner surfaces served as the stationary phase. In capillary GC, the efficiency of separations of volatile analytes could be dramatically increased in comparison to packed columns by decreasing the internal diameter of the tubing to <300 μm. The amount of stationary phase in the columns is low (a few milligrams), and also correspondingly low is the sample capacity of the chromatographic system. The small sample volumes caused problems with their manipulation, vaporization, and final introduction into the mobile phase, and the necessity of large volumes of highly diluted target analytes was another consequence of miniaturization [4]. The continuous and sensitive detection of the small mass flows of analytes in these miniaturized systems was enabled by the "universal" flame ionization detector (McWilliam) and later especially by the GC/MS coupling operated in different modes.

Special Contributions. For the practical and analytical application of the new techniques, a number of related instrumental developments arising in miniaturized GC were performed in the years between 1960 and 1996:

- Production, surface deactivation and coating of metal, glass, and fused-silica capillary tubing with reproducible generation of layers of polymers of defined film thickness, polarity, and high-temperature stability on native and specially prepared chemically modified surfaces
- Coupled column systems containing narrow-bore capillary columns for valveless multidimensional GC [1]
- Mono- (specific) and multidimensional detection by ionization detectors, especially on interfacing with GC/MS, and in the scanning mode [2]
- Sampling techniques for different types of samples and analytical problems [4]

The author's activities on chiral separations with GC, LC, and CE became of great interest in the Institüt as asymmetric syntheses with metallorganic catalysts were extensively investigated. For example, in 1993, a paper by F. Kober et al. [5] was published. In this regard, I also mention a paper on chiral separations of acidic drugs by CE, which was published together with Jaques Crommen, Professor of Pharmacy, at the University of Liège, in whose laboratory I spent a year of research work as awardee of the Alexander v. Humboldt Foundation [6].

The adoption of the principle of miniaturization with the aim of optimization of separation efficiency occurred in liquid chromatography, when columns of high performance with packings of chemically modified small porous particles (i.d. <10 μm), preferably from silica, were introduced by J. J. Kirkland, J. Huber, I. Halász, K. K. Unger, and others. Synthesis and modification of these porous materials with chemically stable and selective layers of small and large molecules of suitable polarity are generally difficult to perform with high reproducibility. The main aim of our studies was to contribute to the chemistry of stationary phases on the surfaces of porous support materials by the development of procedures for polymer coating (as applied earlier for capillary GC). The common silanization procedures for anchoring molecules of the desired "polarity" (including chiral species) to the porous surfaces was used. As examples, the chemistry of synthesis of the so-called polymer-coated stationary phases by catalytic immobilization of C18-substituted polymethyl siloxanes (PMSC18), the highly alkali stable polybutadiene-coated Al_2O_3 [7], and the silica-coated polybutadiene maleic acid (PBDMA) for ion-exchange chromatography (IEC) of the alkali and alkali earth cations [8], are mentioned here to represent the activities of the author's groups in the HPLC field. An application of the PBDMA cation-exchange phase, which allows for fast HPLC separation of all the alkali and alkaline earth cations in about 10 min at high resolution, is shown in Chapter 5, Section 56.I.5, Fig. 1 of Gehrke et al. (Eds.), *Chromatography: A Century of Discovery*, Vol. 64, Elsevier, 2001; see Schomburg, Chapter 5.

The introduction of the flexible and thin-walled fused silica tubing into capillary electrophoresis by J. J. Jorgenson in 1981, with narrow internal diameter (50–100 μm), initiated another spectacular trend in the development of highly efficient and selective analytical separation methods, which were initially based on the electromigration principle. This electromigrative separation technique provoked enthusiasm for methodological work in this field and in our lab. In aqueous and non-aqueous (but dielectric) buffer media of defined pH, large ionic and nonionic molecules can be quickly and efficiently resolved in different CE modes depending on the chemical structure of the analytes and on the composition of the electrolyte serving as separation buffer. In analogy, since 1989, the coating chemistry developed and applied for GC and HPLC along with the surface chemistry in the separation systems of CE became one of the major topics of the author's group.

Examples of developments were CE systems for separations in the capillary zone electrophoresis (CZE), electrokinetic chromatography (EKC), and capillary gel electrophoresis (CGE) modes, which were specially optimized with regard to

- Modification of absolute and relative migration times (selectivities) by charged and uncharged, also chiral buffer additives.

- Separation of large biomolecules in the CGE mode using gels and polymer solutions [9].
- Dynamic and permanent modification of surfaces by charged and uncharged oligomeres for the manipulation of the electroosmotic flow and the suppression of undesirable analyte–wall interactions [10–12]. A fast CGE was achieved by Yin et al. in a gel-filled capillary at extraordinary resolution [7] [see also Schomburg, Chap. 5 in Gehrke et al. (Eds.), *Chromatography: A Century of Discovery*, Vol. 64, Elsevier, 2001]. References [11] and [12] concern chapters in books on capillary electrophoresis, edited by P. Camillieri and Morteza G. Kharledi.

It appears obvious that in both chromatography and capillary electrophoresis some related or analogous instrumental and chemical problems were solved; they are related, with a similar approach of miniaturization, which worked out to be important as a basic feature in CE, but for another reason in CG and LC, i.e., for the effective abduction of the Joule heat. In spite of the principally different separation mechanisms in chromatography and basic capillary electrophoresis, the methodological and instrumental experience with miniaturization in chromatography had a significant and positive effect on the achievable analytical performance in CE. This can also be concluded in view of the recent achievements concerning the EKC and EC modes of capillary electrophoresis, which are hybrids of LC and CE; however, they unite in many regards, the main features of the two separation methods.

REFERENCES (for Gerhard Schomburg)

1. G. Schomburg, H. Husmann, and F. Weeke, Aspects of double column gas chromatography with glass capillary columns with glass capillaries involving intermediate trapping, *J. Chromatogr.* **112**, 205–217 (1975).
2. D. Henneberg, U. Henrichs, and G. Schomburg, Special techniques in the combination of gas chromatography and mass spectrometry, *J. Chromatogr.* **112**, 343–352 (1975).
3. G. Schomburg and E. Ziegler, Progress and experiences with the Mülheim computer system in gas chromatographic data handling, *Chromatographia* **5**, 96–104 (1972).
4. G. Schomburg, H. Behlau, R. Dielmann, F. Weeke, and H. Husmann, Sampling techniques in capillary gas chromatography, *J. Chromatogr.* **142**, 87–102 (1977).
5. F. Kober, K. Angermund, and G. Schomburg, Molecular modeling experiments on chiral recognition in GC with specially derivatized cyclodextrins as selectors, *J. High. Resol. Chromatogr.* **16**, 299–311 (1993).
6. M. Fillet, I. Bechet, G. Schomburg, P. H. Hubert, and J. Crommen, Enantiomeric separation of acidic drugs by capillary electrophoresis using a combination of charged and uncharged β-cyclodextrins as chiral selectors, *J. High Resol. Chromatogr.* **19**, 669–673 (1996).
7. G. Schomburg, U. Bien-Vogelsang, A. Degee, and J. Köhler, Synthesis of stationary phases for reversed phase LC using silanization and polymer coating (polybutadiene coating of alumina), *Chromatographia* **19**, 170–179 (1984).
8. P. Kolla, G. Schomburg, and J. Köhler, Polymer coated cation exchange stationary phases on the basis of silica (PBDMA), *Chromatographia* **23**, 465–472 (1987).

9. H.-F. Yin, J. A. Lux, and G. Schomburg, Production of polyacrylamide gel filled capillaries for capillary gel electrophoresis (CGE). Influence of capillary surface pretreatment on performance and stability, *J. High Resol. Chromatogr.* **13**, 624–627 (1990).

10. M. Gilges, M. H. Kleemiss, and G. Schomburg, CZE separation of basic and acidic proteins using poly(vinylalcohol) coatings in fused silica capillaries, *Anal. Chem.* **66**, 2038–2046 (1994).

11. G. Schomburg, Oligonucleotides, in P. Camillieri (Ed.), *Capillary Electrophoresis, Theory and Practice*, CRC Press, Boca Raton, FL, 1993, pp. 255–310.

12. G. Schomburg, Coated capillaries in high performance capillary electrophoresis, in M. G. Khaledi (Ed.), *High Performance Capillary Electrophoresis, Theory, Techniques and Applications*, Wiley, New York, 1998, Chapter 14, pp. 481–520.

6.C.5. Chromatographic Enantioseparations [1] (William H. Pirkle and Christopher J. Welch)

The explosion of research interest in chirotechnology (the science of making and measuring enantiopure materials), which began in the early 1980s and continues unabated to this day, can in some measure be attributed to the availability of tools for the rapid and reliable quantitation of enantiopurity. Where it once took days or weeks to obtain oftentimes questionable results, accurate and reliable measurements can now be obtained in a matter of minutes. During the 1990s, the technique of chromatographic enantioseparation became the method of choice for analytical determinations of enantiopurity. The availability of a variety of Chiral stationary phases (CSPs) and the existence of automated chromatography equipment have made analytical determinations of enantiopurity almost routine. This method is very widely used, particularly in the pharmaceutical sector, where most new chiral drugs are now manufactured and sold in enantiopure form.

Pirkle-Type CSPs. Among the many types of CSPs that have been developed, the Pirkle-type, or brush-type CSPs, have proved to be among the most useful for many liquid chromatographic enantiomer separations. These CSPs consist of an enantioenriched, small-molecule selector immobilized on an inert chromatographic support, typically silica gel. Separation is achieved when the two enantiomers of the analyte are differentially adsorbed by the CSP. A combination of simultaneous, geometrically constrained, intermolecular interactions utilizing forces such as hydrogen bonding, $\pi-\pi$ attraction, ionic interactions, and steric repulsion can result in diastereomeric adsorbates with differing free energies, the prerequisite for enantioseparation. The design and development of CSPs in the Pirkle laboratories stems from a research program aimed at developing a better understanding of molecular interactions. Pirkle has been a pioneer in demonstrating that chromatography can be an exquisitely sensitive tool for the study of molecular recognition.

ADVANTAGES OF PIRKLE-TYPE CSPs. In general, Pirkle-type CSPs possess a number of advantages relative to other CSP types. Since the selector is a small molecule that is often completely synthetic, a structure that contains no labile or reactive components can usually be developed: In addition, the mode of attachment of the selector to the chromatographic support can be chosen for durability. Brush-type CSPs are typically covalently attached to the chromatographic support. Thus, most brush-type CSPs are chemically robust, and are generally quite longlived. Longevity is useful for an analytical CSP, but truly essential for a preparative CSP, where continuous operation for several years may be required. The chemical robustness of brush-type CSPs results in the ability of these CSPs to be used in a wide variety of mobile phases, which provides greater flexibility in method development, especially when poorly soluble analytes are being investigated.

An additional advantage of brush-type CSPs stems from the fact that the selectors are typically small molecules with molecular weight under 1000. Consequently, the selectors can be very densely arrayed on the chromatographic surface, resulting in a CSP that is highly resistant to sample overload and that has a very high preparative capacity. Finally, most synthetic CSPs are available in either enantiomeric form. Consequently, either elution order (+ before −, or − before +) can be chosen. Elution of the minor enantiomer before the major is generally preferred in analysis, while elution of the desired component before the undesired can greatly increase productivity in preparative HPLC.

Preparative Enantioseparation by HPLC. The use of chromatography for the preparative separation of pharmaceutical enantiomers has become increasingly popular. While often expensive, chiral HPLC offers the tremendous advantage of speed. Consequently, many pharmaceutical companies now use preparative chiral HPLC as an integral part of the drug development process. For preparative separations on the multi-kilogram level, a highly enantioselective CSP is of great importance. All other things being equal, a highly enantioselective CSP will afford the least costly separation. There has been considerable progress in the design of highly enantioselective CSPs, with several reports of enantioselectivities in excess of 100. With this level of enantioselectivity relatively low-tech(nology), inexpensive, separation processes such as batch adsorption can be utilized.

Discovery of New CSPs. The discovery of new CSPs in the Pirkle group has relied heavily on the principle of reciprocity [1–3]. For illustration, see Fig. 5 in Chap. 5 of 2001 Chromatography book [1]. Simply stated, this principle suggests that if a "gloves" CSP is capable of separating enantiomeric hands, then a "hands" CSP should be able to separate enantiomeric gloves. Of course, there are sometimes exceptions; for example, the manner in which the selector is tethered to the support can influence enantioselectivity. Nevertheless, this principle has proven to be quite useful in the discovery and development of CSPs. Factors that are important for convenient preparative enantioseparation include high enantioselectivity, good solubility of the analyte in the mobile phase, elution of the desired component before the undesired one, and the ability of the product enantiopurity to be upgraded via crystallization.

A New Approach to CSP Development. We recently reported a strategy for synthesis and screening for libraries of CSPs, which offers a tool for the rapid assessment of which CSP will work best for preparatively resolving the enantiomers of a given compound. The technology has two parts: (1) a parallel synthesis of libraries of milligram quantities of diverse stationary phases on porous silica particles [4], and (2) screening the resulting libraries for chromatographic performance [5]. This method for synthesis and screening for CSP libraries has a number of advantages when compared to alternative methods used for developing new chromatographic stationary phases. For example, CSP library synthesis on a milligram scale results in tremendous savings in materials and reagents. In addition, the parallel nature of the library synthesis means that tens or hundreds of new CSPs can be prepared within the time required for making one full sized CSP. Furthermore, the CSP library screening technique is simple, rapid, and inexpensive, and can be automated. Another advantage is that after evaluation, the CSP libraries can be washed and reused. Finally, and most importantly, the compound mixture of interest can be directly used in the screening assay without the need for purification, derivatization, immobilization, or formation of conjugates. Since both the CSP and the analyte being evaluated are exactly the same as what will be used in the full-scale chromatographic separation, a successful screening result can be confidently translated into successful chromatography on the large scale without worrying about things such as tether effects or selector immobilization.

REFERENCES (for William H. Pirkle and Christopher J. Welch)

1. W. H. Pirkle and C. J. Welch, Chiral HPLC and Enantioseparation of Pharmaceuticals, in *Chromatography: A Century of Discovery 1900–2000: The Bridge to the Sciences/Technology*, C. W. Gehrke, R. L. Wixom, and E. Bayer (Eds.), Journal of Chromatography Library, Vol. 64, Elsevier Science B.V, Amsterdam, 2001, pp. 441–452.

2. W. H. Pirkle, C. J. Welch, and B. Lamm, Design synthesis and evaluation of an improved enantioselective naproxen selector, *J. Org. Chem.* **57**, 3854–3860 (1992).

3. W. H. Pirkle, C. J. Welch, J. A. Burke, and B. Lamm, Target-directed design of chiral stationary phases, *Anal. Proc.* **29**, 225–226 (1992).

4. C. J. Welch, G.A. Bhat, and M. N. Protopopova, Silica based solid phase synthesis of chiral stationary phases, *Enantiomer* **3**, 463–469 (1998).

5. C. J. Welch, M. N. Protopopova, and G. A. Bhat, Microscale synthesis and screening of chiral stationary phases, *Enantiomer* **3**, 471–476 (1998).

6.C.6. The Enantiomeric Separations Revolution (Daniel W. Armstrong)

Our early work with cyclodextrin [1–3] led me into the area of chiral molecular recognition and enantiomeric separations [4–6]. My main interest was to understand, in a fundamental manner, the ultimate in molecular recognition (i.e., the ability of one molecule to distinguish between enantiomers of another molecule). Developing a highly successful class of chiral stationary phases that would be used worldwide (as they are today) was never a prime consideration. Prior to the early 1980s, enantiomeric separations tended to be avoided or ignored, as they were thought to be tedious, difficult, or impossible to do. I developed the original stable cyclodextrin-bonded phase for

HPLC because (1) using cyclodextrins in the mobile phase was very expensive at that time, unless one recycled the cyclodextrin additive: (2) the cyclodextrin mobile-phase additive approach produced relatively inefficient separations; and (3) I was curious as to the reversed-phase selectivity of cyclodextrin columns (which separated on the basis of inclusion complexation) versus that of the more common alkyl-type reversed-phase stationary phases. Thus, our early work on cyclodextrins consisted of roughly equal parts of theory, mechanism, achiral retention/separations, and enantiomeric separations. While all of these studies received a good deal of attention from academia, industry, and government scientists, it was the enantiomeric separations that seems to have garnered the most attention.

The mechanistic aspect of our cyclodextrin (and other chiral selector research) work overlapped with the practical aspects that were so popular with others. By understanding how and why chiral recognition occurs, one can use it more effectively to achieve the desired separation. By the mid-1980s, we had a basic understanding of the chiral recognition mechanism for cyclodextrins in aqueous solutions [4–6]. Our early modeling and mechanistic work gave rise to a set of rules for separation.

Most intriguing was our discovery that host–guest inclusion complexation was not always necessary or even preferable for chiral recognition with cyclodextrins. In the gas phase and in non-hydrogen-bonding solvents, such as acetonitrile (i.e., the polar organic mode), external adsorption and association produced enantioseparations that could not be obtained under conditions that favored inclusion complexation [7,8]. This greatly enhanced the understanding of these systems and expanded their utility as well.

The five basic classes of chiral stationary phases for HPLC are outlined: (1) ligand exchange, (2) macrocyclic, (3) π-complex, (4) polymeric, and (5) synthetic. These are outlined by Armstrong in Supplement S11 and Chapter 5 of Gehrke et al. Vol. 64, cited above. It should be noted that there can occasionally be some overlap between these classes.

Macrocyclic Stationary Phases. Three main classes of chiral stationary phase (CSPs) utilize macrocyclic compounds as chiral selectors: (1) cyclodextrins, (2) antibiotics, and (3) crown ethers. Altogether, this particular group of chiral selectors is responsible for >95% of all GC enantiomeric separations (i.e., mainly on derivatized cyclodextrins), >95% of all CE enantiomeric separations (i.e., cyclodextrins plus glycopeptide antibiotics), and ~50% of all LC separations reported.

π-Complex Stationary Phases. Chiral stationary phases (CSPs) that interact with racemic analytes through one or more π–π interactions have been used since at least the 1960s in chromatography [9]. Early versions of these columns were sometimes referred to as "charge transfer" stationary phases. The chiral selector in these types of stationary phases contained either a π-donor or π-acceptor moiety [9–12]. If the CSP was a π donor, then the analyte to be resolved must contain a π-acceptor group or vice versa. Other simultaneous interactions must be present as well, if enantioselective separations are to occur. Generally these other interactions consist of hydrogen bonding, steric repulsion and/or dipolar interactions. Separations on π-complex-type CSPs are usually done in the normal-phase mode. This is because $\pi-\pi$ interactions, hydrogen bonding, and dipolar interactions are more pronounced in nonpolar solvents.

Polymeric Stationary Phases. Polymeric chiral selections can be classified in several different ways. There are naturally occurring chiral polymers, synthetic chiral polymers, and hybrid varieties. Although interesting from an academic standpoint, the totally synthetic chiral polymers, have not had a great impact on the LC separation of enantiomers, to date. The principal synthetic polymers used as CPSs are poly(triphenyl-methyl methacrylate) and poly(2-pyridyl diphenylmethyl methacrylate) types [13]. These exist in right- and left-hand helical forms depending on the configuration of the chiral catalyst used in the polymerization reaction [13]. Most compounds resolved on these CPSs are better resolved on one or more of the others CSPs described in this monograph.

Up to the present time, the natural occurring chiral polymers have had a far greater impact on LC enantioseparations than have the synthetic or hybrid polymer CSPs. The natural polymers can be divided into two main types: (1) proteins and (2) carbohydrates:

1. *Proteins.* Bonded protein CSPs have played an important role in the analytical separations of enantiomers [14–16]. They are used in the reversed-phase mode with aqueous buffers or hydroorganic solvent systems. Early versions of these columns suffered from a lack of hardiness and longevity. However the second generation of protein-based CSPs had much improved stability and efficiency. The α_1-acid glycoprotein column has proven to be particularly useful to the pharmaceutical industry for quickly resolving a variety of amine containing compounds.

2. *Carbohydrates.* Cellulose and amylose are among the most common of naturally occurring chiral polymers. They are very poor chiral selectors in their native state. However, when their hydroxy-functional groups are derivatized (particularly with aromatic moieties via ester or carbamate linkages), and they are properly immobilized on a silica support, they become highly effective chiral stationary phases. The initial work on functionalized cellulose-based CSPs was done in Europe [17,18].

By the beginning of the twenty-first century, enantiomeric separations and analyses became essential in a wide variety of areas. These include the pharmaceutical and medicinal sciences (including pharmacokinetics and pharmacodynamics), since it is known that different enantiomers can have different physiological effects and biological disposition [1,19]. The tremendous advances in the LC analysis of enantiomers provided the impetus for the US Food and Drug Administration (FDA) to develop and issue guidelines for the development of new stereoisomeric drugs in 1992 [2]. In the environmental area, many chiral pesticides and herbicides have enantioselective effects and biodegradation rates [3,20]. The food and beverage industry is becoming increasingly concerned with the analysis of enantiomers that can affect flavor, fragrance, and nutrition and can be used to monitor fermentation, age, and even adulteration of products [21,22]. Enantiomeric separations are frequently the most useful way to determine the enantiomeric purity of newly synthesized chiral compounds and in elucidating reaction mechanisms [23–25]. Stereochemical analysis also can be of importance in geochemistry [5,6,24,25], geochronology [26], biochemistry, and in some areas of materials science [7].

Conclusions. While LC has played a leading role in enantioselective separations, it is the combination of all of the often complementary separation techniques (LC, CE, GC, TLC), that has allowed today's scientists to successfully address and solve most problems involving the resolution and analysis of enantiomeric compounds. The other important aspect of the enantiomeric separations work described herein, is that it has greatly enhanced our understanding of molecular recognition. CSPs also were beneficial in the study of chromatographic theory and mechanisms. For example, one can isolate and study chiral versus nonchiral retention mechanisms. Enantiomers can be useful probe molecules since they have the same size, shape, solvent solubility, and other properties. Mechanistic studies involving chiral molecules and/or CSPs will undoubtedly increase in the foreseeable future.

REFERENCES (for Daniel W. Armstrong)

1. E. J. Ariens, E. W. Wuis, and E. F. Veringa, Stereoselectivity of bioactive xenobiotics: A pre-Pasteur attitude in medicinal chemistry, pharmacokinetics and clinical pharmacology, *Biochem. Pharmacol.* **37**, 9–18 (1988).
2. Anonymous, FDA's policy statement for the development of new stereoisomeric drugs, *Chirality* **4**, 338 (1992).
3. D. W. Armstrong, G. L. Reid III, M. L. Hilton, and C.-D. Chang, Relevance of enantiomeric separations in environmental science, *Environ. Pollut.* **79**, 51 (1993).
4. D. W. Armstrong, L. He, T. Yu, J.-T. Lee, and Y.-S. Liu, Enantiomeric impurities in chiral catalysts, auxiliaries, synthons and resolving agents. Part 2, *Tetrahedron Asymmetry* **10**, 37–60 (1999).
5. G. Eglinton and M. Calvin, Chemical fossils, *Sci. Am.* **216**, pp. 32–43 (1967).
6. R. P. Philip, *Chem. Eng. News* **64**(6), 28–43 (1986).
7. K. Robbie, M. J. Brett, and A. Lakhtakia, Chiral sculptured thin films, *Nature* **384**, 616 (1996).
8. V. A. Davankov and S. V. Rogozhin, Ligand chromatography as a novel method for the investigation of mixed complexes: stereoselective effects in amino acid copper (II) complexes, *J. Chromatogr.* **60**, 280 (1971).
9. L. H. Klemm and D. Reed, Optical resolution by molecular complexation chromatography, *J. Chromatogr.* **3**, 364 (1960).
10. F. Mikeš, G. Boshart, and E. Gil-Av, Resolution of optical isomers by high performance liquid chromatography, using coated and bonded chiral charge-transfer complexing agents as stationary phases, *J. Chromatogr.* **122**, 205–221 (1976).
11. F. Mikeš and G. Boshart, Resolution of optical isomers by high-performance liquid chromatography: a comparison of two selector–selectand systems, *J. Chromatogr.* **149**, 455–464 (1978).
12. F. Mikeš and G. Boshart, Binaphthyl-2,2′-diyl hydrogen phosphate. A new chiral atropisomeric selector for the resolution of helicenes using high performance liquid chromatography, *J.C.S., Chem. Commun.* 173–174 (1978).

13. Y. Okamoto, S. Honda, I. Okamoto, H. Yuki, S. Murata, R. Noyori, and H. Takaya, Novel packing material for optical resolution: (+)-poly(triphenylmethyl methacrylate) coated on macroporous silica gel, *J. Am. Chem. Soc.* **103**, 6971 (1981).

14. J. Hermansson, Direct liquid chromatographic resolution of racemic drugs using a1-acid glycoprotein as the chiral stationary phase, *J. Chromatogr.* **269**, 71 (1983).

15. S. Allenmark, B. Bomgren, and H. Boren, Direct liquid chromatographic separation of enantiomers on immobilized protein stationary phases III. Optical resolution of a series of N-aroyl D, L-amino acids by high performance liquid chromatography on bovine serum albumin covalently bonded to silica, *J. Chromatogr.* **269**, 63 (1983).

16. J. Haginaka. C. Seyama, and N. Kanasugi, Ovoglycoprotein-bonded HPLC stationary phases for chiral recognition, *Anal. Chem.* **67**, 2579 (1995).

17. H. Hakli, M. Mintas, and A. Mannschreck, Preparative separations of enantiomeric diaziridines by liquid chromatography, *Chem. Ber.* **112**, 2028 (1979).

18. K. R. Lindner and A. Mannschreck, Separation of enantiomers by high-performance liquid chromatography on triacetylcellulose, *J. Chromatogr. A* **193**, 308 (1980).

19. E. J. Ariens, Bias in pharmacokinetics and clinical pharmacology, *Clin. Pharmacol. Ther.* **42**, 361 (1987).

20. A. M. Krstulovíc (Ed.), *Chiral Separations by HPLC*, Ellis Horwood, Chichester, 1989.

21. D. W. Armstrong, C.-D. Chang, and W. Y. Li, Relevance of enantiomeric separations in food and beverage analyses, *J. Agric. Food Chem.* **38**, 1674 (1990).

22. K. H. Ekborg-Ott and D. W. Armstrong, Stereochemical Analyses of Food Components, in S. Ahuja (Ed.), *Chiral Separations,* American Chemical Society, Washington, DC, 1997, Chapter 9, p. 201.

23. D. W. Armstrong, T. J. Lee, and L. W. Chang, Enantiomeric impurities in chiral catalysts, auxiliaries and synthons used in enantioselective synthesis, *Tetrahedron Asymmetry* **9**, 2043–2064 (1998).

24. D. W. Armstrong, Y. Tang, and J. Zukowski, Resolution of enantiomeric hydrocarbon biomarkers of geochemical importance, *Anal. Chem.* **63**, 2858 (1991).

25. D. W. Armstrong, E. Y. Zhou, J. Zukowski, and B. Kosmowska-Ceranowicz, Enantiomeric composition and prevalence of some bicyclic monoterpenoids in amber, *Chirality* **8**, 39 (1996).

26. J. L. Bada, in G. C. Barrett (Ed.), *Chemistry and Biochemistry of Amino Acids*, Chapman and Hall, New York, NY, 1985, p. 6.

6.C.7. The Invention of Sephadex® [1–3] (Jerker O. Porath)

We all understood that to discover or invent adsorbents that could separate the components in a peptide or protein mixture into discrete zones on columns by elution (linear chromatography) would be an "open sesame." Such adsorbents were badly needed. Per Flodin and I had packed our electrophoresis columns with starch grains to suppress convection in the buffer media. To me the enlightenment came one day about 4 years after I had joined Tiselius. When testing the efficiency of bed packing, I discovered that some dyes (e.g., DNP–amino acids) migrated as discrete zones, and, of course, that was a certain sign of isotherm linearity. My curiosity further increased after an accident when Noris Siliprandi, an Italian research guest, absentmindedly forgot to turn on the current overnight. The next morning his assisting wife eluted the substances from the column and found that a strange unexpected kind of separation had occurred. I had certain suspicions about what had happened, undertook systematic studies, and discovered that test substances often migrated in order of their molecular size within a volume less than that of the column. Some of them moved with stronger retention because they were linearly or nonlinearly adsorbed. Unfortunately, the eluates were contaminated by extracts from the starch. I deemed the results not to be recommended as a suitable fractionation procedure; I did not publish. At that time premature publication was considered risky.

I will now digress to what happened before my time at the Institute of Biochemistry in Uppsala; in the 1940s, Tiselius gave one of his students, Björn Ingelman, the task to undertake a study of a slimy product that had become a great nuisance in a sugar factory in southern Sweden. Sugarbeet extracts were contaminated by some unknown substance that made crystallization of sugar very difficult. Ingelman identified the mucous substance as dextran. As a blood plasma expander, the dextran later became the most important product of Pharmacia, at that time a small company, located in Stockholm. Pharmacia moved to Uppsala to come closer to our Institute. Ingelman made a comprehensive study on dextran for his doctoral thesis. He crosslinked dextran and obtained a lump of gel that, as a curiosity, was placed on a remote shelf of the factory lab, where it remained for several years. Eventually this gel came into my possession, thanks to P. Flodin. The introduction of crosslinked dextran has been described elsewhere [1,2]. I will only make some brief comments here. Already in my first experiments with cross-linked dextran, I observed both molecular sieving and linear adsorption, just as I had done on starch about 2 years earlier. The crosslinked dextran was stable, and no leakage was observed. It appeared that we had found the chromatographic medium we had been looking for. We also tried other hydrogels, among them crosslinked polyvinyl alcohol, but decided to continue to work with the dextran gels. Our publication was delayed to give Pharmacia time to develop the necessary knowledge in technology and to build a factory.

The new separation method using dextran gels appealed to me for the following reasons:

- Solutes migrated as compact zones analogous to those obtained under the influence of linear adsorption.
- We had a much simpler alternative to the ultracentrifuge (the complicated machine designed by Theodor Svedberg) for estimation of molecular size, and it could be used for smaller molecules.
- It could be applied on an analytical or preparative scale and had a potential for industrial applications.

From the beginning, P. Flodin and I had a clear perception of the molecular mechanism behind molecular sieving, but we never formulated a full-fledged theory. I refer to a review on gel filtration theory by T. C. Laurent [3]. My interest in size-exclusion chromatography (SEC) soon faded, but only temporarily. I tried to combine precipitation and molecular sieving in a salt concentration gradient on Sephadex columns. The method was published under the name of "zone precipitation," later to be called "reverse gradient salting-out chromatography" [2]. About 35 years later, in 1998, the method was independently rediscovered under an even fancier name: "on-column precipitation–redissolution chromatography."

Agar was introduced for chromatography by Poulsen and agarose by S. Hjertén [4]. With this matrix SEC could be extended to large molecules. However, agar and agarose derivatives have both advantages and shortcomings. We solved most of the problems by crosslinking the gels. Bed compaction at moderate pressures was practically eliminated. This was an important step forward in our attempts to improve hydrogels for chromatography.

A series of systematic studies by Jan-Christer Janson, Torgny Laas, and myself [4] showed that the rigidity of agarose products depends strongly on the length of the cross-linker used to stabilize the gel. The patent right to exploit this discovery was sold to Pharmacia in an effort to increase our chance of receiving financial support from The Swedish Board of Technical Development. Royalties caused mixed blessings. Financial help from elsewhere dried up, and we had to pay back the major portion of the royalties to the Board. However, Pharmacia prospered and their products found widespread use under such trade names as Sepharose and Superose.

Crosslinking of dilute agarose with a high concentration of divinylsulfone (DVS) extends the molecular sieving range upward, but no DVS–agarose products ever appeared on the market, few further studies were carried out, and no optimization trials were made. The knowledge of DVS–agarose went into oblivion. This unfortunate situation is a consequence of the industrial takeover of almost all research on adsorption media for chromatography. Using DVS-crosslinked $0.5-1.0\%$ agarose, SEC can be extended to particles of molecular mass exceeding 10^6 Da, to small viruses such as tobacco mosaic virus.

Hydrophobic interaction (HI) and electron donor-acceptor (EDA) selectivity represent different "dimensions" in separation technology. This is clearly demonstrated by experiments with composite columns consisting of tandem coupled beds of hydrophobic and EDA–adsorbent. Aromatic substances can be selectively extracted from a complex mixture. A clean-cut separation of serum proteins into albumin and immunoglobulin fractions can be achieved on such tandem columns. We have synthesized and studied quite a few aromatic and heteroaromatic thioagaroses [5]. Sven Oscarsson and I introduced 3-(2-pyridylmethylenethio)-2 hydroxypropylagarose, which is now on the market. Aromaticity is not necessary for donor or acceptor properties, as shown by the "T-gel" with the ligand $CH_2-CH_2-SO_2-CH_2-CH_2-S-CH_2-CH_2OH$.

Strategies and Future Prospects [6]. The diversified supply of different adsorbents makes it exceedingly difficult to decide what kind of strategy should be chosen to solve a particular protein fractionation problem. We may distinguish between three kinds of chromatographic strategies to deal with such problems:

1. To maximize the separation in a single bed.
2. To separate the substances according to their affinities for a number of selected adsorbents: differential affinity chromatography.
3. A combination of strategies 1 and 2.

Almost all chromatographic applications have been made in single beds, while only a few have been performed by differential chromatography and, as far as I know, none by the combined use of strategies 1 and 2.

Cascade-mode multiaffinity chromatography (CASMAC) of human serum exemplifies a particular mode of differential chromatography [6–8]. There are many problems waiting in the future that require improved chromatographic techniques. Further development of strategies beyond strategies 1 and 2 above can be anticipated. Displacement and spacer displacement chromatography offer extremely powerful ultimate purification methods for protein pharmaceuticals of natural origin. Synthetic proteins and peptides for medical use will also require extraordinarily effective chromatographic methods. It should be possible to use all the adsorbents that I have described in micro- and sub-micro-scale chromatography. My interest is now going in the opposite direction, toward macro- and megascale, with other demands on the matrix. A new kind of matrix based on crosslinked polyethyleneimine and agarose has recently been synthesized.

Eventually, adsorbents based on this new matrix may find use for concentration of low-grade ore extracts and mine-tailing leachates, as well as for environmental cleanup processes such as removal of aromatic contaminants (with EDA adsorbents) and toxic metal ions (with immobilized metal-affinity "IMA"-adsorbents) from industrial and natural waters. Only the first steps in such a pretentious program have been taken and practical applications have to wait far into the next century.

REFERENCES (for Jerker O. Porath)

1. J. Porath and P. Flodin, Gel filtration: A method for desalting and group separation, *Nature*, **183**, 1657 (1959).
2. J. Porath, The twelfth Hopkins memorial lecture. Molecular-sieving and non-ionic adsorption on polysaccharide gels, *Biochem. Soc. Trans.* **7**, 1197 (1979).
3. T. C. Laurent, History of a theory. Chromatography classic, *J. Chromatogr.* **633**, 1 (1993).
4. J. Porath, T. Löös, and J. C. Janson, Agar derivatives for chromatography, electrophoresis and gel-bound enzymes. II. Rigid agarose gels cross-linked with divinyl sulphone (DVS), *J. Pg Chromatogr.* **103**, 49 (1975).
5. J. Porath and S. Oscarsson, A new kind of thiophilic electron-donor-acceptor adsorbent, *Makromol. Chem. Macromol. Symp.* **17**, 359 (1988).
6. J. Porath and P. Hansen, Cascade-mode multiaffinity chromatography. Fractionation of human serum proteins, *J. Chromatogr.* **550**, 751 (1991).
7. L. Kägedal, Immobilized metal ion affinity chromatography, in J.-C. Janson and L. Rydén (Eds.), in *Protein Purification*, 2nd Ed., Wiley-Liss, New York, NY, 1988. p. 311.
8. R. G. Pearson, Hard and soft acids, HSAB, part I. Fundamental principles, *J. Chem. Ed.* **45**, 581 (1968).

6.C.8. Sephadex Derivatives for Extraction and Sample Preparation (Jan B. Sjövall)

The work with complex matrices such as plasma and feces made it clear that appropriate sample preparation procedures were essential for a successful analysis of complex mixtures of bile acids and steroids in biological materials. Chromatography with conventional two-phase solvent systems were not practical for this purpose because of the need for solvent equilibration. With a graduate student, Ernst Nyström, I began to study how stationary phases could be prepared by chemical modification of crosslinked dextran gels. Methylated Sephadex was prepared that permitted molecular sieving in organic solvents, but it was also clear that partitioning between the mobile phase and the solvent–gel phase took place. Depending on the solvent polarity, reversed- or straight-phase partition separations were obtained. In straight-phase systems, separations of lipids, protected peptides, and other lipophilic compounds were amplified by the combined action of partitioning and size exclusion. High column efficiencies could be obtained by recycling chromatography or with capillary columns made with superfine particles (17–23 μm). Sterols differing by a CH_2 group in the side chain could be separated with a separation factor of 1.017 using recycling chromatography in heptane/chloroform/methanol systems. Reusable capillary columns with an injection port could easily be prepared in 1.5-mm-i.d. Teflon tubing and used with a moving-chain flame ionization detector constructed by Eero Haathi in Turku, with whom we had a most stimulating collaboration [1]. Preparative columns extending through two floors and sometimes containing 2 kb of gel were also useful (see Supplement S11 online at http:/www.chemweb.com/preprint/ and Chap. 5 in Gehrke et al., Vol. 64, 2001.).

It soon became clear that a wider range of substituted Sephadexes were needed for sample preparation purposes. Less polar derivatives were required for establishment of better reversed-phase systems, and ion exchangers of variable polarity were also desirable for group separations of biological extracts, in which metabolites occurred in different forms of conjugation. A postdoctoral fellow, Jim Ellingboe, observed that long-chain olefin oxides were commercially available in bulk quantities from Ashland Chemical Company. These were used to synthesize hydrophobic derivatives of Sephadex LH-20 in a BF_3-catalyzed reaction [2,3]. We found that reversed-phase separation of cholesterol esters was possible on a capillary column of hydroxyalkoxypropyl–Sephadex G-25 [4]. In a similar way, chloro- and bromohydroxypropyl derivatives could be prepared with epichloro- and epibromohydrins, respectively. These derivatives, whose polarities could be modified by appropriate hydroxyalkylations, served as intermediates in the preparation of ion exchangers and other stationary phases for ligand exchange purposes [3]. A variety of these were synthesized, of which those substituted with diethylamino, triethylamino, and sulfonic acid groups were the most useful ones [5]. The neutral hydrophobic, hydroxyalkyl derivatives could be used for reversed- or straight-phase partition chromatography, the mechanism depending on the solvent used. Extreme reversed- and straight-phase conditions were obtained with aqueous methanol and heptane, respectively [4]. Ethylene chloride or chloroform could be used as modifiers to adjust the polarity to the desired level in either case. When the polarities of the solvent and the gel–solvent phases were about the same, size exclusion was the predominant separation mechanism [1]. Our and other laboratories applied the hydrophobic Sephadex gels to the separation of a wide range of lipid soluble compounds, and the separations achieved were in most cases superior to those previously described. Nonpolar compounds such as triglycerides and polyisoprenoids were readily separated according to carbon number. Conditions were mild, adsorptive losses were not observed, and it was possible to separate labile compounds, such as trimethylsilyl ethers of 1,2-diacylglycerides in a reversed-phase system prior to GC/MS analysis.

After our development of the hydrophobic gel phases, reversed-phase high-performance chromatography, as we know it today, began to develop. Our soft gels did not allow the use of high pressures, and separation times in high-efficiency columns were very long. It was therefore evident that, if not synthesized on a solid matrix, these gels would find their main applications in low-pressure open-column systems, mostly for the fractionation of crude biological extracts prior to analysis with faster high-resolution methods such as GC/MS. With some exceptions this has also become their major use. Strategies were developed in the early 1970s for the preparation and group fractionation of biological samples for multicomponent analysis (metabolic profiling) of steroids, bile acids, and metabolites of other lipophilic compounds in urine, plasma, feces, and tissues by GC/MS [5].

Our sample preparation procedures are based on passage of the sample through an appropriate sequence of lipophilic, neutral, and ion-exchanging gel beds. The strategy is to retain either the analytes or the interfering material (digital chromatography). Groups of unconjugated and conjugated metabolites differing in charge and acidity are sequentially eluted by stepwise displacement from the ion-exchanging beds. Selective gels can be prepared for group isolation of compounds with a common structural feature, such as synthetic steroids carrying an ethynyl group (retained on a lipophilic sulfonic acid gel in silver form). However, in applications involving multicomponent analyses, passage of the sample through a sequence of beds, each with a desired property and separation mechanism, is more flexible than the use of beds with multiple functional groups.

REFERENCES (for Jan B. Sjövall)

1. J. Sjövall, E. Nyström, and E. Haahti, Liquid chromatography on lipophilic Sephadex: Column and detection techniques, *Adv. Chromatogr.* **6**, 119–170 (1968).
2. J. Ellingboe, E. Nyström, and J. Sjövall, Liquid chromatography on lipophilic-hydrophobic Sephadex derivatives, *J. Lipid Res.* **11**, 266–273 (1970).
3. J. Ellingboe, B. Almé, and J. Sjövall, Introduction of specific groups into polysaccharide supports for liquid chromatography, *Acta Chem. Scand.* **24**, 463–467 (1970).
4. C. W. Gehrke, R. L. Wixom, and E. Bayer (Eds.), *Chromatography: A New Discipline of Science*, Elsevier, Amsterdam, 2001.
5. J. Sjövall and M. Axelson, Sample work-up by column techniques, *J. Pharm. Biomed. Anal.* **2**, 265–280 (1984).

6.D. SPECIAL TOPICS

For Section 6.D on special topics in chromatography, highlights were prepared in abstract form by the Editors from the main cited paper, book, or invited review noted in each section heading (with references included from the authors at the end of each topic).

6.D.1. Advanced Chromatographic and Electromigration Methods in Biosciences (Z. Deyl, I. Miksik, F. Tagliaro, and E. Tesarová)

Advanced Chromatographic and Electromigration Methods in Biosciences, edited by Z. Deyl et al. [1], deals with chromatographic and electrophoretic methods applied for the separation (quantitation and identification) of biologically relevant compounds. Individual separation modes are dealt with to an extent, which follows their applicability for biomedical purposes: liquid chromatography and electromigration methods are highlighted.

The chapters list the literature covering the 1987–1997 period, and have been written by specialists in a particular area and with an emphasis on applications to the biomedical field. This implies that theoretical and instrumental aspects are kept to a minimum, which allows the reader to understand the text. Considerable attention is paid to method selection, detection and derivatization procedures, and troubleshooting. The majority of examples given represent the analyses of typical naturally occurring mixtures. Adequate attention is paid to the role of the biological matrix and sample pretreatment, and special attention is given to forensic, toxicological, and clinical applications. The book also has an extensive Index of Compounds Separated [1].

As in chromatography, considerable present research is now focusing on an assembly of macrocyclic building blocks to produce supramolecular structures with desired properties, including ligand-exchange processes, cyclodextrin phase arrays, enantioselective and molecular recognition as given in 13 papers on supramolecular chemistry at the 35th Midwest Regional Meeting of the American Chemical Society, Oct. 25–27, 2000 in St. Louis, MO [2].

REFERENCES (for Z. Deyl, I. Miksik, F. Tagliaro, and E. Tesarová)

1. Z. Deyl, I. Miksik, F. Tagliaro, and E. Tesarová, Advanced Chromatographic and Electromigration Methods in Biosciences, J. Chromatogr. Library, Elsevier Science, The Netherlands, **60**, (1998).
2. 35th Midwest Regional meeting of the American Chemical Society, October 25 to 27, 2000, Sponsored by the St. Louis, MO, ACS Section, 13 abstract and oral papers.

6.D.2. Monolithic Columns [1] (Nabuo Tanaka)

Packing materials for HPLC have advanced from irregular to spherical, large to small, plain adsorbents to chemically bonded materials, and from low efficiency to high efficiency. The development of high-efficiency columns in the 1970s and 1980s was followed by the improvement of the chemistry of reversed-phase materials, leading to the present state-of-the-art ODS materials without any secondary retention effects.

The maximum obtainable plate numbers of popular commercial columns have plateaued for a few decades at around 10,000–30,000 plates per column. Interest in increasing the absolute efficiency of HPLC has been stimulated by the development of

CE, capillary electrochromatography (CEC), and ultra-high-pressure liquid chromato-graphy (UHPLC), which allows the use of very small particles generating plate numbers of 100,000–200,000 or more.

Monolithic columns possess desirable features for future separation media. Organic polymer-based materials will make significant contributions owing to their compatibility with biological macromolecules, although they have often been used as alternatives to silica-based materials in HPLC. Anybody can make monolithic columns, because one can circumvent the difficult steps in column preparation, namely, particle sizing and column packing. This will allow many researchers to participate in the development. Micro-HPLC using long monolithic capillary columns may also become a serious contender with GC, CEC, or UHPLC for attaining ultra-high efficiency.

This issue of the *Journal of High Resolution Chromatography* [1] presents a series of papers on monolithic stationary phases (F. Svec et al.), microfabricated monolith columns for liquid chromatography (F. E. Regnier), miniaturized molecularly imprinted continu-ous polymer rods (T. Takeuchi and J. Matsui), affinity chromatography (K. Amatschek), and enantiomeric separations of acidic and neutral compounds by capillary electrochro-matography with β-cyclotextrin-bonded polyacrylamide gels (Bohui Xiong et al.).

REFERENCE (for Nabuo Tanaka)

1. N. Tanaka, Monolithic columns, *J. High Resolut. Chromatogr.* **23**, 1 (2000).

6.D.3. Perfusion Chromatography [1] (Fred E. Regnier)

Dr. Regnier presents a review on perfusion chromatography as a technique based on fluid dynamics for reducing stagnant mobile-phase mass transfer in liquid chromatography. This is achieved by supports with large pores that allow mobile phase to flow through the particles.

The resolving power of liquid chromatographic systems has improved dramatically, with much of the success attributable to the minimization of band spreading through the design and production of new packing materials [2,3]. Theory indicates that band spread-ing is predominantly caused by intraparticle diffusion; longitudinal diffusion, solute dif-fusion in the mobile phase, eddy diffusion, and adsorption–desorption kinetics play a less significant role [4–6]. Diffusion through stagnant pools of mobile phase in particles is a slow process, particularly in the case of macromolecules [7]. Although this problem may be diminished by reducing particle size and, therefore, diffusion path length, it is not practical to use porous particles of less than 3–5 μm.

More recent studies have shown that there is another solution to the problem of stag-nant mobile-phase mass transfer. A new chromatographic packing material (POROS) has been developed, which has particle transecting pores of 6000–8000 Å diameter [8] to allow liquid flow through the sorbent. The advantage of this perfusion process is that intraparticle flow convectively transports solutes to the stationary phase in the particle interior. The convective transport of proteins into 10- and 20-μm particles at·high mobile-phase velocity was found to be an order of magnitude more rapid than diffusive transport. POROS-based supports separated proteins in less than a minute at mobile-phase velocities 1–30 times higher than those used in high-performance liquid chromatography [9].

This process of reducing intraparticle mass transfer by perfusing liquid through a support particle has been called *perfusion chromatography* [8–10]. Because perfusion chromatography is based on fluid mechanics, it is possible to enhance the kinetics of mass transfer without altering the separation mechanism.

The perfusible packing materials in the studies described above were fabricated in a multistep copolymerization of styrene and divinylbenzene. Electron microscopy shows that these particles have essentially a bimodal pore network in which the 6000–8000-Å through-pores are interconnected by 800–1500 Å diffusive pores. Solutes are convectively transported into the interior of the particle by the 6000–8000-Å through-pores and then diffuse into the 800–1500-Å pores. The term *diffusive pore* will be used to describe these smaller interconnecting pores.

Perfusion chromatography is a new technique based on fluid mechanics that decreased the separation of time of proteins by an order of magnitude or more. Rapid separations in all the conventional modes of chromatography, except size exclusion, have been achieved using the perfusion regime. The technique will be greatest use in two areas: (1) an analytical environment where large numbers of routine analyses are required and (2) a process environment where enhanced intraparticle mass transport will increase throughput and productivity.

REFERENCES (for Fred E. Regnier)

1. F. E. Regnier, Perfusion chromatography, *Nature* **350**, 634–635 (1991).
2. F. E. Regnier and K. M. Gooding, *Chromatography*, 5th edn (E. Heftmann, Ed.), Ch. 14, Elsevier, Amsterdam, 1991.
3. F. E. Regnier, HPLC of biological macromolecules—the 1st decade, *Chromatographia* **24**, 241–251 (1987).
4. J. C. Giddings, *Dynamics of Chromatography*, Dekker, New York, 1965.
5. J. N. Done and J. H. Knox, Performance of packings in high speed liquid chromatography II. Zipax. Effect of particle size, *J. Chromatogr.* **10**, 606–612 (1972).
6. C. Horvath and H. J. Lin, Movement and band spreading of unsorbed solutes in liquid chromatography, *J. Chromatogr.* **126**, 401–420 (1976).
7. L. R. Snyder, Column efficiencies in liquid adsorption chromatography: Past, present and future, *J. Chromatogr. Sci.* **7**, 352–360 (1969).
8. N. Afeyan, N. F. Gordon, I. Mazsaroff, L. Varady, S. P. Fulton, Y. B. Yang, and F. E. Regnier, Flow-through particles for the high-performance liquid chromatographic separation of bio-molecules: Perfusion chromatography, *J. Chromatogr.* **519**, 1–29 (1990).
9. N. Afeyan, S. P. Fulton, and F. E. Regnier, Perfusion chromatography packing materials for proteins and peptides, *J. Chromatogr.* **544**, 267–279 (1991).
10. N. Afeyan, R. Dean, and F. E. Regnier, U.S. Pat. Applic., Ser. No. 595661 (1989).

6.D.4. Chromatographic Methods for Biological Materials [1] (Toshihiko Hanai)

A strategy for chromatographic process selection for biological materials has been developed [1]. The characterization of solutes under different chromatographic conditions is important in solving purification problems in various fields of chemistry. The chromatographic behavior reflects the relative difference in the nature of solutes. Provided that solutes can be characterized by their physicochemical properties, the chromatographic conditions may be optimized and the nature of solutes can be derived from in the chromatographic behavior (retention times).

Modern liquid chromatography was first developed in the latter part of the 1900s, as high performance liquid chromatography, and its main applications since have been in biorelated fields. The development of new packing materials and a variety of applications allow the classification of separation systems based on the physicochemical properties of the analytes and packing materials. Biomedical investigations have been simplified by the development of new packing materials; urine and serum samples can now be injected directly, polar compounds can be analyzed in the reversed-phase mode on carbon packing materials instead of using ion-exchange liquid chromatography, and so on. Protein–drug binding constants can be measured with high precision. In addition, the development of computational chemical calculation permits optimization of the separation conditions, based on the molecular properties of the analytes. The main developments over the late 1990s are summarized in Ref. 1.

Organic polymer packing materials are chemically stable and are suitable for use in an automated instrument. The theoretical plate number (reflecting the separation power of the packing materials) is usually not sufficient, and the handling is still not easy compared to inorganic packing materials. However, organic polymer packing materials are particularly suitable for applications involving biorelated compounds (biopolymers, owing to their chemical inertness and the easy modification of their surface for affinity chromatography).

REFERENCE (for Toshihiko Hanai)

1. T. Hanai, Sélection of chromatographic methods for biological materials, in Z. Deyl, I. Mikšik, F. Tagliaro, and E. Tesarová (Eds.), *Advanced Chromatographic and Electromigration Methods in Biosciences*, Elsevier Science B.V., Amsterdam, 1998.

6.D.5. The Potential Role of Separations [1] (F. E. Regnier)

Advances in life science have triggered a revolution in healthcare, biotechnology, and agriculture. Separations have been essential tools in this revolution. "Liquid chromatography and electrophoresis have played a critical role in protein purification, peptide sequencing, amino acid analysis, and DNA sequencing: all of which had a major impact in advancing our understanding of protein and DNA structure," F. Regnier states. But what of the future? What part will separation systems play in the continuing evolution of biochemistry, biotechnology, and molecular biology, if any? Two of the major ongoing endeavors in miniaturized analytical systems described are biodot arrays and microfluidic systems.

Regnier's review [1] considers the current state and possible future of separation methods in the rapidly developing field of bioscience. For example, drug development by combinatorial methods allow the synthesis of more than 10^3 compounds per day and it is clear that the analysis of such large numbers of samples will need a radically new approach in which the technology developed for the etching of integrated electronic circuits is likely to play a major role. An alternative possibility is the development of multiple analyses with parallel bundles of open tubular columns. There are many practical problems to be overcome but it is probably that chromatographic and electrophoresis methods coupled with mass spectrometry will figure prominently until such time as miniaturization at a molecular level using nanotubes becomes a reality.

REFERENCE (for F. E. Regnier)

1. F. E. Regnier, The evolution of analysis in life science research and molecular medicine: The potential role of separations, *Chromatographia Supplemental I*, **49**, S56–63 (1999).

6.D.6. SilicaROD™—a New Challenge in Fast High-Performance Liquid Chromatography Separations [1] (Karin Cabrera et al.)

High-performance liquid chromatography (HPLC) has become one of the most widely used methods for the analysis of compound mixtures in industry, especially for the quality control of products. Nowadays productivity is the major and dominant upcoming issue; specifically, the goal is to drastically reduce the analysis time and cost per analysis.

The heart of each HPLC method is the column, which enables the resolution of compounds according to the selectivity and column performance. The quality of such a column, i.e., the separation performance, is determined mainly by the particle size and size distribution and the quality of the packing of the particles [2]. In contrast to such conventional materials, in SilicaROD™ columns have packing that consists of one rod of continuous monolithic porous silica [1]. The rod is obtained by the hydrolysis and polycondensation of alkoxysilanes and possesses large through-pores, which are comparable to the interstitial voids of a particle-packed column and mesopores of the same size and volume as in porous particles providing comparable specific surface areas for the chromatograhic interaction. Thus the SilicaROD column shows similar chromatographic properties with respect to retention and selectivity as HPLC columns filled with particles of the same specific surface area and pore diameter [3,4].

In contrast to regular columns, SilicaROD columns exhibit 90% interstitial porosity as compared to 80% in case of packed columns, due to the presence of large through-pores [4]. Further, they can be operated at higher flow rates with lower backpressures than can conventional columns, which is a distinct advantage over particle-packed columns. In addition, the column efficiency is maintained at high flow rates, because of the flat plate height vs. linear velocity dependency [4].

The most important feature of a SilicaROD column is its higher porosity over conventional particle-filled columns and the high mechanical stability of the rod, which allows work to be carried out at higher flow rates, resulting in faster separations.

The SilicaROD columns represent a novel generation of HPLC columns, that show the same excellent performance and selectivity of particle-filled columns but provide the unique opportunity to perform much faster separations. This is due to the morphological structure of the SilicaROD: a continuous monolithic column bed made of porous silica, a bimodal pore size distribution composed of through- and mesopores. A higher porosity as in HPLC columns packed with particles permits high flow rates of low backpressures without loosing efficiency. This opens up new ways of optimizing separation methods with respect to speed. SilicaROD columns enables flow gradients as a tool to reduce analysis time down to a few minutes without any modification of standard HPLC instrumentation.

REFERENCES (for Karin Cabrera et al.)

1. K. Cabrera, G. Wieland, D. Lubda, K. Nakanishi, N. Soga, H. Minakuchi, and K. K. Unger, Silica ROD™—A new challenge in fast high-performance liquid chromatography separations, *Trends Anal. Chem.* **17**(1), 50–53 (1998).

2. L. R. Snyder and J. J. Kirkland, *Introduction to Modern Liquid Chromatography* (2nd Edn.), John Wiley and Sons, New York, 1979.

3. H. Minakuchi, K. Nakanishi, N. Soga, N. Ishizuka, and N. Tanaka, Octadecylsilylated porous silica rods as separation media for reversed-phase liquid chromatography, *Anal. Chem.* **68**, 3498 (1996).

4. H. Minakuchi, K. Nakanishi, N. Soga, N. Ishizuka, and N. Tanaka, Effect of skeleton size on the performance of octadecylsilylated continuous porous silica columns in reversed-phase liquid chromatography, *J. Chromatogr. A* **762**, 135 (1997).

6.D.7. The History of Sephadex® [1] (Jan Christer Janson)

"We wish to report a simple and rapid method for the fractionation of water-soluble substances." This quote is the opening sentence of the now-classical paper by J. Porath and P. Flodin [2], who, in 1959, introduced gel filtration as a practical separation technique. In retrospect, it must be regarded as one of the most significant advances made in the methodology of biochemistry. It had an enormous impact, not only scientifically, where the speed and simplicity of the new method contrasted strongly with most of the separation techniques then available, but also commercially. For Pharmacia AB, Uppsala, Sweden, the company that developed, manufactured, and marketed the first gel filtration medium, Sephadex, it meant a pioneering effort that laid the foundation for the development of a large number of original products that are now indispensable to bioscience and biotechnology.

The introduction of the gel filtration technique was preceded by a combination of events, which together make an extraordinary example of how scientific tradition, the prepared mind of a scientist, and the close links between academia and industry are all utilized in an optimal way. The beginning, or rather the roots, of the success story of Sephadex can be traced back to 1925. In this year, Arne Tiselius, at the age of 23, became the personal assistant to Professor Theodor Svedberg at the Department of Physical Chemistry, University of Uppsala. The year before, Svedberg had performed the first successful ultracentrifugation experiment, demonstrating for the first time the molecular weight monodispersity of a protein (hemoglobin). On Svedberg's suggestion Tiselius developed the moving-boundary method of electrophoresis of proteins. His results not only supported Svedberg's findings of proteins as homogeneous entities but also revealed the complexity of natural protein mixtures, such as whole serum, which for the first time was separated into its well-known components—albumin and, as named by Tiselius, α-, β-, and γ-globulins. In 1937 Tiselius became the first Professor of Biochemistry at Uppsala University [1].

In early 1941, at the age of 23, Björn Ingelman, a chemist, graduated from the Faculty of Natural Sciences in Uppsala joined Tiselius as a postgraduate student. Tiselius asked Ingelman to study the proteins and polysaccharides in sugar beet, a project financed by the Swedish sugar industry. One spinoff was the possibility of commercial exploitation of the sugarbeet pectin to substitute for the normally imported citrus pectin, which was in short supply during World War II. It was, however, soon found that the sugarbeet pectin was of adequate quality. At about the same time Ingelman's studies serendipitously led him to another polysaccharide of extremely high molecular weight and composed solely of glucose, namely, dextran. How this discovery was made and how it was developed into a plasma volume expander and commercialized by Pharmacia, is in its details outside the scope of this article and has been reviewed [3]. However, it is by no means insignificant because gel filtration on Sephadex was developed as a direct consequence of Ingelman's continued work on dextran.

In 1954, P. Flodin left Tiselius' laboratory to take up a position at Pharmacia, and Porath continued the development of column zone electrophoresis in vertical columns, now with cellulose as the main packing material. The two colleagues, however, kept in close contact and met regularly to discuss scientific and other matters. Porath was not entirely happy with the cellulose powder, which displayed a variety of adsorption effects, and he tried several alternative convection suppressors. On the suggestion by P. Flodin in late Autumn 1956, he also tried a powder obtained by grinding a block polymerisate of dextran crosslinked with epichlorohydrin.

Soon after, in early 1957, J. Porath could report to P. Flodin that he had observed the same sieving properties in the columns packed with the particles of crosslinked dextran as they had experienced some years earlier in the starch columns. The crosslinked dextran particles had physical properties much better than those of starch, and their porosity could be controlled with accuracy, and the two scientists immediately realized that their discovery had important implications, both scientifically and commercially [1].

The first, and now classical, paper was published by J. Porath and P. Flodin in *Nature* in June 1959 [2]. It was entitled "Gel filtration: A method for desalting and group separation." This was also the first time the proposed name for the technique was used. The choice was preceded by lengthy discussions. Tentatively, J. Porath and P. Flodin had used mainly the terms "molecular sieving" and "restricted diffusion" to describe the technique, but these were considered too general. The term "gel filtration" was suggested by Arne Tiselius [4]. It was immediately accepted because it was concise, emphasizing the use of a gel and indicating the main mechanism—separation according to molecular size.

The first paper was soon followed by others. In 1959, six more papers were published, three of which were from Porath's laboratory and two from Pharmacia. Porath's papers dealt with the separation of enzymes in snake venoms [5] and of pituitary harmones [6,7]. The papers from Pharmacia described the purification of pepsin [8] and the fractionation of low-molecular-weight dextrans [9]. During the following 2 years, more systematic studies were reported by Porath [10], Gelotte [11], Granath and Flodin [12], and Flodin [4]. Both Porath and Gelotte observed adsorption phenomena of mainly aromatic substances to the more tightly crosslinked Sephadex types.

These gels consist of hydrophilic chains that are crosslinked. They are devoid of ionic groups; the polar character is due almost entirely to the high content of hydroxyl groups. Although water-insoluble, the gels are nevertheless capable of considerable swelling. The fractionation depends primarily on differences in molecular size, although phenomena have been observed that indicate the influence of other factors.

In 1959, a patent application was filed describing a number of ion-exchanger derivatives of Sephadex. During 1960–1962, a whole series of products were introduced based on Sephadex G-25 and Sephadex G-50. The first ion exchangers were based on diethylaminoethyl (DEAE), carboxymethyl (CM), and sulfoethyl (SE) groups, and were available in coarse, medium, and fine grades obtained from block polymerizates.

By introducing hydroxypropyl groups into Sephadex G-25, one obtains a lipophilic derivative that has the ability to swell in both water and polar organic solvents. This product is called "Sephadex LH-20" (LH stands for Lipophilic and Hydrophilic, and the number 20 indicates that its approximte solvent regain is 2.0), and it has become a routine separation medium for many kinds of steroids. Sephadex LH-20 was introduced in 1966. Ten years later, in 1976, the more porous-type Sephadex LH-60 was launched on the market.

REFERENCES (for Jan Christer Janson)

1. C. Janson, The History of the Development of Sephadex, Pharmacia AB, Biotechnology, Uppsala, Sweden; and Biochemical Separation Center, Uppsala University Biomedical Center, Uppsala, Sweden, *Chromatographia* **23**, 361–369 (1987).

2. J. Porath and P. Flodin, Gel filtration: A method for desalting and group separation, *Nature* **183**, 1657 (1959).

3. A. Grönwall and B. Ingelman, The introduction of dextran as a plasma substitute, *Vox Sang* **47**, 96–99 (1984).

4. P. Flodin, Methodological aspects of gel filtration with special reference to desalting operations, *J. Chromatogr.* **5**, 103 (1961).

5. W. Björk and J. Porath, Fractionation of snake venom by the gel-filtration method, *Acta Chem. Scand.* **13**, 1256 (1959).

6. J. Porath, Fractionation of polypeptides and proteins on dextran gels, *Clin. Chim. Acta* **4**, 776–778 (1950).

7. E. B. Lindner, A. Elmquist, and J. Porath, Gel filtration as a method for purification of protein-bound peptides exemplified by oxytocin and vasopressin, *Nature* **184**, 1565 (1959).

8. B. Gelotte and A.-B. Krantz, Purification of pepsin by gel filtration, *Acta Chem. Scand.* **13**, 2127 (1959).

9. P. Flodin and K. Granath, Symposium über Makromoleküle in Wiesbaden, (1959), Verlag Chemie Weinheim.

10. J. Porath, Gel filtration of proteins, peptides and amino acids, *Biochim. Biophys. Acta* **39**, 193 (1960).

11. B. Gelotte, Studies on gel filtration: Sorption properties of the bed material Sephadex, *J. Chromatogr.* **3**, 103 (1960).

12. K. Granath and P. Flodin, Fractionation of dextran by the gel filtration method, *Die Makromolekulare Chemie*, **48**, 160 (1961).

6.D.8. High-Speed LC of Proteins and Peptides [1] (T. Issaeva, A. Kurganov, and Klaus K. Unger)

Since the very first high-performance liquid chromatographic separation by J. F. K. Huber in 1964 [2] performed with particles of a mean size of $\sim$40 μm, the diameter of HPLC packings has steadily reduced to 5 or even 3 μm. The main benefit of the reduction in particle size is an increase in the column efficiency and a decrease in the analysis time. As is well known, the efficiency of a chromatographic column, expressed as the number of the theoretical plates N, relates to the length of the column L and to the height of the theoretical plate H: $N = L/H$ [2]. Under an optimal linear flow rate u_{opt} for a well-packed chromatographic column, the height of the theoretical plate H is $\sim$2d_p. Therefore, the number of the theoretical plates is inversely proportional to the particle size [2], $N \approx L/2d_p$ [1].

A new generation of non-porous silica reversed-phase (RP) packings, commercially available from Micra Scientific, was tested for separations of peptides and proteins by means of the gradient HPLC. Extremely high-speed separations were achieved using conventional chromatographic equipment; six proteins could be completely separated within 6 s. Tryptic digest peptides could be resolved in more than 40 components within 2–3 min. The effects of the experimental parameters, such as temperature and flow rate, were investigated.

The results obtained in this work clearly show the great potential of nonporous adsorbents in the acceleration of the chromatographic analysis. The lack of the corresponding chromatographic equipment, such as fast and sensitive detectors, high-speed integrators, and high-pressure pumps, prevented the invention of these packings in the routine chromatographic analysis in the past. Nowadays, these problems have been overcome. Very-high-speed separations of proteins and peptides can be realized using commercially available chromatographic equipment and commercially available chromatographic columns packed with nonporous micrometer-sized particles. Application of these adsorbents in microcolumn and capillary HPLC promises an even further decrease of analysis times [1].

REFERENCES (for T. Issaeva)

1. T. Issaeva, A. Kurganov, and K. Unger, Super-high-speed liquid chromatography of proteins and peptides on non-porous Micra NPS-RP packings, *J. Chromatogr. A* **846**, 13–23 (1999).
2. L. C. Snyder and J. J. Kirkland (Eds.) *Introduction to Modern Liquid Chromatography*, Wiley-Interscience, New York, 1974.

6.D.9. Separation and Analysis of Peptides and Proteins [1] (C. K. Larive et al.)

High-performance liquid chromatography continues to be the method of choice for protein and peptide analysis. Reviews published in the late 1990's include reviews of equipment [2] and theory [3]. New packing materials useful for protein chromatography have recently been reviewed by Leonard [4]. Reversed-phase liquid chromatography is a popular choice for the separation of peptides and proteins. In general, most articles are focused on new column support materials used to improve the speed of protein and peptide separations. These include perfusion stationary phases, which contain very large pores and nonporous materials consisting of very-small-diameter particles ($\sim$2 μm). Porous stationary phases have been reviewed by Rodrigues [5]. The use of nonporous sorbents for protein separations has been reviewed by Lee [6]. Both types of chomatographic media can be used for fast separations. However, they also have limitations. Perfusion chromatographic media is less stable and the small particle size nonporous stationary phase can lead to high backpressures.

This review focuses on selected applications of the separation and analysis of peptides and proteins published during 1997–1998. Specific topic areas covered include high-performance liquid chromatography (HPLC), ultrafiltration, capillary electrophoresis (CE), affinity-based methods for protein isolation, separation, and mass spectrometry (MS) [1].

REFERENCES (for C. K. Larive)

1. C. K. Larive, S. M. Lunte, Min Zhong, M. D. Perkins, G. S.Wilson, G. Gokulrangan, T. Williams, F. Afroz, C. Schöneich, T. S. Derrick, C. R. Middaugh, and S. Bogdanowich-Knipp, Separation and analysis of peptides and proteins, *Anal. Chem.* **71**(12), 389R–423R (1999).
2. W. R. LaCourse and C. O. Dasenbrock, Column liquid chromatography: Equipment and instrumentation, *Anal. Chem.* **70**, 37R–52R (1998).
3. J. G. Dorsey and W. T. Cooper, Liquid chromatography: Theory and methodology, *Anal. Chem.* **70**, 591R–644R (1998).
4. M. Leonard, New packing materials for protein chromatography, *J. Chromatogr. B* **699**, 3–27 (1997).
5. A. E. Rodrigues, Permeable packings and perfusion chromatography in protein separation, *J. Chromatogr. B* **699**, 47–61 (1997).
6. W.-C. Lee, Protein separation using non-porous sorbents, *J. Chromatogr. B* **699**, 29–45 (1997).

6.D.10. A Special Report on HPLC 2000 [1] (Ronald E. Majors)

A special summation report is presented by R. E. Majors on HPLC 2000 covering highlights of the 24th International Symposium held in Seattle, WA, June 24–30, 2000 on high-performance liquid phase separations and related techniques.

The topics covered include (1) separations in the life sciences, (2) HPLC column technology, (3) capillary electrochromatography, (4) capillary electrophoresis, (5) nanotechnology, (6) sample preparation and processing, (7) spectroscopy investigations of liquid–phase separations, and (8) HPLC 2001.

REFERENCES (for Ronald E. Majors)

1. R. E. Majors, A special report on HPLC 2000, *LC-GC North America* **18**(11), 1122–1134 (2000).

6.D.11. More on Advances in Chromatography Technology [1] (Ronald E. Majors)

Highlighted in this 2000 issue of the journal *LC-GC North America* include:

- "GC ovens—a hot topic"—**J. V. Hinshaw**
- "Milestones in chromatography—A. A. Zhukhovitskii—a Russian pioneer of gas chromatography"—**L. S. Ettre and V. G. Berezkin**
- "Directions in discovery—fast, efficient separations in drug discovery—LC-MS analysis using column switching and rapid gradients"—**S. R. Needham**
- "Hybrid organic–inorganic particle technology: Breaking through traditional barriers of HPLC separations"—**Yung-Fong Cheng et al**.
- "Evaluating the isolation and quantification of sterols in seed oils by solid-phase extraction and capillary gas–liquid chromatography"—**Bryan Ham et al**.

A. A. ZHUKHOVITSKII. The Editors highly recommend a close reading by chromatographers of these articles by well-known scientists presented in the November 2000 issue of LC-GC. First, a special report is given on Topics in HPLC 2000 by Ronald Majors [1] (see Section 6.D.10, above). This is followed by a discussion on the activities of A. A. Zhukhovitskii (1908–1991), a Russian pioneer of gas chromatography. In this paper, Ettre and Berezkin [2] describe the principles of the special techniques that he developed: his life, Chromathermography, the heat dynamic method, chromatography without carrier gas, chromarheography, chromadistillation, and Zhukhovitskii—the man, with cited references.

GC OVENS. A story unfolds of the principles developed by Zhukhovitskii on adsoprtion–desorption measurements in the 1940s and 1950s; these then relate to the leading edge of CG ovens by John V. Hinshaw [3], and how GC ovens have undergone an evolutionary development to the present highly efficient temperature-programmed ovens.

HYBRID ORGANIC–INORGANIC PARTICLE TECHNOLOGY. The heart of a chromatographic separation is in the column itself that determines the important characteristics of retention and selectivity of the analyte. Yung-Fong Cheng and colleagues [4] introduced a family of reversed-phase columns that contain packings based on hybrid particles. The porous spherical particles in the hybrid organic–inorganic columns are synthesized using a mixture of two organosilanes: one that forms silica units and another that forms methylsiloxane units. The resulting hybrid particles contain methylsiloxane groups throughout their internal and external structures. The hybrid organic–inorganic particles combine the best properties of silica—high efficiency and excellent mechanical strength—with the best properties of organic polymers—a wide pH stability range and reduced silanol effects. In this article, the authors describe reversed-phase HPLC packings based on hybrid particles and demonstrate the packings' retention and selectivity characteristics. They also show that these media have long lifetimes at elevated temperatures under conditions at which conventional bonded silica fails rapidly. The benefits of operating at elevated temperatures are increased efficiency and reduced backpressure. When combined with short columns containing 2.5-μm hybrid particles, these conditions enable fast, high-resolution gradient separations, which are useful for high-throughput analyses. Cheng has demonstrated that a 20-mm-long column containing 2.5-μm hybrid organic–inorganic particles enable high-speed gradient separations with analysis times as short as 1 min.

FAST EFFICIENT SEPARATIONS IN DRUG DISCOVERY. S. R. Needham, Laboratory Director of the LC-MS Analysis group, Alturas Analytics, Inc., discusses fast, efficient separations in drug discovery: LC-MS analysis using column switching and rapid gradients [5]. He reports a dual-column LC-MS-MS method developed to rapidly analyze drugs and metabolites in complex matrices. This system achieves baseline separation of drugs and isomeric metabolites. Currently, the system configured to inject every 2.0 min. Future modifications to the autosampler will allow injections every 1.5 min. With the use of higher flow rates and shorter columns, the run time of this system could approach 1.0 min/per sample and yet retain complete chromatographic separation of the solutes. Diverse classes of drugs and metabolites can be analyzed routinely using this LC-MS-MS method [5].

ISOLATION BY SPE AND QUANTIFICATION OF STEROLS BY CGC. A study evaluating the isolation and quantification of sterols in seed oils by solid-phase extraction and capillary gas–liquid chromatography was reported by B. Ham and colleagues of Thionville Laboratories, New Orleans, LA [6]. They found the use of C18 reversed-phase SPE columns to be an acceptable alternative to the traditional TLC approach. The amount of solvent needed is greatly reduced, as is the time involved in separating the unsaponifiable compounds from the sterol fraction. The sterols recovered by SPE in all samples tested were free of interfering peaks. In conclusion, the two tests performed—the comparison of SPE with an established TLC method and the comparison of the six seed oil sterols obtained by SPE with literature values—both demonstrated that using SPE in place of TLC is a convenient, accurate, time-saving, and cost-effective alternative to the traditional method for a wide variety of seed oil sterol fractions.

REFERENCES (for Ronald E. Majors, Section 6.D.11)

1. R. E. Majors, A special report on HPLC 2000, *LC-GC North America* **18**(11), 1122–1134 (2000).
2. L. S. Ettre and V. G. Berezkin, A. A. Zhukhovitskii—A Russian pioneer of gas chromatography, *LC-GC North America* **18**(11), 1148–1155 (2000).
3. J. V. Hinshaw, GC Ovens—A hot topic, *LC-GC North America* **18**(11), 1142–1147 (2000).
4. Y.-F. Cheng, T. H. Walter, Z. Lu, P. Iraneta, B. A. Alden, C. Gendreau, U. D. Neue, J. M. Grassi, J. L. Carmody, J. E. O'Gara, and R. P. Fisk, Hybrid organic–inorganic particle technology: Breaking through traditional barriers of HPLC separations, *LC-GC North America* **18**(11), 1162–1172 (2000).
5. S. R. Needham, Fast, efficient separations in drug discovery—LC-MS analysis using column switching and rapid gradients, *LC-GC North America* **18**(11), 1156–1161 (2000).
6. B. Ham, B. Butler, and P. Thionville, Evaluating the isolation and quantification of sterols in seed oils by solid-phase extraction and capillary gas–liquid chromatography, *LC-GC North America* **18**(11), 1174–1181 (2000).

6.E. CHROMATOGRAPHY COLUMN VALIDATION [1]

The development of the chromatographic method can often be time-consuming and difficult. The traditional approach for this task typically involves selecting a favorite or frequently used combination of solvent, pH, buffer, and column as a starting point, then making subsequent adjustments in mobile composition until a satisfactory method emerges. This design, however, is laborious and often does not yield the best method. A more systematic approach that utilizes good experimental design and follows the workflow described in this outline can be used to efficiently and effectively develop high-quality reversed-phase chromatographic methods. This method development work-flow includes method scouting, where selectivity factors such as pH, organic modifier, and column chemistry are evaluated to determine which experimental parameters are most effective in manipulating selectivity to achieve the required resolution. The best separation that results from the scouting experiment is then optimized or fine-tuned to obtain the final desired result. The final method may be validated to ensure that it meets the requirements for its intended use.

Desirable Information for Method Development. To design a sound method for a scouting experiment, it is important to collect as much information about the chemical nature of the sample and analytes as possible, such as

1. Sample solubility
2. Number of analytes
   How many peaks of interest are to be separated?
3. Chemical structure(s)
4. Functional groups
   How do the analytes differ?

Are there ionizable species?

What are the pK_a values?

How will pH influence the chromatography?

5. Detection

What type of detection is required or possible?

Are the analytes of interest UV-absorbing (λ_{max})?

6. Concentration range and quantitative requirements

What detection levels are necessary?

What quantitation limits are needed?

7. Sample matrix effects

Are there matrix effects that would interfere with the separation or detection?

Intended Use of the Method. Another consideration is to understand or define the goal or intended use of the method. A clear understanding of the intended use will help to define the performance parameters that must be met as well as the key parameters to use to validate the method. For example, a method may need to separate an analyte of interest from all the components in the sample, or the method may need to separate not only the analyte of interest from the other components but also all the other components from each other. Other considerations may include: how much throughput and resolution are needed, and what tailing factors are acceptable.

Method Optimization. There are several parameters that can be manipulated to fine-tune or optimize a separation. Gradient slope is a physical parameter that can be manipulated with this in mind. Another optimization parameter is temperature. Every chemical process is influenced by temperature and temperature can provide dramatic selectivity differences.

Method Validation. This is a process that establishes whether performance characteristics of the analytical method meet the requirements for the intended application. The typical chromatographic method characteristics evaluated during the validation process are specificity, linearity, range, limit of detection, limit of quantitation, accuracy, precision, and robustness.

REFERENCE (for Section 6.E)

1. Selections from Waters Corp., *The Science of What's Possible. A Guide to Systematic Method Development and Validation*; available online at www.waters.com/methods (2008).

6.F. DEVELOPMENTS IN LC COLUMN TECHNOLOGY DURING 2006–2008 (Ronald E. Majors)

HPLC Column Technology—State of the Art Ron E. Majors presents an in-depth overview on HPLC column technology—state of the art to 2008 in LC-GC [1]. He enlisted 10 experts and pioneers in the field from industry and academia to contribute their technical knowledge.

It has often been stated (or maybe overstated) that the column is the heart of the chromatograph. Without the proper choice of column and appropriate operating conditions, method of development and optimization of the high-performance liquid chromatographic (HPLC) separation can be frustrating and unrewarding experiences. Since the beginning of modern liquid chromatography, column technology has been a driving force in moving separations forward. Today, the driving forces for new column configurations and phases are the increased need for high-throughput applications, for high-sensitivity assays, and to characterize complex samples such as peptide digests and natural products.

In the last several years, advances are still being made in column technology with smaller porous particles (1–2 μm in diameter), ultra-high-pressure HPLC, high-temperature ($\leq 200°C$) columns, nanocolumns with diameters under 100 μm and rapid separation columns enabling high-resolution separations in seconds. LC-on-a-Chip experimentation is now driving columns to smaller and smaller dimensions but making LC-MS interfacing even easier. Polymer- and silica-based monoliths have seen major improvements with better reproducibility, a variety of stationary phases, and commercial availability. New particle designs such as superficially porous particles for high-speed applications have come on the scene. Improvements in applications-specific columns such as those for chiral separations, sensitive biological samples, and very polar compounds are being shown every year. The area of multidimensional LC and comprehensive LCxLC has become a reality in the tackling of complex samples.

In time for the HPLC 2008 Symposium held in Baltimore this year, I have assembled a special edition of *LC-GC North America* to highlight the state-of-the-art in HPLC column technology. Experts and pioneers in the field of HPLC column technology from industry and academia were asked to contribute their technical knowledge. In this issue, we have presented an overview of column advances in the last 2 years (Majors), followed by a look at high throughput in high-pressure LC (Rozing), polymeric monolithic columns (Svec and Krenkova) and silica-based monolithic columns (Cabrera), high temperature HPLC (Yang), chiral chromatography columns (Beesley), enhanced stability stationary phases (Silva and Collins), and rounded out with a treatise on hydrophilic interaction chromatography (McCalley). The contributors were asked to provide an update on the phase and column technology in their respective areas with a focus on advances made in recent years. Focus was directed primarily to the most recent advances. These papers are an excellent source of information on the most recent developments in LC column technology.

REFERENCE (for Ronald E. Majors, Section 6.F)

1. R. E. Majors. Recent developments in LC column technology, *Supplement to LC-GC North America* **24**(54), April 2006, and **26**(54), April 2008.

7

CHROMATOGRAPHY— ADVANCES AND APPLICATIONS IN ENVIRONMENTAL, SPACE, BIOLOGICAL, AND MEDICAL SCIENCES

Charles W. Gehrke

Department of Biochemistry and the Agricultural Experiment Station Chemical Laboratories, College of Agriculture, Food and Natural Resources, University of Missouri, Columbia

David S. Hage

Department of Chemistry, University of Nebraska, Lincoln

CHAPTER OUTLINE

Chromatography: A Science of Discovery. Edited by Robert L. Wixom and Charles W. Gehrke
Copyright © 2010 John Wiley & Sons, Inc.

7.A. EARLY YEARS OF AUTOMATED CHEMISTRY [1] (Charles W. Gehrke)

Analytical chemistry and biochemistry are changing disciplines; the 20th century marks a period where revolutions have occurred in analytical chemistry that will have a dramatic impact through the coming years of the 21st century. These changes have been brought about to a large extent by new analytical methods in chromatography. We are now in a decade of chromatographies and hyphenated techniques, interfaced with high- and low-resolution mass spectrometry and computers for fast data reduction. Some of our most important environmental problems have been solved with this array of instrumentation combined with sensitive and selective analytical chromatographic methods and detectors.

The world of analytical chemistry has changed immeasurably since the 1950s. The analytical laboratory of today is far different from that in which most of us were trained; the diverse types of chromatographic methods available to us today have greatly altered our approaches to the problems of analysis and constitute important successes in every discipline of the analytical and biological sciences.

As Manager of the Experiment Station Chemical Laboratories (ESCL) (from 1950 to 1987), starting in 1950, my mission was to help professors and graduate students in the college with the use of chemistry in their research programs. I had just completed 7 years of teaching all aspects of chemistry at Missouri Valley College, Marshall, Missouri, and thus looked forward to my new position as Associate Professor and the challenges of research support in chemistry to agriculture. Deans John Longwell and Sam Shirkey of the College of Agriculture, University of Missouri, Columbia, Missouri, stated in 1950 that my responsibilities as State Chemist were to streamline and update the analysis programs for N, P, K, Ca, Mg, and sampling in fertilizers and limestones; of course, this program brought substantial funding of $500,000 to 1 million dollars per year to the College. I also had the responsibility of teaching analytical chemistry and biochemistry to the graduate students in the College of Agriculture. In a few years with a broad knowledge in analytical chemistry and the help of 12 graduate students and 20 staff members, the analyses for these elements were quickly changed from manual to automated methods using the equipment of the Technicon Corporation. As a result, 8 automated methods, called "Missouri Methods" were adopted by AOAC International (formerly Association of Analytical Chemists) as "Official Methods". We could now quantitatively analyze 30 to 60 samples per hour, whereas formerly this same work required days.

Figure 7.1 shows a photo of Gehrke as centennial President of the AOAC in 1984 with keynote speaker, Nobel Laureate Linus Pauling. Figure 7.2 shows Gehrke's early chromatography team at the University of Missouri.

Figure 7.1. Photo of Drs. Linus Pauling (left) and Charles W. Gehrke (right) at 1984 meeting of the International Association of the AOAC. Linus Pauling was the keynote speaker and C. W. Gehrke was the Centennial President. [Reproduced from C. W. Gehrke, R. W. Wixom, and E. Bayer (Eds.), *Chromatography: A Century of Discovery*, Vol. 64, Elsevier, Amsterdam, 2001 and Supplement S12 on the Internet (http://www.chemweb.com/preprint), with permission from Elsevier, Amsterdam.]

Figure 7.2. University of Missouri chromatography team. [Reproduced from C. W. Gehrke, R. W. Wixom, and E. Bayer (Eds.), *Chromatography: A Century of Discovery*, Vol. 64, Elsevier, Amsterdam, 2001 and Supplement S12 on the Internet (http://www.chemweb.com/preprint), with permission from Elsevier, Amsterdam.]

7.B. CHROMATOGRAPHY IN ENVIRONMENTAL ANALYSIS SINCE THE 1960s

7.B.1. The Analytical Biochemistry Laboratories (ABC) & Environmental Analysis (Charles W. Gehrke and Lyle D. Johnson) [1]

ABC Labs. In 1968, I founded the Analytical Biochemistry Laboratories (ABC Labs), a for profit corporation, with two of my graduate students—Jim Ussary of my staff and David Stalling of the National Fishery Laboratory in Columbia, Missouri. The University was only mildly receptive. Technicon Corporation of Tarrytown, NY, wanted me to set up an automated laboratory for testing of agricultural feeds, fertilizers, and amino acids. This was a natural, as my job in the Experiment Station Chemical Laboratories (ESCL) was the same for the College of Agriculture and its staff. The Graduate School offered help and a site in the University Research Park near the Reactor facility. I said I would think about it. After due consideration, I declined as I could not accept the work and avoid a conflict of interest. We then raised $200,000 of private money, bought 70 acres of land on the outskirts of Columbia and started the ABC Co. In 2000, we occupied three additional large laboratories in the University Reactor for pharmaceutical studies and the distribution of short-lived radioactive isotopes to pharmaceutical companies.

Jim Ussary was our first CEO of ABC from 1968 to 1980. The three founders were still members of the Board of Directors in 2001; however, the work of our company is now 80% directed to pharmaceutical analysis and bio-analytical chemistry. Over 200 scientists are now employed in Columbia, Missouri. In the early years, our work was centered on the analysis of pesticides and aquatic toxicology services. Lyle Johnson, one of my first graduate students, directs the analytical environmental and field studies programs of ABC with 35 employees.

Introduction to Environmental Analyses. Over the past few decades, developments in analytical chemistry have increased our understanding of life and environmental effects on the basis of molecular chemistry. We have been able to detect and quantitate trace amounts of compounds affecting the very basics of life processes. The existence and persistence of some industrial and agricultural chemicals since 1960 had caused concern in the reproduction and longevity of higher food chain animals. The discovery and innovation of progressive analytical procedures allowed scientists and consequently government regulators to develop chemistries and to implement enforcement guidance, harmonious with nature and beneficial to mankind and wildlife environment.

Two significant developments in analytical separation techniques had a large impact on the environmental regulatory process. J. E. Lovelock, in 1958, observed a physical characterization of noble gases ionized by radioactive beta emitters [2]. This characteristic was quenched by electrophilic compounds passing through the gaseous plasma, which led to the development of the argon-ionization detector, followed by the electron-capture detector (ECD). The latter is highly sensitive to halogen-containing compounds and many other organic species (see also Lovelock in Ref. 1, Chap. 5). For more details on the advances in detectors for gas chromatography, and on the maturation of high performance liquid chromatography as a separation tool applied to environmental analyses, see Chap. 9 of this current book.

The development of the Lovelock ECD for GC [2] was such a significant event, which through its capacity to detect pesticides in just about everything, in all kinds of matrices, set the scene for Rachel Carson in 1962 and the environmental movement. The GC analyses that followed showed that polychlorinated biphenyls, excellent electrical insulating material in transformers, though resistant to biodegradation, found its way into the environmental food chain, affecting the reproductive function of higher food chain birds, most notably the American Bald Eagle. Other agro-chemicals, such as DDT, chlordane, heptachlor, aldrin, dieldrin, endrin, BHC, lindane, methoxychlor, and toxaphene, were also suspected of contributing to adverse environmental effects. The Lovelock detector made it practical to detect these chlorinated species selectively over many other organic molecules. This capability led to screening and monitoring programs sponsored by governmental agencies, academic research groups, and industrial organizations for wildlife, soil, water, air and foodstuff, giving a total biosystem perspective of chemical exposure. These early years of cause and effect led government regulators to understand the long-term effects of persistent chemicals to body functions from chronic exposure.

7.B.2. The Belgian Dioxin Crisis (Charles W. Gehrke) [1]

Thirty-eight years after the start of the environmental movement, 1999, in Belgium, will be remembered as the year of the 'dioxin crisis'. This is a very important example of an environmental problem and its solution involving contamination of chicken feed with PCBs, PCDDs and PCDFs. At the Research Institute for Chromatography (RIC), Pat Sandra relates how he used matrix solid phase dispersion clean up and CGC–μECD and CGC–MS for fast screening of samples and the solution of the dioxin crisis in Belgium.

The complete story of this 1999 crisis is presented in Ref. 1, Gehrke et al., 2001, Chap. 5 by P. Sandra [1], and further discussed in Chap. 9 of this current text.

7.C. AMINO ACID ANALYSIS BY GAS–LIQUID AND ION-EXCHANGE CHROMATOGRAPHY—30 YEARS (Charles W. Gehrke)

This section presents a general review of our three-volume treatise on gas–liquid chromatography of amino acids, published in 1987 [3]. Our investigations on the gas chromatography of the 20 protein amino acids and many nonprotein amino acids, presented in Chapter 1 of Ref. 3 (see also Ref. 4), began with the critical first step to obtain accurate measurement of the amino acid content of proteinaceous substances: sample preparation and protein hydrolysis. The main sources of variance between the amino acid content of an HCl hydrolysate and the protein from which it was derived have been published [5], and results are presented from studies that compared high-temperature, short-time hydrolysis (145°C, 4 h) with the more traditional (110°C, 24 h) procedure for a diverse set of sample types. The results demonstrate the feasibility of the higher temperature and shorter hydrolysis time for many applications. The results of a prehydrolysis oxidation for analysis of cystine as cysteic acid and methionine as methionine sulfone for the diverse samples were evaluated, and amino acid values extrapolated from multiple hydrolysis times at 145°C were compared with multiple hydrolyses at 110°C, illustrating a means for obtaining values for amino acids such as isoleucine, valine, threonine, and serine, which are more accurate than those from a single hydrolysate, and more rapid than the typical 110°C multiple hydrolysate technique [5]. We also evaluated interlaboratory variations in sample preparation, which revealed that although differences resulting from hydrolysate preparation by two different laboratories can be minimized, those variations are greater than the chromatographic or analytical variability. Our studies further demonstrated that use of glass tubes with Teflon-lined screwcaps as hydrolysis vessels compared favorably with sealed glass ampules, and that 4 h hydrolysis at 145°C with these tubes, after careful exclusion of air, is a rapid, precise, and practical method for protein hydrolysis. The importance of careful sample preparation to the achievement of accurate data is stressed; this chapter provides both a review of HCl hydrolysis of proteins and practical information on preparation of hydrolysates [5] (Fig. 7.3).

The GC analysis of amino acids as the *N*-trifluoroacetyl (*N*-TFA) *n*-butyl esters—the established method developed principally in our laboratories—provides an effective and

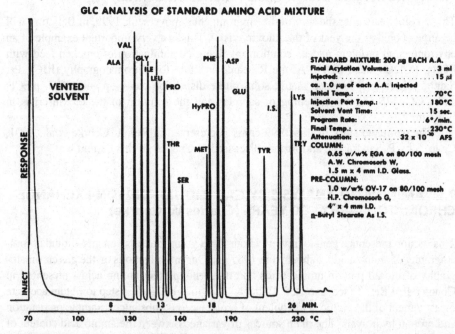

Figure 7.3. GLC analysis of standard amino acid mixture. [Reproduced from C. W. Gehrke, R. W. Wixom, and E. Bayer (Eds.), *Chromatography: A Century of Discovery*, Vol. 64, Elsevier, Amsterdam, 2001 and Supplement S12 on the Internet (http://www.chemweb.com/preprint), with permission from Elsevier, Amsterdam.]

reliable means of amino acid determination that is applicable to a very wide range of analytical needs. My research group, graduate students, and colleagues during 1962–1975 established the fundamentals of quantitative derivatization and the conditions of chromatographic separation, and defined the interactions of the amino acid derivatives with the stationary and support phases. Our studies and continued refinements since 1974 have resulted in a precise and accurate, reliable, straightforward method for amino acid measurement [4,5].

We conducted an extensive array of the applications of GC of amino acid analysis on a wide range of sample matrices, from pine needle extracts to erythrocytes. The *Experimental* section developed [3, Vol. 1, Chap. 1] provides a thorough description of our quantitative analytical procedures, including preparation of ethylene glycol adipate (EGA) and silicone mixed-phase chromatographic columns. The EGA column, which is used to separate and quantitate all the protein amino acids, except histidine, arginine, and cystine, is composed of 0.65 w/w% stabilized grade EGA on 80/100 mesh acid–washed Chromosorb® W, 1.5 m × 4 mm-i.d. glass. For quantitation of histidine, arginine, and cystine, the silicone mixed phase of 1.0 w/w% OV-7 and 0.75 w/w% SP-2401 on 100/120 mesh Gas-Chrom® Q (1.5 m × 4 mm-i.d. glass) performed extremely well. We also described the preparation and use of ion-exchange resins for sample cleanup, and complete sample derivatization to the N-TFA n-butyl esters. The amino acids are esterified by reaction with n-butanol · 3 N HCl for 15 min at 100°C, and the excess n-butanol · 3 N HCl is removed under vacuum at 60°C. Any remaining moisture is removed azeotropically with dichloromethane; then the amino acid esters are trifluoroacylated by reaction with trifluoroacetic anhydride (TFAA) at 150°C for 5 min in the presence of dichloromethane as solvent. Immediately following the *Experimental* section are valuable comments on various parts of the method [4], which provide guidance on the use of the entire technique, from sample preparation to chromatography to quantitation. Of particular value is a comparison of GLC and IEC results of hydrolysates of diverse matrices. This extensive comparison of an array of sample types showed that the values obtained by the two techniques were generally in close agreement.

Both GLC and IEC analyses of multiple hydrolysates were performed to evaluate the reproducibility of hydrolysate preparation and to compare GLC and IEC analyses of the same hydrolysates. The total amino acids found in the same hydrolysates were essentially identical by both GLC and IEC. As the sets of three hydrolysates were prepared at the same time under identical conditions, differences between the GLC and IEC analyses of the same hydrolysates might be expected to exceed the differences between identically prepared hydrolysates. However, the slight differences in the amounts of certain amino acids present in the different hydrolysates can be observed, emphasizing that variations do arise as a result of the hydrolysis itself, even under preparation conditions most conducive to reproducibility [3,4].

As the sulfur-containing amino acids are of particular interest in nutrition, cystine and methionine analyses are discussed in detail. The quantitative determination of amino acids in addition to cystine and methionine in preoxidized hydrolysates by IEC is described [3], and a rapid oxidation–hydrolysis procedure is presented that allows accurate analysis of cystine, methionine, lysine, and nine other amino acids in feedstuffs and other biological matrices. Floyd Kaiser has subsequently used the N-TFA n-butyl ester method for more than 17 years on a routine basis in our corporate laboratory (Analytical Biochemistry Laboratories, Columbia, MO); and his observations on the analysis of an extremely wide range of sample types over this timespan are presented as *Experiences of a Commercial Laboratory*, providing valuable practical information on amino acid analysis by GC [3, Vol. 1, Chap. 2, pp. 53–55].

The average recovery of cystine from a wide range of matrices without the use of performic acid was 55.5% as compared to results obtained with performic acid oxidation. Similarly, methionine is preferably analyzed as methionine sulfone.

Interlaboratory evaluation of 145°C, 4 h hydrolysis, in which one laboratory used sealed ampules and the other laboratory used Teflon-lined screwcap tubes, demonstrated excellent agreement of amino acid values. In summary, we found that the hydrolysis of a range of different protein-containing matrices at 145°C for 4 h in glass tubes with Teflon-lined screwcaps after vacuum removal of air, nitrogen, purge, and sonication performed as well as sealed glass ampules at both 145°C, 4 h and 110°C, 24 h hydrolysis conditions.

The analysis of amino acids as the *N*-TFA *n*-butyl esters is an established technique that offers much to scientists concerned with the determination of amino acids. The method offers excellent precision, accuracy, selectivity, and is an economical complementary technique to the elegant Stein–Moore ion-exchange method (see Figs. 7.4–7.7).

We also provided both a detailed account and historical perspective on development of GC amino acid analysis and describe the solution of problems encountered as the methods evolved [3]. The *N*-TFA *n*-butyl ester and trimethylsilyl (TMS) derivatives are discussed, including reaction conditions, chromatographic separations, mass spectrometric (MS) identification of both classes of derivatives, interactions of the arginine, histidine, and cystine derivatives with the liquid phase and support materials, and application of the methods in [3, Vol. 1, Chap. 3]. The acylation of arginine posed a problem in early studies; the successful solution of this problem paved the way for a high-temperature acylation procedure, which is now widely used with numerous acylating reagents [6]. Likewise, esterification of the amino acids was investigated in detail, resulting in a direct esterification procedure that quickly and reproducibly converts the amino acids to *n*-butyl esters. This approach has also been widely used to form various amino acid esters [7].

The early development of GLC analysis of iodine- and sulfur-containing amino acids as the TMS derivatives is described, with the finding that bis(trimethylsilyl)-trifluoroacetamide (BSTFA), a silylating reagent that we invented and patented, is an effective silylating reagent to form amino acid derivatives [3, Vol. 1, Chap. 3; 8]. Our studies on the derivatization of the protein amino acids with our silylation reagent, BSTFA, led the conversion of the amino acids to volatile derivatives in a single reaction step. Although certain amino acids tend to form multiple derivatives that contain varying numbers of TMS groups, high-temperature, long-reaction-time derivatization permits quantitative analysis of the amino acids as the TMS derivatives. Our studies on the GLC of the TMS amino acids resulted in the development of a 6-m column of 10% OV-11 on Supelcoport® for separation of the TMS derivatives.

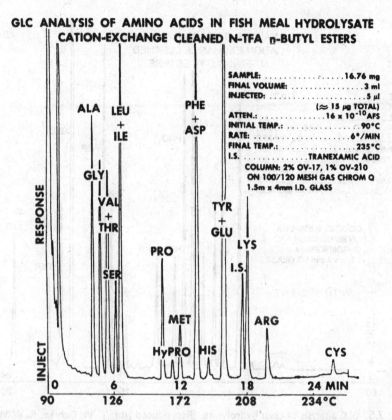

GLC ANALYSIS OF AMINO ACIDS IN FISH MEAL HYDROLYSATE CATION-EXCHANGE CLEANED N-TFA n-BUTYL ESTERS

SAMPLE: ....16.76 mg
FINAL VOLUME: ....3 ml
INJECTED: ....5 µl
($\approx$ 15 µg TOTAL)
ATTEN.: ....16 x 10^{-10} AFS
INITIAL TEMP.: ....90°C
RATE: ....6°/MIN
FINAL TEMP.: ....235°C
I.S. ....TRANEXAMIC ACID
COLUMN: 2% OV-17, 1% OV-210
ON 100/120 MESH GAS CHROM Q
1.5m x 4mm I.D. GLASS

Figure 7.4. GLC analysis of amino acids in fishmeal hydrolysate. [Reproduced from C. W. Gehrke, R. W. Wixom, and E. Bayer (Eds.), *Chromatography: A Century of Discovery*, Vol. 64, Elsevier, Amsterdam, 2001 and Supplement S12 on the Internet (http://www.chemweb.com/preprint), with permission from Elsevier, Amsterdam.]

The development of a chromatographic column system for the *N*-TFA *n*-butyl esters came about from the realization that the derivatives of arginine, histidine, and cystine were not reproducibly eluted from columns with polyester liquid phases, although this type of column was excellent for analysis of the other protein amino acids. We developed a siloxane mixed-phase column specifically for these three amino acids; the final system was an ethylene glycol adipate (EGA) column for 17 amino acids and the mixed-phase column for the remaining three [6,8]. Our summary points out, that the foundation of a successful amino acid analysis by GC is composed of two elements: (a) reproducible and quantitative conversion of amino acids to suitable volatile derivatives, and (b) separation and quantitative elution of the derivatives by the gas chromatographic column (see Fig. 7.8 for photo from our Mass Spectrometry Laboratory).

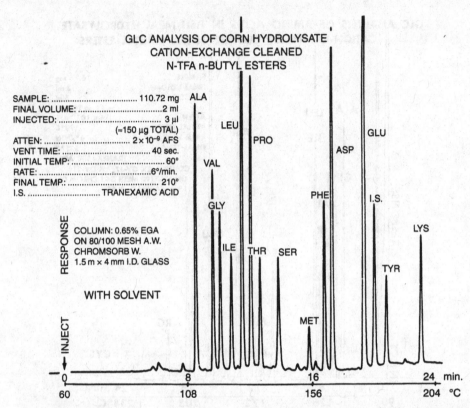

GLC ANALYSIS OF CORN HYDROLYSATE
CATION-EXCHANGE CLEANED
N-TFA n-BUTYL ESTERS

SAMPLE: .... 110.72 mg
FINAL VOLUME: ....2 ml
INJECTED: .... 3 µl
　　　　　　　　　　　　　(≈150 µg TOTAL)
ATTEN: .... 2 × 10⁻⁹ AFS
VENT TIME: .... 40 sec.
INITIAL TEMP: ....60°
RATE: ....6°/min.
FINAL TEMP.: .... 210°
I.S. .... TRANEXAMIC ACID

COLUMN: 0.65% EGA
ON 80/100 MESH A.W.
CHROMSORB W.
1.5 m × 4 mm I.D. GLASS

WITH SOLVENT

Figure 7.5. GLC analysis of corn hydrolysate. [Reproduced from C. W. Gehrke, R. W. Wixom, and E. Bayer (Eds.), *Chromatography: A Century of Discovery*, Vol. 64, Elsevier, Amsterdam, 2001 and Supplement S12 on the Internet (http://www.chemweb.com/preprint), with permission from Elsevier, Amsterdam.]

Research by Other Investigators in the Late 1970s and 1980s. The following paragraphs summarize research conducted by 29 investigators, who are contributors to our three-volume treatise on amino acid analysis by gas–liquid chromatography [3]. S. L. MacKenzie of the Plant Biotechnology Institute, NRC, Canada, a leader in the successful development of the N-heptafluorobutyryl (N-HFB) isobutyl ester derivatives, describes the rationale for his extensive work, and presents in detail the derivatization, separation, and applications of this derivative [3, Vol. 1, Chap. 4] (see also Refs. 9–11). His chapter contains a section entitled *Important Comments*, pointing out that derivatization is the most crucial factor in reproducible analysis of amino acids by GLC.

　　　Noting that there are some 50 diseases known to be due to anomalies of amino acid metabolism, J. Desgres and P. Padieu of the National Center for Mass Spectrometry and the Laboratory of Medical Biochemistry at the University of Dijon, France, describe the development of the N-HFB isobutyl derivatives for the clinical analysis of amino acids,

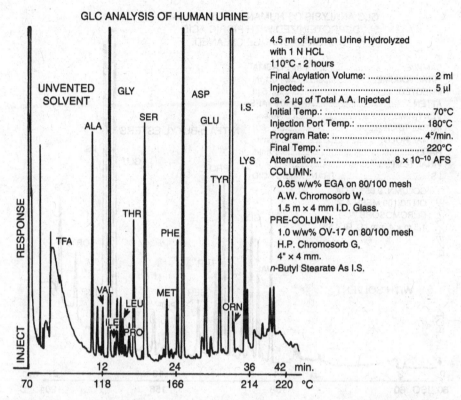

GLC ANALYSIS OF HUMAN URINE

4.5 ml of Human Urine Hydrolyzed
with 1 N HCL
110°C - 2 hours
Final Acylation Volume: .... 2 ml
Injected: .... 5 μl
ca. 2 μg of Total A.A. Injected
Initial Temp.: .... 70°C
Injection Port Temp.: .... 180°C
Program Rate: .... 4°/min.
Final Temp.: .... 220°C
Attenuation.: .... 8 × 10⁻¹⁰ AFS
COLUMN:
 0.65 w/w% EGA on 80/100 mesh
 A.W. Chromosorb W,
 1.5 m × 4 mm I.D. Glass.
PRE-COLUMN:
 1.0 w/w% OV-17 on 80/100 mesh
 H.P. Chromosorb G,
 4" × 4 mm.
n-Butyl Stearate As I.S.

Figure 7.6. GLC analysis of human urine. [Reproduced from C. W. Gehrke, R. W. Wixom, and E. Bayer (Eds.), *Chromatography: A Century of Discovery*, Vol. 64, Elsevier, Amsterdam, 2001 and Supplement S12 on the Internet (http://www.chemweb.com/preprint), with permission from Elsevier, Amsterdam.]

and the adaptation of the method to the routine clinical analysis of amino acids in [3, Vol. 1, Chap. 5; 12]. The experimental protocol is applied to the analysis of normal and pathologic physiological fluids, including phenylketonuria, maple syrup disease, idiopathic glycinemia, cystathionase deficiency, cystathionase synthetase deficiency, and renal absorption disorders. After use of their method for more than 6 years in a clinical laboratory, they conclude that GLC is perfectly suited for the daily analysis of more than 30 amino acids, and emphasize the importance of GC/MS in elucidating metabolic disorders.

I. M. Moodie of the National Research Institute for Nutritional Diseases and the Metabolic Unit, Tygerberg Hospital, South Africa, describes [3, Vol. 1, Chap. 6; 13,14] the development of an efficient GLC method specifically to routinely produce accurate determination of protein amino acids in fishery products and presents his choice of suitable derivative and columns, modification of derivative preparation, and sample preparation techniques for packed-column analysis of the *N*-HFB isobutyl esters.

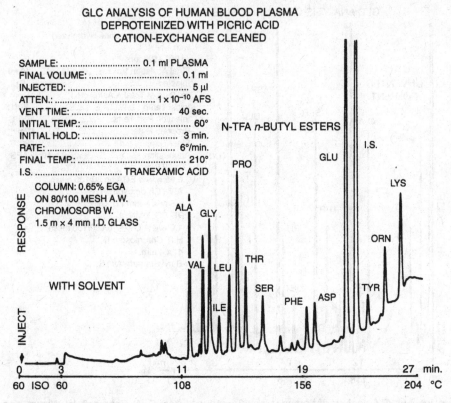

GLC ANALYSIS OF HUMAN BLOOD PLASMA
DEPROTEINIZED WITH PICRIC ACID
CATION-EXCHANGE CLEANED

SAMPLE: .... 0.1 ml PLASMA
FINAL VOLUME: .... 0.1 ml
INJECTED: .... 5 µl
ATTEN.: .... 1×10^{-10} AFS
VENT TIME: .... 40 sec.
INITIAL TEMP.: .... 60°
INITIAL HOLD: .... 3 min.
RATE: .... 6°/min.
FINAL TEMP.: .... 210°
I.S. .... TRANEXAMIC ACID

COLUMN: 0.65% EGA
ON 80/100 MESH A.W.
CHROMOSORB W.
1.5 m x 4 mm I.D. GLASS

Figure 7.7. GLC analysis of human blood plasma. [Reproduced from C. W. Gehrke, R. W. Wixom, and E. Bayer (Eds.), *Chromatography: A Century of Discovery*, Vol. 64, Elsevier, Amsterdam, 2001 and Supplement S12 on the Internet (http://www.chemweb.com/preprint), with permission from Elsevier, Amsterdam.]

Development of the chiral diamide phases for resolution of amino acid enantiomers is the subject of a chapter in our treatise [3, Vol. 2, Chap. 1], by **E. Gil-Av, R. Charles, and S.-C. Chang** of the Weizmann Institute of Science, Israel. Being pioneers in research on the separation of enantiomers, Gil-Av et al. [15,16] discuss the evolution of optically active phases from the α-amino acid derivative phases (e.g., N-TFA-L-Ile-lauroyl ester), to dipeptide phases (e.g., N-TFA–L-Val–L-Val–O-cyclohexyl) to the still more efficient and versatile diamide phases of the formula $R_1CONHCH(R_2)CONHR_3$. Gil-Av et al. describe the synthesis and purification of the diamide phases and the determination of their optical purity, and the influence of structural features of the diamides on resolution and thermal stability is discussed in detail. In these extensive studies, the structure of R_1, R_2, and R_3 in the above formula were varied with R_1 and R_3 representing various n-alkyl, branched alkyl, and alicyclic groups, and R_2 representing various aliphatic and aromatic groups. Chain lengthening of R_1 and R_3 groups produced the desired increase in thermal

Figure 7.8. Photo of the University of Missouri, Columbia, Mass Spectrometry Laboratory. Shown (left to right) are Dr. Klaus Gerhardt, organic chemist, Mr. Roy Rice, mass spectrometrist, and Dr. Charles W. Gehrke. [Reproduced from C. W. Gehrke, R. W. Wixom, and E. Bayer (Eds.), *Chromatography: A Century of Discovery*, Vol. 64, Elsevier, Amsterdam, 2001 and Supplement S12 on the Internet (http://www.chemweb.com/preprint), with permission from Elsevier, Amsterdam.]

stability, yielding phases operable at 200°C and above. The sequel to Gil-Av's research is described in Chapter S-9 De (Supplement available online at http://www.chemweb.com/preprint).

The separation of amino acid enantiomers using chiral polysiloxane liquid phases and quantitative analysis by the novel approach of "enantiomeric labeling" was presented by **E. Bayer, G. Nicholson, and H. Frank** of the University of Tübingen, Germany in our 1987 treatise [3, Vol. 2, Chap. 2; 17,18]. They synthesized a novel type of chiral stationary phase, which possesses sufficiently high thermal stability to enable GC separation of the enantiomers of all the protein amino acids within about 30 min. The stationary phase, subsequently named Chirasil-Val®, is a polysiloxane functionalized with carboxyl alkyl groups to which valine-*t*-butylamide residues are coupled through an amide linkage. Frank et al. [17,18], describe the criteria that the stationary phase must fulfill, the steps taken to synthesize the required stationary phase, the preparation of capillary columns with Chirasil-Val stationary phases, and GC conditions needed to optimize the separations.

The concept of the use of D-amino acids as internal standards for the quantitative analysis of the L-amino acids is first and elegantly described by H. Frank et al. [19]; they point out that prior to analyses of amino acids in biological fluids, a several-step procedure is usually required to separate the amino acids from high-molecular-weight material, carbohydrate derivatives, fatty acids, oligopeptides, biogenic amines, and other compounds, and that recovery of the amino acids during this process must be either quantitative or of a constant and known percentage. Also, variations can occur during derivatization and in the response of the detector, and discriminatory effects in the injector can arise when using the split technique of sample injection.

When the optical antipodes are used as internal standards, not only are volumetric losses during workup, derivatization, and GC compensated; differences in chemical behavior, such as recovery from ion-exchange columns during cleanup, yield of derivatization, and stability of the derivatives are also compensated. With "enantiomeric labeling," the internal standard suffers exactly the same fate as its optical antipode. Frank et al. [19], illustrate the use of "enantiomeric labeling" for analyses of blood serum and gelatin, present in the mathematical expressions for determining the concentration of each amino acid in samples. Sensitivity, precision, and accuracy of the enantiomeric labeling technique are discussed, with accuracy of determination of synthetic samples surpassing ion-exchange chromatography (IEC). Thus GC analysis of amino acids by "enantiomeric labeling" is a powerful technique for a wide variety of applications. For situations where many analyses are required, Bayer also has introduced a new automated derivatization robot [20].

The analysis of amino acid racemization and the separation of amino acid enantiomers is presented by H. Frank of the University of Tübingen, Germany [3, Vol. 2, Chap. 3; 20,21]. The author presents examples of enantiomers of free amino acids and amino acid analogues that exhibit different biological properties. A tragic illustration is the glutamic acid derivative, thalidomide, a sedative drug used in the 1950s that induced malformations of newborns. Later, the L-enantiomer of the metabolite was found to be teratogenic; whereas, the D-antipode is not (see Chap. S-9 De [3]). The stereochemical analysis of amino acids is important for studies on the biochemical, pharmacological, or toxicological properties of enantiomers, as well as its utilization in structure elucidation, peptide synthesis, racemization monitoring, and pharmaceutical research.

For a meaningful determination of lower percentages of D-amino acids in proteins, Frank [3, Vol. 2, Chap. 3; 20,21] resorted to deuterium/hydrogen exchange and GC/MS, which enables the determination of 0.01% of D-enantiomer originally present in a protein, and found an even smaller quantity of D-enantiomers by use of an added standard compound as a reference peak. By using a series of dilutions, ideally D-enantiomers could be measured when present at 0.01–0.001% of the L-enantiomer. The methods for determination of optical purity of amino acids have been greatly refined, making previously intractable problems now appear relatively straightforward [3].

A. Darbre, of Kings College, London, one of the first investigators to use GC for determination of amino acids, includes an interesting perspective [3, Vol. 2, Chap. 4; 22,23]. He describes studies of the N-TFA amyl and methyl esters, including column evaluations, various methods of esterification, trifluoroacylation, and quantitative studies using gas density balance and flame ionization detectors. His investigation includes use of radioactive amino acids to study the breakdown of some derivatives during chromatography by monitoring the column effluent for radioactivity. Darbre describes his methods of solid sample injection, in situ precolumn derivatization, and quantitative studies of biological samples using the N-TFA amino acid methyl esters. He also discussed GC separation of acylated dipeptide methyl esters, pointing out the more recent advances in MS techniques for peptide analysis. Darbre's most recent work has centered on the N-HFB isobutyl esters, a practical derivatization method, and its application to analysis of blood plasma and other biological samples. The use of electron capture detection (ECD), alkali flame ionization detection (AFID), and flame ionization detection (FID) are compared for relative molar response (RMR) values of the amino acid derivatives and the analysis of biological samples by capillary columns with FID, AFID, and ECD. Darbre foresees the next stage of development to include work at even higher levels of sensitivity (nanomole to picomole amounts), and perhaps eventually the use of aseptic conditions to eliminate the problems of background contamination.

The N-TFA n-propyl esters have been the subject of an in-depth study by G. Gamerith, at the University of Graz, Austria [3, Vol. 2, Chap. 5; 24,25] for medical diagnostic applications. Gamerith thoroughly discusses his effective amino acid analysis method for screening individuals with potential inborn errors of metabolism and demonstrates the application of the method for quantitative screening of urine specimens of patients with mental retardation.

Cyril Ponnamperuma of NASA and the University of Maryland elegantly presents the role of GC in the study of cosmochemistry [3, Vol. 2, Chap. 6; 26,27]. One of the most fundamental questions of all science is the "origin of life." How did it first begin? Since the laws of physics and chemistry are universal laws, it may be legitimate to extrapolate from the circumstances on earth to elsewhere in the universe, and our present-day telescopes reveal that there are 10^{23} stars, which, like our own sun, can provide the photochemical basis for plant and animal life. Ponnamperuma presents the central role of GC and interfaced MS spanning two decades of study on chemical evolution, prebiotic chemistry, and his search for life molecules on the moon and planets, ancient sediments, and the extraterrestrial Murchison meteorite. The detection and quantitation of minute traces of amino acids in complex mixtures was made possible by highly sensitive and selective GC techniques that were first developed at the University of Missouri by Gehrke, Kuo, and Zumwalt [3, Vol. 2, Chap. 6].

Our experiments in 1984, with the methods of HPLC followed by capillary GC, interfaced with MS presented for the first time unambiguous evidence for the existence of extraterrestrial nucleobases in the Murchison meteorite. The nucleobases uracil, thymine, cystosine, adenine, and xanthine were also found in the primordial soup generated by an electric discharge. GC/MS has thus provided the chemist with a powerful research tool for the study of organic cosmochemistry.

W. A. König, of the University of Hamburg, Germany, describes the analysis of amino acids and N-methylamino acids by capillary GC of diastereomeric derivatives on nonchiral phases and the use of isocyanates as reagents for enantiomer separation on chiral phases in [3, Vol. 3, Chap. 1; 28,29]. König discusses practical applications of configurational analysis of amino acids, and presents in detail the use of diastereomeric derivatives. Factors that affect the precision of determination of the proportion of a mixture of enantiomers are described, including optical purity of the chiral reagent and reaction kinetics of enantiomers in the formation of diastereomeric derivatives.

I. Abe and S. Kuramoto [3, Vol. 3, Chap. 2; 30,31] address in depth two topics: (1) a comparison of the resolution of various D-amino acid derivatives on optically active columns and (2) the "age dating" of sediments by determination of the extent of racemization of amino acids in fossil shells. The influence of the ester group on chromatographic resolution was evaluated by synthesis and analysis of the methyl, ethyl, n-propyl, isopropyl, and neopentyl esters, and the effect of the acyl group was investigated by preparation and analysis of the TFA, PFPF, and HFB derivatives. A total of nine different derivatives and the different substituent groups were investigated in terms of retention times, D and L peak-to-peak resolution, and elution order on both glass and fused-silica Chirasil-Val columns. This systematic study provides detailed information on the separation of amino acid enantiomers; and his greatest difficulty was in the separation of the lower-molecular-weight amino acids (glycine, threonine, isoleucine, alloisoleucine), which elute within a very narrow range.

I. Abe and S. Kuramoto [3] also describe the application of enantiomeric analysis to age dating of sediments of the Osaka Plain. The rate of amino acid racemization was dependent on temperature, pH, and ionic strength. Temperature especially affects the rate of racemization; for instance, racemization half-lives at $0°C$ are about 100 times longer than at $25°C$. The expressions for calculation of the age of fossils from the enantiomer ratios are then presented. Sediments between 50 and 64.5 m in depth were dated at 27,000–28,000 and 29,000–30,000 years before the present; this illustrates the potential for age dating from D/L ratios of several amino acids.

G. Odham and G. Bengtsson, of the University of Lund, Sweden present [3, Vol. 3, Chap. 4; 32,33], present an excellent description of the GC microanalysis of amino acids using electron capture detection (ECD) and MS with emphasis on micro environmental applications, especially in studies on the interactions between microorganisms and other cells. Their research has focused on situations in which the sample size is limited to microliters or micrograms, and the amino acid concentrations are in the ppb or ppt (parts per billion or trillion) range; this combined limitation of both very low sample amount and very low concentrations within the sample creates special analytical demands, which the authors clearly address. They describe microscale sample preparations, including hydrolysis (for proteins and peptides), extraction (for free amino acids), ion-exchange cleanup, and derivatization, providing much practical information on each procedure. For capillary GC separation, their use of splitless and on-column injections is described, as limited sample quantity prohibited use of the split technique. The authors point out that both splitless and on-column injections permit quantitative analysis of amino acids, while split injection is unreliable at least for the higher-boiling-point amino acids.

The oxazolidinones are the subject of another chapter in our treatise [3, Vol. 3, Chap. 5] (see also Refs. 34 and 35). P. Hušek, of the Endocrinology Institute, Prague, Czechoslovakia, has developed a novel derivative; he considers alternatives to the esterification–acylation approach to amino acid derivatization and presents possibilities for the simultaneous derivatization of β-amino and carboxyl groups. He investigated the use of perhalogenated acetones, i.e., 1,3-dichloro-1,1,3,3-tetrafluoroacetone (DCTFA), for forming cyclic derivatives thus developing a "single-medium, two-reagent" reaction approach for amino acid derivatization. Amino acids that do not contain side-chain reactive groups can be analyzed after cyclization only, while those with reactive side-chain groups are analyzed after treatment with an acylating agent. Hušek clearly describes the derivatization procedures and the chromatographic analysis after cyclization and acylation. He also discusses applications of his analytical method to analysis of physiological fluids; recommends a practical procedure for isolation of free amino acids from 50 to 100 μL of serum or urine, and presents a rapid procedure for the determination of thyroid hormonal compounds, diiodothyronine, triiodothyronine, and thyroxine, and their deamination and decarboxylation products.

Capillary GC, invented by **M. J. E. Golay**, was first described in 1957. At the beginning, capillary columns were made mainly of stainless steel or other metals, but with the introduction of fused-silica columns in 1979 by Dandeneau and Zerenner, fused silica has largely replaced all other column materials. Our study took advantage of the fact that two different derivative types have been thoroughly investigated and are now well established for packed column amino acid analysis: the N-TFA n-butyl esters developed by Gehrke et al. [3, Vol. 1, Chaps. 1–3] (see also Refs. 3 and 4), and the N-HFB isobutyl derivatives primarily developed by MacKenzie and coworkers [3, Vol. 1, Chap. 4; 9,10].

C. W. Gehrke and R. W. Zumwalt have demonstrated [3, Vol. 3, Chap. 7; 36] that fused-silica bonded phase columns are a very effective means for separating and quantitatively measuring amino acids in protein hydrolysates as both the N-TFA n-butyl and N-HFB isobutyl derivatives. Comparisons are presented of columns with differing liquid-phase polarities, providing information on the effect of polarity on separation of particular amino acids, and evaluation of split, splitless, and on-column injection methods revealed the on-column injection technique to be most reproducible. Fused-silica capillary columns with chemically bonded phases of 0.25 μm film thickness [J&W Durabond DB-1, methyl silicone, and DB-5, bonded phenyl (5%) methylsilicone] were used to separate and quantitate the amino acid derivatives. All the protein amino acid N-TFA n-butyl esters were separated on a short (20 m × 0.25-mm-i.d.) DB-5 column, while only the derivatives of lysine and tyrosine were not well resolved on a DB-1 column of the same length and diameter. Conversely, the DB-5 column gave better separation of all the protein amino acid N-HFB isobutyl esters than did the DB-1 column, with the DB-5 column providing an improved lysine/tyrosine separation. A number of analytical applications on diverse sample types are presented in which data from fused-silica capillary GC analyses are compared to results obtained form IEC.

Amino acid analysis of commercial animal feeds, feed ingredients, lysozyme, β-lactoglobulin, and ribonuclease by capillary GLC as both the N-TFA n-butyl esters and the N-HFB isobutyl esters were in agreement with the IEC results [36]. The comparison clearly demonstrates that the capillary GC method is a powerful tool for the measurement of amino acids. We describe the uniting of established methods of amino acid derivatization that were developed by Kaiser et al. in 1974 [4] and MacKenzie in 1974 [9], with the more recent development of fused-silica immobilized phase capillary columns into a high-resolution, precise, and *accurate* method for measurement of amino acids.

A representative chromatogram by ion exchange chromatography of a complex physiological fluid for amino acids is presented in Figure 7.9. This separation was made in the Experiment Station Chemical Laboratories, University of Missouri.

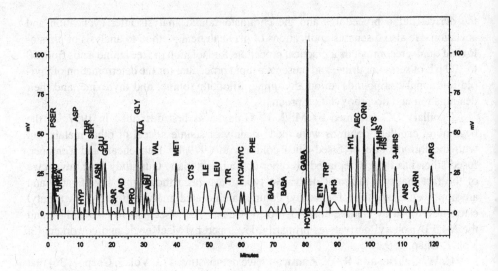

For peak identification, see Abbreviations.

Abbreviations:			
PSER	O-Phosphoserine	LEU	Leucine
TAU	Taurine	TYR	Tyrosine
PETN	O-Phosphoethanolamine	HYC	Cystathionine
UREA	Urea	AHYC	Allocystathionine
ASP	Aspartic acid	PHE	Phenylalanine
HYP	Hydroxyproline	BALA	β-Alanine
THR	Threonine	BABA	β-Aminoisobutyric acid
SER	Serine	GABA	γ-Amino-n-butyric acid
ASN	Asparagine	HCYS	Homocystine
GLU	Glutamic Acid	ETN	Ethanolamine
GLN	Glutamine	TRP	Tryptophan
SAR	Sarcosine	NH3	Ammonia
AAD	α-Aminoadipic acid	HYL	δ-Hydroxylysine
PRO	Proline	AEC	S-(2-Aminoethyl)-L-cysteine
GLY	Glycine	ORN	Ornithine
ALA	Alanine	LYS	Lysine
CIT	Citrulline	1-MHIS	1-Methylhistidine
ABU	α-Amino-n-butyric acid	HIS	Histidine
VAL	Valine	3-MHIS	3-Methylhistidine
MET	Methionine	ANS	Anserine
CYS	Cystine	CARN	Carnosine
ILE	Isoleucine	ARG	Arginine

Figure 7.9. Ion-exchange chromatogram of amino acids in a complex physiological fluid. Chromatogram of a complex physiological fluid, obtained on a Hitachi L8800 Amino Acid Analyzer. This figure shows the recording at 570 nm; the parallel chromatogram recorded at 440 nm is not shown. The analysis time was 2 h. (Courtesy of Dr. T. Mawhinney of the Experiment Station Chem. Lab., Univ. Missouri.)

7.D. CHROMATOGRAPHY IN SPACE SCIENCES—GLC AND IEC OF APOLLO MOON SAMPLES [1] (Charles W. Gehrke)

The lunar samples from Apollo flights 11, 12, 14–17 provided the students of chemical evolution with an opportunity of examining extraterrestrial materials for evidence of early prebiological chemistry in the solar system [37]. Our search was directed to water-extractable compounds with emphasis on amino acids. Gas chromatography, ion-exchange chromatography and gas chromatography combined with mass spectrometry were used for the analysis. The characterization of carbon compounds indigenous to the lunar surface is of particular interest as these investigations could result in findings which would advance our knowledge of the processes of chemical evolution. The Apollo missions have provided us with the requisite extraterrestrial material for study. Figures 7.10 and 7.11 give a photographic glimpse of this important time in our scientific history.

Our search for water-extractable organic compounds, with emphasis on amino acids, in Apollo 17 fines, and a summary of the analysis for amino acids in samples from Apollo flights 11, 12, 14–17, are presented in our 1975 publication [37]. The chromatography and procedural techniques are also described in detail in Ref. 1, Chap. 4 (Section D). Along with a discussion on methodological strategies and examples of chromatographic data from Apollo flights 11, 12, 14–17, is a section on the controversy of contamination

Figure 7.10. Photo of Drs. Robert W. Zumwalt (left), Charles W. Gehrke (center), and Dr. David Stalling (right) at NASA, Ames Research Center, California, with Charles W. Gehrke holding Apollo 11 sample (1969). [Reproduced from C. W. Gehrke, R. W. Wixom, and E. Bayer (Eds.), *Chromatography: A Century of Discovery*, Vol. 64, Elsevier, Amsterdam, 2001 and Supplement S12 on the Internet (http://www.chemweb.com/preprint), with permission from Elsevier, Amsterdam.]

Figure 7.11. Photo of GLC chromatograph for NASA, the four-column Varian 2100 biomedical gas chromatograph used in the Apollo 11, 12, 14–17 studies. It is now housed in the Smithsonian Institution, Washington, DC. A second instrument is housed in the Boone Country Historical Society, Columbia, MO. Charles W. Gehrke (right), Robert W. Zumwalt (center), and Kenneth C. Kuo (left) are shown in the photo. [Reproduced from C. W. Gehrke, R. W. Wixom, and E. Bayer (Eds.), *Chromatography: A Century of Discovery*, Vol. 64, Elsevier, Amsterdam, 2001 and Supplement S12 on the Internet (http://www.chemweb.com/preprint), with permission from Elsevier, Amsterdam.]

of moon rock and the National Academy of Science experiment that followed with Apollo 14 to address the contamination issues [1]. The chromatographic methods and strategies which were developed and used in investigations on the lunar regolith samples are also directly applicable to the search for life molecules (amino acids and genetic code) in other space specimens such as the Mars meteorites, and in the Mars samples being brought to the Earth.

7.E. CHROMATOGRAPHY OF NUCLEOSIDES—CANCER BIOMARKERS AND STRUCTURAL CHARACTERIZATION (Charles W. Gehrke)

In the 1960s, we were developing charcoal adsorption cleanup of samples and GC methods for the genetic bases and modified nucleosides in preparation for the search for these molecules in the returned moon samples by NASA. Thus, it was a small step to applications and to problems in nucleic acid biochemistry. Our studies in nucleic acid biochemistry were directed toward a better understanding of how the chemical structure of nucleic acids are correlated with their unique biological functions. This information can then be used to gain a deeper insight into how normal cells regulate their metabolic activities, allows speculation on how they evolved to their respective biological role(s), and permits potentially a correlation of the altered structures of nucleic acids in abnormal or diseased states to biological function. An understanding of how cells behave normally and in the diseased state provides the basis for the development of rational therapeutics and improved diagnostic tools.

Studies have been and are being undertaken in many laboratories (1980–2000) on nucleic acid metabolites as "cancer markers," and of chemical carcinogens and mutagens adducted to nucleic acids for assessment of human exposure to environmental insults. Methodological limitations have hampered the advancement and exploitation of using modified nucleosides and their signals in routine tests in clinical chemistry or as important determinant life molecules in biochemical research. The development of high-resolution chromatographic methods for qualitative identification and quantitative measurement of an array of nucleosides along with chemical information on nucleic acid components has been a challenge to analytical biochemists since the beginning of the 1960s. Information on modification in nucleosides may help decipher the functions of tRNAs, mRNA, rRNA, and DNA. HPLC has evolved into a powerful research tool since the mid-1980s to solve structure problems and in biochemical analysis.

Since the late 1970s, our laboratory has made major progress on reversed-phase high-performance liquid chromatography and UV–photodiode array detection (RPLC-UV) nucleoside analysis and has developed comprehensive chromatographic methods and quantitative enzymatic RNA hydrolysis procedures. In total, 67 known nucleosides can be identified, and 31 ribonucleosides and 6 deoxynucleosides can be quantified directly in a single chromatographic run from an enzymatic hydrolysate of RNAs, DNAs, and physiological fluids. In collaborative efforts with scientists across the world, we have applied these methods in a number of interesting investigations. Next, we briefly introduce the reversed-phase high-performance liquid chromatography and UV–photodiode array detection (RPLC-UV) for ribo- and deoxynucleosides and place emphasis on research in the areas of biomarkers of cancer and normal metabolism and in structural characterization.

7.E.1. Reversed-Phase Liquid Chromatography (RPLC)

Information on HPLC instrumentation; chromatographic parameters for high-resolution, high-speed, and high-sensitivity separation of nucleosides; analytical and semipreparative enzymatic hydrolysis of nucleic acids; and cleanup procedures (ultrafiltration procedure and phenylboronate gel column cleanup) for ribonucleosides in physiological fluids are presented in our papers [38–40].

It is a challenge to the analytical biochemist to simultaneously separate and measure such a large number of nucleosides in a complex biological matrix. One of the major problems for nucleoside chromatography is to obtain the needed reference molecules so that the information for the essential qualitative and quantitative analytical references can be established. Only about 20 modified ribonucleosides can be obtained through commercial sources; thus we have standardized the chromatographic retention times, obtained RPLC–UV spectra, and determined the molar response factors for a large number of ribonucleosides (scientists in their respective laboratories need to standardize and calibrate their analytical system for modified nucleoside analysis in a broad range of biological matrices). To overcome this limitation, we selected three unfractionated

TABLE 7.1. Nomenclature of Ribonucleosides and Index Numbers

IUPAC Name	Letter Symbol	Index Number
Adenosines		
Adenosine	A	4
2'-*O*-Methyladenosine	Am	61
1-Methyladenosine	m^1A	21
1-Methyl-2'-*O*-methyladenosine	m^1Am	
2-Methyladenosine	m^2A	66
2-Thioadenosine	s^2A	
2-Methylthioadenosine	ms^2A	
3-Methyladenosine	m^3A	
1,3-Dimethyladenosine	m^1m^3A	
5'-Methylthioadenosine	$ms^{5'}A$	
1,N^6-Dimethyladenosine	m^1m^6A	
N^6-(*N*-Formyl-α-aminoacyl)adenosine	f^6A	
N^6-Methyladenosine	m^6A	67
N^6-Methyl-2-methylthioadenosine	ms^2m^6A	
N^6,N^6-Dimethyladenosine	m_2^6A	74
N^6-Methyl-2'-*O*-methyladenosine	m^6Am	71
2-Hydroxyadenosine	o^2A (isoG)	
N^6-Carbamoyladenosine	nc^6A	
N^6-Threoninocarbonyladenosine	tc^6A (t^6A)	63
N^6-Methyl-N^6-threoninocarbonyladenosine	mtc^6A (mt^6A)	70
N^6-Threoninocarbonyl-2-methylthioadenosine	ms^2tc^6A (ms^2t^6A)	72
N^6-Glycinocarbonyladenosine	gc^6A (g^6A)	50
N^6-Methyl-N^6-glycinocarbonyladenosine	mgc^6A (mg^6A)	
N^6-(Δ^2-Isopentenyl)adenosine	i^6A	78
N^6-(Δ^2-Isopentenyl)-2-methylthioadenosine	ms^2i^6A	80
N^6-(*cis*-4-Hydroxyisopentenyl)adenosine	cis oi^6A	
N^6-(4-Hydroxyisopentenyl)-2-methylthioadenosine	ms^2oi^6A	79
9-(2'-*O*-Ribosyl-β-D-ribofuranosyl)adenine	rA	
Inosine	I	29
1-Methylinosine	m^1I	43
2-Methylinosine	m^2I	
7-Methylinosine	m^7I	16
9-β-D-Ribofuranosylpurine (Nebularine)	neb	
7-β-D-Ribofuranosylhypoxanthine		
Cytidines		
Cytidine	C	1
2'-*O*-Methylcytidine	Cm	27
2-Lysinocytidine (Lysidine)	k^2C?	
2-Thiocytidine	s^2C	20
3-Methylcytidine	m^3C	18
N^4-Methylcytidine	m^4C	22
N^4-Methyl-2'-*O*-methylcytidine	m^4Cm	
N^4-Hydroxymethylcytidine	om^4C	

(Continued)

TABLE 7.1. *Continued*

IUPAC Name	Letter Symbol	Index Number
N^4-Methyl-2-thio-2'-O-methylcytidine	m^4s^2Cm	
N^4-Acetylcytidine	ac^4C	48
5-Methylcytidine	m^5C	23
5-Methyl-2'-O-methylcytidine	m^5Cm	
5-Hydroxymethylcytidine	om^5C	12

Guanosines

Guanosine	G	3
2'-O-Methylguanosine	Gm	45
1-Methylguanosine	m^1G	46
N^2-Methylguanosine	m^2G	49
3-Methylguanosine	m^3G	
7-Methylguanosine	m^7G	28
N^2,N^2-Dimethylguanosine	m_2G	57
N^2,N^2-Dimethyl-2'-O-methylguanosine	m_2Gm	
$N^2,N^2,7$-Methyltrimethylguanosine	m_2m^7G	
Queuosine	Q	40
ß-D-Mannosylqueuosine	$_{man}Q$	41
ß-D-Galactosylqueuosine	$_{gal}Q$	42
Xanthosine	X	32
1-Methylxanthosine	m^1X	32
2-Methylxanthosine	m^7X	32

Uridines

Uridine	U	2
2-Thiouridine	s^2U	33
2-Thio-2'-O-methyluridine	s^2Um	
2-Selenouridine	Se^2U	
3-(3-Amino-3-carboxypropyl)uridine	acp^3U, (nbt^3U)	32
3-Methyluridine	m^3U	37
4-Thiouridine	s^4U	36
2,4-Dithiouridine	s^2s^4U	
4-Thiouridine disulphide	$(s^4U)_2$	
5-(β-D-Ribofuranosyl)uracil (Pseudouridine)	Ψ	6
5-(2'-O-Methyl-ß-D-ribofuranosyl)uracil, (2'-O-Methylpseudouridine)	Ψm	39
5-(β-D-Ribofuranosyl)-N^1-methyluracil, (1-Methylpseudouridine)	$m^1Ψ$	17
5-(2'-O-Methyl-ß-D-ribofuranosyl)-N^1-methyluracil, (1-Methyl-2'-O-methylpseudouridine)	$m^1Ψm$	
5,6-Dihydrouridine	hU (D)	5
5-Methyl-5,6-dihydrouridine	m^5hU (m^5D)	
5-Methyluridine	$m^5U(T)$	30
5-Methyl-2'-O-Methyluridine	m^5Um (Tm)	53
5-Methyl-2-thiouridine	m^5s^2U (s^2T)	52

(Continued)

TABLE 7.1. *Continued*

IUPAC Name	Letter Symbol	Index Number
5-Hydroxyuridine	o^5U	11
5-Carboxyhydroxymethyluridine	com^5U	
5-Carboxymethyluridine	cm^5U	7
5-Carboxymethyl-2-thiouridine	cm^5s^2U	
5-Methoxyuridine	mo^5U	34
5-Methoxy-2-thiouridine	mo^5s^2U	55
5-Aminomethyluridine	nm^5U	
5-Aminomethyl-2-thiouridine	nm^5s^2U	
5-Methylaminomethyluridine	mnm^5U	9
5-Methylaminomethyl-2'-O-methyluridine	mnm^5Um	
5-Methylaminomethyl-2-thiouridine	mnm^5s^2U	25
5-Methylaminomethyl-2-selenouridine	mnm^5Se^2U	
5-Carboxymethylaminomethyluridine	$cmnm^5U$	8
5-Carboxymethylaminomethyl-2'-O-methyluridine	$cmnm^5Um$	
5-Carboxymethylaminomethyl-2-thiouridine	$cmnm^5s^2U$	24
5-Carbamoylmethyluridine	ncm^5U	14
5-Carbamoylmethyl-2'-O-methyluridine	ncm^5Um	
5-Carbamoylmethyl-2-thiouridine	ncm^5s^2U	
5-Methoxycarbonylmethyluridine	mcm^5U	44
5-Methoxycarbonylmethyl-2-thiouridine	mcm^5s^2U	60
5-Methylcarboxymethoxyuridine	$mcmo^5U$	54
5-Methylcarboxymethoxy-2-thiouridine	$mcmo^5s^2U$	68
6-Carboxyuridine (Oridine)	c^6U (O)	
Hydroxywybutosine	$Y_{OH,oyW}$	75
Wybutosine	$Y_{t,yW}$	76
Wyosine	Y, W	77

tRNAs: *E. coli*, brewer's yeast, and calf-liver as reference sources of the nucleosides. Each of these tRNAs contain unique as well as common nucleosides and provide an array of modified nucleosides that are often encountered by researchers. Some minor differences in the modified nucleoside profile may be observed in these three tRNAs from different sources, especially for *E. coli* tRNAs; this problem can be resolved by using a reliable supplier or by standardization of a selected lot of tRNAs obtained in large quantity and of good homogeneity. The nucleoside peaks are identified by an assigned index number, which essentially corresponds to their respective elution order. Table 7.1 lists the IUPAC names, symbols, and index numbers of the nucleosides that have been determined by RPLC-UV. Other ribonucleosides that have not yet been characterized by RPLC-UV are also included in this table. A total of 67 ribonucleosides have been chromatographically and spectrometrically characterized. Table 7.2 presents the stepwise gradients for two solvents and time to achieve the separation. The high-resolution HPLC chromatograms at 254 nm of reference nucleosides from unfractionated *E.coli* tRNAs, rat liver tRNAs, and yeast tRNAs are shown in Figures 7.12–7.14.

TABLE 7.2. High-Resolution RPLC Elution Sequence of Ribonucleosides

hU, 4.4	ψ, 4.6	cm^5U, 5.0	$cmnm^5U$, 5.1	mnm^5U, 5.4	C, 5.9	cmo^5U, 6.1	o^5U, 6.4	om^5C, 6.7	ncm^5U, 7.2	U, 8.4	m^7I, 8.6	$m^1\psi$, 9.2	m^3C, 9.4
om^5U, 9.5	s^2C, 9.6	m^1A, 10.6	m^5C, 11.4	m^4C, 11.9	$cmnm^5s^2U$, 12.1	mnm^5s^2U, 12.6	2,5-PCNR, 14.6	4,3-PCNR, 14.8	Cm, 15.5	m^7G, 18.0			
I, 18.5	m^5U, 19.8	G, 19.0	acp^3U, 20.1	X, 20.4	s^2U, 21.1	mo^5U, 21.6	s^4U, 24.0	m^3U, 24.5	Um, 25.3	Ψm, 26.9	Q, 27.4	^{man}Q, 28.1	^{gal}Q, 28.7
mcm^5U, 29.8	Gm, 30.4	m^1G, 31.0	ac^4C, 32.0	m^2G, 32.4	A, 32.6	g^6A, 33.5	m^5s^2U, 34.4	m^5Um, 34.8	$mcmo^5U$, 35.0	mo^5s^2U, 35.8	m^1I, 29.4	rA, 37.0	
m^2m^2G, 37.8	mcm^5s^2U, 39.7	Am, 40.1	t^6A, 41.6	m^2A, 45.2	m^6A, 46.2	$mcmo^5s^2U$, 47.4	mt^6A, 48.3	m^6Am, 49.4	ms^2t^6A, 51.9	m^6m^6A, 56.3			
Y_{OH}, 59.0	Y_w, 65.0	cis io⁶A, 70.0	i⁶A	cis ms²io⁶A, ms²i⁶A, 84.6									

Total 70 molecules

2,5-PCNR is 2-pyridone-5-carboxamide-N^1-ribofuranoside
4,3-PCNR is 4-pyridone-3-carboxamide-N^1-ribofuranoside

Ribonucleosides with elution times not assigned: m^1Am, ms^2A, m^3A, m^1m^3A, ms^2m^6A, o^2Am, nc^6A, mg^6A, m^6Cm, m^5Cm, om^4C, m^2m^7G, m^2I, m^1X, m^7X, $m^1\Psi m$, s^2Um, Se^2U, s^2s^4U, $(s^4U)_2$, m^5hU, com^5U, cm^5s^2U, nm^5s^2U, mnm^5Se^2U, mnm^5Um, mmm^5Um, $cmnm^5Um$, ncm^5Um, c^6U, 2-Ribosylguanine, Lysidine, and 2,4-Diaminopyrimidine nucleoside; **total 34 molecules**

HPLC OF NUCLEOSIDES IN UNFRACTIONATED E. coli tRNAs

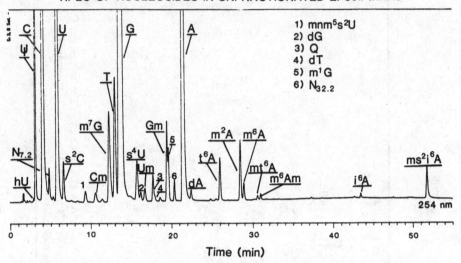

Figure 7.12. HPLC of nucleosides in unfractionated *E. coli* tRNAs. [Reproduced from C. W. Gehrke, R. W. Wixom, and E. Bayer (Eds.), *Chromatography: A Century of Discovery*, Vol. 64, Elsevier, Amsterdam, 2001 and Supplement S12 on the Internet (http://www.chemweb.com/preprint), with permission from Elsevier, Amsterdam.]

HPLC OF NUCLEOSIDES IN UNFRACTIONATED RAT LIVER tRNAs

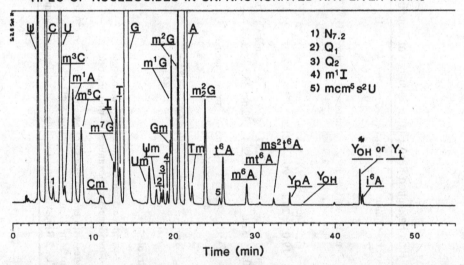

Figure 7.13. HPLC of nucleosides in unfractionated rat liver tRNAs. [Reproduced from C. W. Gehrke, R. W. Wixom, and E. Bayer (Eds.), *Chromatography: A Century of Discovery*, Vol. 64, Elsevier, Amsterdam, 2001 and Supplement S12 on the Internet (http://www.chemweb.com/preprint), with permission from Elsevier, Amsterdam.]

HPLC OF NUCLEOSIDES IN UNFRACTIONATED BREWER'S YEAST tRNAs

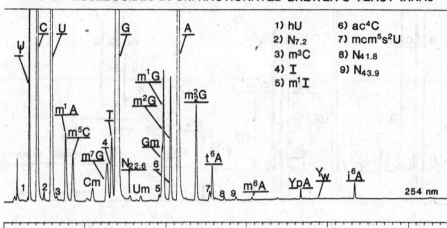

1) hU
2) $N_{7.2}$
3) m^3C
4) I
5) m^1I
6) ac^4C
7) mcm^5s^2U
8) $N_{41.8}$
9) $N_{43.9}$

254 nm

Time (min)

Figure 7.14. HPLC of nucleosides in unfractionated Brewer's yeast tRNAs. [Reproduced from C. W. Gehrke, R. W. Wixom, and E. Bayer (Eds.), *Chromatography: A Century of Discovery*, Vol. 64, Elsevier, Amsterdam, 2001 and Supplement S12 on the Internet (http://www.chemweb.com/preprint), with permission from Elsevier, Amsterdam.]

7.E.2. RPLC-UV Methods for Total Nucleoside Composition of RNAs and DNAs

We have developed nucleoside chromatography protocols for a broad array of RNAs and DNAs, and they have been applied extensively. In general, RNA nucleoside chromatography requires emphasis on resolution and flexibility, and for DNA the emphasis is on accuracy and speed. There is also an increasing need in biochemical analysis for high sensitivity. With new instrumentation and column technology, an increased LC-UV sensitivity of more than 10-fold can be achieved so that low picograms of nucleosides can be quantitated routinely. Unfractionated tRNA constitutes one of the most complicated mixtures of biopolymers known, and high resolution is required for this analysis (see Figs. 7.12–7.14).

Isoacceptor tRNAs are usually available only in very small amounts (less than a few micrograms); however, an advantage with single-species tRNAs is that they are less complicated in composition. For analysis of single-species tRNAs, intermediate-resolution and higher-sensitivity (high-speed) protocols are generally used. An accurate identification and quantitation of the total nucleoside composition is very important in providing supplementary and confirmatory information in support of tRNA sequence studies. Figure 7.15 shows the separation of nucleosides in tRNA^{Leu} from bovine serum. The quantitative results from five isoacceptor tRNAs obtained by RPLC-UV have been compared to sequence analysis results [41]. The lower m^7G value is indicative

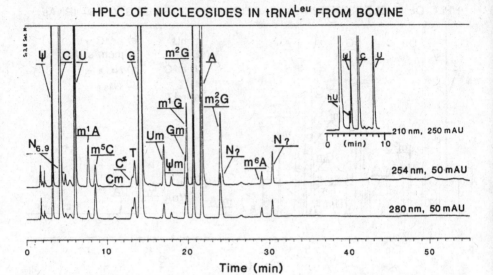

Figure 7.15. HPLC of nucleosides in tRNA^{Leu} from bovine serum. [Reproduced from C. W. Gehrke, R. W. Wixom, and E. Bayer (Eds.), *Chromatography: A Century of Discovery*, Vol. 64, Elsevier, Amsterdam, 2001 and Supplement S12 on the Internet (http://www.chemweb.com/ preprint), with permission from Elsevier, Amsterdam.]

of the instability of m^{7}G at alkaline pH during enzymatic hydrolysis. It is of interest that differences of one residue number for uridine in tRNA^{Phe} and uridine and guanosine values in tRNA^{Val} were observed from the two analytical methods analytical methods. We present the mol% values of all the nucleosides in four bovine isoaccepting tRNAs in Chap. 1 of Ref. 42. These four tRNAs were isolated by Gerard Keith's group at the Institute for Molecular and Cellular Biology (IBMC) in Strasbourg, France. Their sequences were not yet determined at the time of HPLC analysis. These mammalian tRNAs have considerable more modifications than do the tRNAs from *Escherichia coli* (*E. coli*), and two unknown modified nucleosides were observed in tRNA^{Leu}.

Ribosomal RNA Nucleoside Analysis. Ribosomal RNA (rRNA) is a high-molecular-weight RNA. In *E. coli* the 70S rRNA has a molecular weight of 2.75×10^6 amu (atomic mass units), and the small subunits, 16S rRNA and 23S rRNA, have 1542 and 4718 residues, respectively. Only 10 methylated nucleosides have been reported in the 16S and 23S rRNAs. To accomplish the chromatography of rRNA for composition analysis, it was necessary to separate and measure one modified nucleoside residue in approximately 5000 nucleotides. This demands a high column capacity so that a large amount of sample (≥ 100 µg) can be injected without loss of resolution. Both of the high-resolution and high-speed chromatographic protocols described for tRNA nucleoside composition analysis have the adequate capacity to meet this requirement for rRNA analysis. Some deoxyribonucleosides were found in the enzymatic hydrolysates of the tRNA samples. However, their presence does not interfere with the measurement of any known modified ribonucleoside. This separation demonstrates the high selectivity of the RPLC so that the respective deoxy- and ribonucleoside are differentiated. RPLC showed qualitative and quantitative differences of modification in both 16S rRNA and 23S rRNA as compared to the literature values. In 16S rRNA we found one additional residue of Ψ, m^5C, and m^2G. Two nucleosides, Gm and m^4Cm, were not found. From 23S rRNA, four additional Ψ, two of m^4C, one of m^5C, two of m^2G, and one of m^2A were found by RPLC. A number of other modifications are in good agreement with the literature values.

Messenger RNA Nucleoside Analysis. Messenger RNAs from viral and eukaryotic cells contain a unique structure known as "caps," which consist of an inverted 7-methylguanosine (m^7Guo) linked to the penultimate nucleoside through a $5'-5'$ triphosphate bridge. These mRNAs usually have a very low amount of internal nucleoside modification ($< \frac{1}{1,000}$). A highly selective RPLC-UV separation using a microscale anion-exchange column was developed for isolation of the cap structures to enhance the resolution and sensitivity of the separation and measurement. The complete method is reported in Ref. 43.

Deoxynucleosides in DNAs. Determination of the molar composition of the major and modified deoxynucleosides in high-molecular-weight DNAs requires a high degree of accuracy and sensitivity. Modified nucleosides in DNA, such as 5-methyldeoxycytidine (m^5dC), 6-methyl-deoxyadenosine (m^6dA), and 4-methyldeoxycytidine (m^4dC), are normally present at 0.1–2 mol% level. Several separation systems were developed and used in our laboratory. The method that we use is dependent on the sample matrices (i.e., presence of RNA, deoxyinosine (dI), inosine (I), nucleobases, and other UV peaks) and amount of DNA sample available. An optimum amount of DNA is 10 μg. The best chromatographic system for the separation of deoxynucleosides is a two-buffer, single-ramp gradient, using a 150 × 4.6-mm Supelcosil LC-18S column. With this column a complete separation can be achieved in less than 15 min. Dual-wavelength quantitation and high-quality data reduction software are essential for the analysis. The deoxynucleoside reference compounds obtained from commercial sources do not have the required purity to obtain the accuracy pair ratio [i.e., $(dC + m^5dC)/dG = 1.000$, and $dT/(dA + m^6dA) = 1.000$] from high-molecular-weight DNAs. Quantitation of the nucleoside composition of a large number of isolated DNA oligomer fragments and synthesized oligomers requires a high sensitivity. In this case, a 5-cm or 3-cm regular bore (3.9–4.6 mm) with 3 or 5 μm particle size columns used in an isocratic separation mode provides the separation in less than 10 min with a fivefold increase in sensitivity. The published HPLC protocols for quantitation of major and modified nucleosides in DNA [39,44] include the precision and linearity of the method. The high-resolution separation of ribo and deoxynucleosides is presented in Figure 7.16.

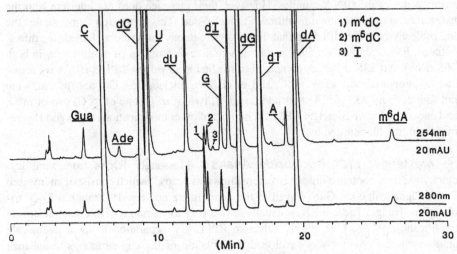

HIGH RESOLUTION HPLC OF RIBO- AND DEOXYNUCLEOSIDES

Figure 7.16. High-resolution HPLC of ribonucleosides and dexoynucleosides. [Reproduced from C. W. Gehrke, R. W. Wixom, and E. Bayer (Eds.), *Chromatography: A Century of Discovery*, Vol. 64, Elsevier, Amsterdam, 2001 and Supplement S12 on the Internet (http://www.chemweb.com/preprint), with permission from Elsevier, Amsterdam.]

7.E.3. Modified Nucleosides in Normal Metabolism and as Cancer Markers

Background. Ernest Borek stated, in 1985, the hope of finding some unique metabolic products or components of malignant cells circulating in body fluids, which can be measured is as old as modern biochemistry [45]. Morton K. Schwartz of the Sloan Kettering Institute coined the term "tumor marker" for such a product. Before we consider whether we have fulfilled such a hope, we ought to define what a "tumor marker" should be. The requirements for an effective "tumor marker" are manifold:

- It should be specific for a malignancy.
- It should provide a minimum of false-positives and false-negatives.
- It should indicate the extensiveness of the malignancy.
- It should preferably diminish or hopefully disappear after effective therapy.

At an international conference held in Vienna under the auspices of the Society for Early Detection of Cancer (1971), it was reported that there were close to 90 reported different tumor markers. Unfortunately, none of these putative "tumor markers" even partly meet the qualifications that we have set above [31].

Most of the tumor markers in use today are proteins. Proteins are the peripheral end products of the molecular mechanisms of every cell. A mammalian cell is endowed with the capability of producing perhaps 10,000 or more different proteins. Unless we chance upon a protein that is causal of a malignancy, or that is universally aberrantly expressed in malignant tissues, looking for protein products to qualify as tumor markers to meet the abovementioned requirements may be hopeless [47,45].

Promising markers are in tRNAs of tumor tissue. The finding of aberrant tRNA-methylating enzymes in tumor tissue prompted the study of the tRNAs themselves. Surprisingly, only a few of the tRNAs in the malignant tumor were found to be different in structure from those in the normal tissue counterpart [46,48]. Guy Dirheimer of Strasbourg isolated 18 different tumor-specific tRNAs and found modification on them [49] different from those in normal counterparts. On the other hand, we have determined with the aid of Japanese colleagues that the primary sequence is the same [50]. Perhaps to enable it to perform its many functions, tRNA is endowed with an extraordinarily complex structure. Its primary sequence consists of about 80 of the four major bases found in other RNAs: adenine, cytosine, guanine and uracil. In addition to these major bases, tRNA contains a large variety of modified bases that are unique to it. The number of modified bases increase with the complexity of the organism. Thus, for example, *Escherichia coli* tRNA may contain only two or three; yeast tRNA may contain five or six, and mammalian tRNA may contain modified bases representing as much as 20% of the total. It has also been shown that the tRNA methyltransferases are abnormally hyperactive in every malignant tissue [48]. Ernest Borek found that the level of excretion of the nucleosides in urine when followed before, during, and after therapy in a malignancy responded well to chemotherapy; and that within 5 days of commencement of therapy in six patients with Burkitt's lymphoma, the excretion levels in urine returned to and remained normal as long as the subjects were in remission.

There have been reports since early 1950 that cancer patients excrete elevated levels of methylated purines and pyrimidines as well as other modified nucleosides [46]. Ample evidence indicated that increased tRNA methylase activity in neoplastic cells was a common and consistent finding, and that increased urinary excretion of modified bases from cancer patients and tumor-bearing animals had also been reported. Methylation of the bases in tRNA had been found to occur after the macromolecule is formed, and of particular interest, these methylated compounds were not reincorporated into the tRNA molecule but thought to be excreted intact. It has been suggested that the high turnover of a subpopulation of tRNA is the major reason for increased excretion of modified nucleosides by cancer patients [48]. The measurement of modified ribonucleosides in body fluids as biological markers of cancer resulted largely from the studies of tRNAs by the late Ernest Borek of Columbia University. In 1974, Gordon Zubrod, then director of National Cancer Institute (NCI), appreciated the possible value of this concept and led to a series of NCI contracts with the University of Missouri to develop high-resolution quantitative chromatographic methods of modified nucleosides for use in tumor marker studies. In the course of our investigation, we developed gas chromatographic and RPLC-UV methods for measuring modified nucleosides, and later the methods were further improved for measuring urinary and serum nucleosides. The RPLC-UV method is far better than the GC method for the highly water-soluble nucleosides, and hence, was used in analysis for a majority of the clinical studies. The method is presented in detail in Refs. 38–41 and 51.

Our Chromatographic Methods. Our research on tRNA catabolites in urine and serum has centered on analysis of the modified nucleosides following isolation of the nucleosides by boronate gel affinity chromatography. Advancements in the isolation, identification, and measurement of modified nucleosides has been striking, and are now providing greater insights into the value of modified nucleosides as potential tumor markers in following the course of cancer and treatment. Numerous research groups in the United States, Europe, and Japan have studied modified nucleosides and their potential relationships to cancer (see Fig. 7.17 and also the comprehensive review by Zumwalt et al. [51]). Trewyn and Grever [52] have provided another excellent review of urinary nucleosides in patients with leukemia; they reviewed the available literature and discussed laboratory analyses, including methods, reference values, and multivariate analyses; and clinical studies covering nonmalignant disease and infection, acute leukemia (childhood and adult), and chronic leukemias. They concluded that measurement of urinary nucleoside excretion offers a potential tool for monitoring disease activity in patients with acute lymphocytic leukemia (ALL), chronic mylenogenous leukemia (CML), and perhaps chronic lymphocytic leukemia (CLL). They also point out that additional work is necessary in following serial determinations of nucleosides at frequent intervals in patients with different types of leukemia to assess the true value of these compounds as an accurate monitor of disease activity within the individual

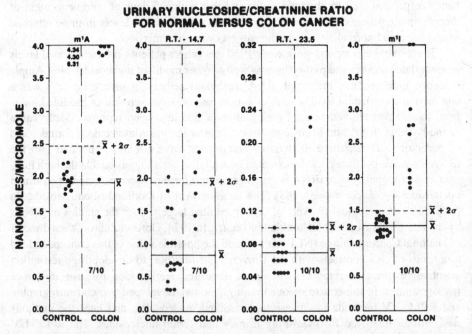

Figure 7.17. Urinary nucleoside/creatinine ratio for normal versus colon cancer. [Reproduced from C. W. Gehrke, R. W. Wixom, and E. Bayer (Eds.), *Chromatography: A Century of Discovery*, Vol. 64, Elsevier, Amsterdam, 2001 and Supplement S12 on the Internet (http://www.chemweb.com/preprint), with permission from Elsevier, Amsterdam.]

patient. In the mid-1980s, we initiated investigations to study the correlation of the levels of serum-modified ribonucleosides with the clinical status of the patient. Longitudinal serum samples were collected from leukemia, lymphoma, and lung cancer patients. Four modified nucleosides were selected—Ψ, m^2m^2G, t^6A, and m^1I—to study the correlation between their levels in serum and the course of the disease. Serum pseudouridine levels showed a direct relationship to total RNA turnover. N^2N^2-dimethylguanosine and N^6-threoinocarbonyl-adenosine, which are found only in tRNA, showed that their concentrations in serum reflect the state of tRNA catabolism. 1-Methylinosine is a very interesting modified nucleoside; the concentration of serum m^1I in normal population is quite high (65 ± 21 nmol/mL) and is one of the commonly elevated nucleosides found in cancer patients [53].

The origin of serum 1-methylinosine is not completely clear at this time. It can be accounted for partially from direct tRNA turnover and deamination of m^1A by adenosine deaminase in serum. We also studied longitudinal collected normal human serum samples and found that the four target nucleosides levels in serum are constant during one day (7:30 a.m., 12:00 noon, and 5:00 p.m.), and over 14 days. For cancer patients we plotted the ratio of each nucleoside to the average concentration found from 94 normal subjects in percent. The results of the longitudinal studies from one selected leukemia patient, one lymphoma patient, and one lung cancer patient are presented in Figures 7.18–7.20. The patient description and correlation of clinical status and modified nucleoside levels are discussed below.

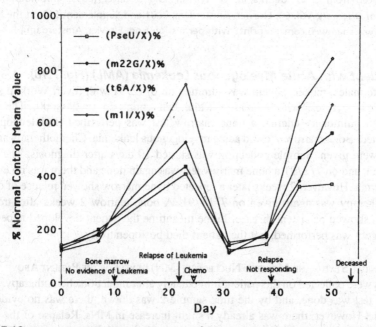

Figure 7.18. Serum modified nucleoside levels in serum from a leukemia (AML) patient. [Reproduced from C. W. Gehrke, R. W. Wixom, and E. Bayer (Eds.), *Chromatography: A Century of Discovery*, Vol. 64, Elsevier, Amsterdam, 2001 and Supplement S12 on the Internet (http://www.chemweb.com/preprint), with permission from Elsevier, Amsterdam.]

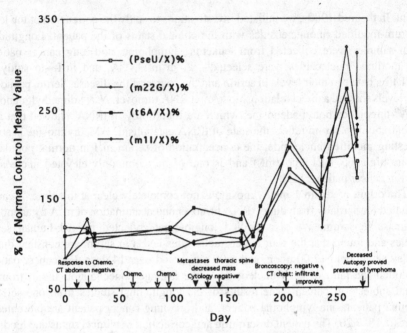

Figure 7.19. Serum modified nucleoside levels in a patient with non-Hodgkin lymphoma. [Reproduced from C. W. Gehrke, R. W. Wixom, and E. Bayer (Eds.), *Chromatography: A Century of Discovery*, Vol. 64, Elsevier, Amsterdam, 2001 and Supplement S12 on the Internet (http://www.chemweb.com/preprint), with permission from Elsevier, Amsterdam.]

Patient with Acute Myelogenous Leukemia (AML) (Fig. 7.18).

A 31-year-old white male, smoker, patient was admitted on 6/12/91 with fever. Workup showed severe leukocytoses, with increased white cell count up to 244,000. The patient went into pulmonary edema, a bone marrow test was performed, and leukophoresis was started. Bone marrow showed acute myelogenous leukemia. Chemotherapy and antibiotics were given. Sample collection was started 10 days after diagnosis. The patient recovered, and on 7/2/91 a bone marrow test was again done and there was no evidence of leukemia. However, 2 weeks later a repeated bone marrow showed relapse of disease. Chemotherapy was again given on 7/18/91. A bone marrow 2 weeks after treatment (day 38) showed persistent disease. In the meantime the patient developed appendicitis and surgery was performed, but the patient died postoperatively.

CLINICAL STATUS AND MODIFIED NUCLEOSIDE (MN) LEVELS OF THE PATIENT ABOVE. Serum samples were collected on this patient immediately after induction chemotherapy. A bone marrow test was done, and by the time sample 2 was drawn, there was no evidence of leukemia. However, there was already a slight increase in MNs. Relapse of the disease was clinically suspected and confirmed by bone marrow and there was a marked increase in all MNs. Reinduction chemotherapy was given, and correspondingly there was a decrease in MNs. A bone marrow test performed on day 38 of the study showed presence

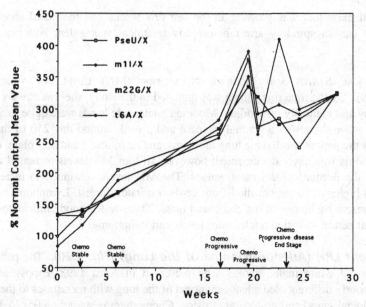

Figure 7.20. Serum modified nucleoside levels in a patient with adenocarcinoma of the lung. [Reproduced from C. W. Gehrke, R. W. Wixom, and E. Bayer (Eds.), *Chromatography: A Century of Discovery*, Vol. 64, Elsevier, Amsterdam, 2001 and Supplement S12 on the Internet (http://www.chemweb.com/preprint), with permission from Elsevier, Amsterdam.]

of leukemic cells at a time where MN levels were also increasing. After day 38, a good clinical correlation of the level of markers increased with clinical deterioration of the patient (see Fig. 7.18).

Patient MP-Y with Non-Hodgkin Lymphoma (Fig. 7.19). This was a 60-year-old white female smoker. Patient presented on 6/01/91 with weakness in the lower extremities, and a MRI showed evidence of cord impression. Biopsy showed non-Hodgkins's lymphoma, large cell type. Postoperatively the patient received radiation therapy, and was started on chemotherapy after assessment of disease. Physical examination revealed axillary lymph node adenopathy and CAT scan of the chest showed chest wall disease and pleural effusion. CAT scan of the abdomen showed metastasis to the lumbar spine. Sample collection started on 9/20/91 while the patient was on chemotherapy. Evaluation of her disease by physical examination indicated a response (decreased size of axillary lymph nodes) and improved neurologic status, and CAT scan showed decreased pleural fluid. CAT scan of abdomen showed no evidence of disease. The patient continued the same regimen of chemotherapy. In February (day 140) the patient went into respiratory distress, and increased pleural effusion was detected. However, cytology or bronchoscopy found no evidence of lymphoma. Chemotherapy was continued as soon as the patient recovered; shortly after the patient was again admitted to hospital with fever. The patient also complained of a chest wall mass that was growing in the last few weeks and then died shortly after this from cardiorespiratory arrest before any treatment was given. Autopsy showed lymphoma.

CLINICAL STATUS AND MODIFIED NUCLEOSIDE (MN) LEVELS OF THE PATIENT (FIG. 7.19). By the time this patient was included in the study, she was already on chemotherapy and clinically responding. Modified nucleoside levels were quite steady until day 175, when there was a gradual increase and a peak around day 250 of the study. Clinically the patient was having lung problems, and an infiltrate and new plural effusion by lymphoma was never documented; however, all four MN levels increased to 250–350% of the normal control mean values. The MN levels continued to increase and remained higher. The patient died from cardiorespiratory arrest. Lymphoma was later proved present by biopsy in the chest wall mass. There was a good clinical correlation of the four serum modified nucleosides levels and lymphoma.

Patient LRH (Adenocarcinoma of the Lung) (Fig. 7.20). The patient was a 58-year-old white male, smoker who presented in April 1993. Biopsy and MRI showed poorly differentiated adenocarcinoma of the lung with metastases to the adrenal gland (adrenal mass) and no pleural effusion. Chemotherapy started on 05/26/93. This patient was not responding to chemotherapy and his clinical status was gradually deteriorating. By October '93 (week 23), the patient was in the end stage of the disease.

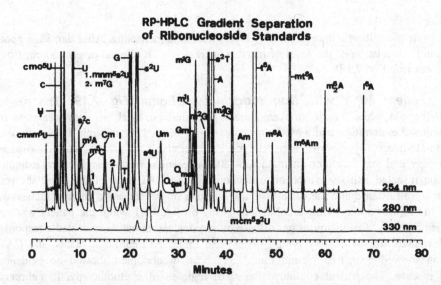

Figure 7.21. RP-HPLC gradient separation of ribonucleoside standards. [Reproduced from C. W. Gehrke, R. W. Wixom, and E. Bayer (Eds.), *Chromatography: A Century of Discovery*, Vol. 64, Elsevier, Amsterdam, 2001 and Supplement S12 on the Internet (http://www. chemweb.com/preprint), with permission from Elsevier, Amsterdam.]

CLINICAL STATUS AND MODIFIED NUCLEOSIDE (MN) LEVELS OF THE PATIENT ABOVE. This patient was not responding to treatments. His clinical status was continually deteriorating. The progression of disease correlated with increased levels of MNs. When the disease reached the end stage, the levels of all four MNs were >300% higher than the normal control mean values. In this case the modified nucleoside levels clearly correlated with the progress of the disease (see Fig. 7.20).

7.E.4. Preparative Isolation of Unknown Nucleosides in Nucleic Acids for Structural Characterization

Knowledge of the chemical structure of nucleoside modifications in nucleic acids is essential for increasing our understanding of their chemical structure and biological function relationships. Transfer ribonucleic acid (tRNA) is one of the most heterogeneous biopolymers known. It not only has a variety of functions within the cell but also contains a much larger proportion of modified nucleosides than do other nucleic acids; more than 60 modified nucleosides have been characterized by HPLC-UV in tRNAs. With our newly developed RPLC-UV nucleoside chromatography methodology [38], providing enhanced resolution and sensitivity, modified nucleosides were detected and identified [54–57]. As tRNA research investigations are conducted on more complex

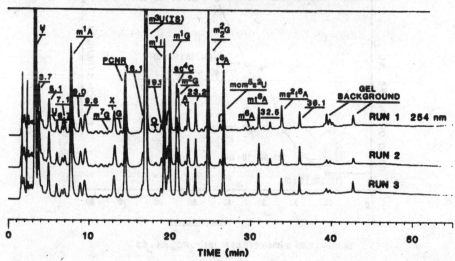

Figure 7.22. Reproducibility of HPLC analysis of ribonucleosides in human urine from three independent runs. [Reproduced from C. W. Gehrke, R. W. Wixom, and E. Bayer (Eds.), *Chromatography: A Century of Discovery*, Vol. 64, Elsevier, Amsterdam, 2001 and Supplement S12 on the Internet (http://www.chemweb.com/preprint), with permission from Elsevier, Amsterdam.]

organisms, it is highly likely that additional modified nucleosides will be discovered as we have observed many unidentified nucleosides in urine and serum. The information on modified nucleosides in human tRNAs is limited, and further investigations should be conducted.

For purification of specific tRNAs, various types of chromatographic and electro-phoretic procedures have been used. Because of the complexity of the initial mixture, the first purification step is generally not for high selectivity but for high capacity. We therefore used the countercurrent distribution (CCD) method, which is mild and serves as a first preparation step with a high capacity. This CCD method was adapted from Holley and Merrill [58] and by Dirheimer and Ebel [49]. This technique permits separ-ation of quantities of tRNAs as high as 5 to 6 g.

In our paper [38], we introduced standard RPLC-UV methodologies for the analysis of nucleosides and nucleoside composition of RNAs, detailed the chromatographic pro-tocols, developed the "nucleoside columns," and gave the essential requirements needed

ELUTION GRADIENT FOR
HIGH-RESOLUTION HPLC OF RIBONUCLEOSIDES

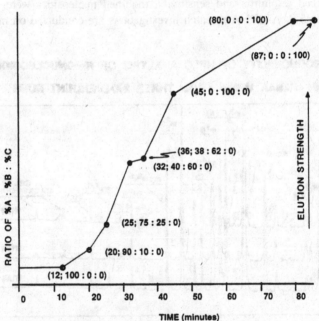

Buffer A: 2.5% Methanol / 0.01 M $NH_4H_2PO_4$, pH = 5.3

Buffer B: 20.0% Methanol / 0.01 M $NH_4H_2PO_4$, pH = 5.1

Buffer C: 35.0% Acetonitrile / 0.01 M $NH_4H_2PO_4$, pH = 4.9

Figure 7.23. Elution gradient for high-resolution HPLC of ribonucleosides. [Reproduced from C. W. Gehrke, R. W. Wixom, and E. Bayer (Eds.), *Chromatography: A Century of Discovery*, Vol. 64, Elsevier, Amsterdam, 2001 and Supplement S12 on the Internet (http://www.chemweb.com/preprint), with permission from Elsevier, Amsterdam.]

in the HPLC instrumentation. Three optimized systems with particular emphasis placed on resolution, speed, and sensitivity were described. In addition, three unfractionated tRNAs were selected: *E. coli*, yeast, and calf liver as sources of "reference nucleosides" to establish the performance of the chromatography (also a quantitative enzymatic hydrolysis protocol to release exotically modified nucleosides from tRNAs was described). In Refs. 40 and 42, we addressed the analytical characterization of nucleosides in nucleic acids, and the chromatography and modification of nucleosides from the perspective of additional chromatographic methodologies for isolation of the nucleic acids, quantitative enzymatic hydrolysis, high-resolution preparative HPLC, and affinity chromatography to obtain the pure single-species nucleosides for UV absorption spectroscopy and interfaced mass spectrometry identification. In addition, we described experiments on the determination of the structure–spectrum relationships, composition, and conformation using an array of advanced analytical techniques of HPLC-UV, FT-IR, NMR, and MS, as well as structure–RPLC retention relationships. In these studies, a consortium of scientists from different institutions joined their expertise and present a comprehensive discussion of the isolation and analytical–structural characterization of tRNAs, oligonucleotides, and nucleosides in RNA and DNA [42]. Two modified nucleosides, A* and G* in yeast initiator tRNA (initiator tRNA^{Met}) at positions 64 and 65 in the T–Ψ stem, were identified as an unmodified guanosine at position 65, and for A* as O-β-D-ribofuranosyl-(1″- 2′)-adenosine at position 64. We elucidated that the final structure for A* at position 64 in yeast initiator tRNA^{Met} was an O-β-D-ribofuranosyl-(1″-2′)-adenosine-5″-phosphate linked by a 3′5′-phosphodiester bond to G at the position 65 [56].

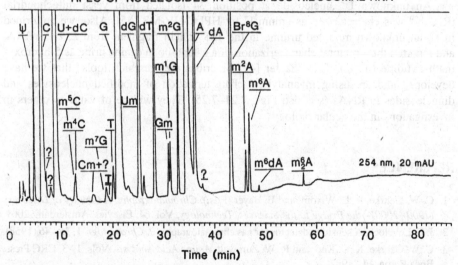

HPLC OF NUCLEOSIDES IN E.COLI 23S RIBOSOMAL RNA

Figure 7.24. HPLC of nucleosides in *E. coli* 23S ribosomal RNA. [Reproduced from C. W. Gehrke, R. W. Wixom, and E. Bayer (Eds.), *Chromatography: A Century of Discovery*, Vol. 64, Elsevier, Amsterdam, 2001 and Supplement S12 on the Internet (http://www.chemweb.com/preprint), with permission from Elsevier, Amsterdam.]

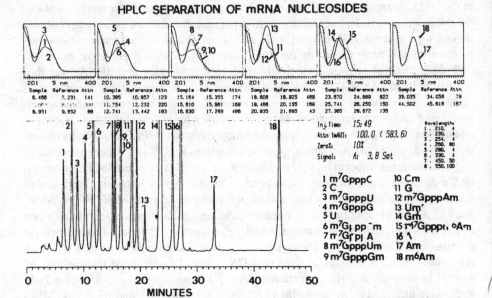

Figure 7.25. HPLC separation of mRNA nucleosides. [Reproduced from C. W. Gehrke, R. W. Wixom, and E. Bayer (Eds.), *Chromatography: A Century of Discovery*, Vol. 64, Elsevier, Amsterdam, 2001 and Supplement S12 on the Internet (http://www.chemweb.com/preprint), with permission from Elsevier, Amsterdam.]

An unknown U* nucleoside at position 34 isolated from yeast mitochondrial tRNA^{Leu} was characterized as cmnm^{5}U by HPLC-UV-MS [55]. Also, we confirmed m^{3}U, an unknown modified uridine in the 16S colicin fragment from *E. coli* rRNA, and reported the structural characterization of a catabolite in canine urine as 5-hydroxy-methycytidine (o^{m5}C). This chapter [42] describes the "research tools" that we have developed and are using in analytical characterization of modified nucleosides and dinucleosides in RNAs (see also Figs. 7.21–7.25). They will be of value to others in investigations in molecular biology.

REFERENCES

1. C. W. Gehrke, R. L. Wixom, and E. Bayer (Eds.), *Chromatography: A Century of Discovery (1900–2000)—the Bridge to the Sciences/Technology*, Vol. 64, Elsevier, Amsterdam, 2001.

2. J. E. Lovelock, A sensitive detector for gas chromatography, *J. Chromatogr.* **1**, 35–46 (1958).

3. C. W. Gehrke, K. C. Kuo, and R. W. Zumwalt, *Amino Acid Analysis*, Vols. 1–3, CRC Press, Boca Raton, FL, 1987.

4. F. E. Kaiser, C. W. Gehrke, R. W. Zumwalt, and K.C. Kuo, Amino acid analysis: Hydrolysis, ion-exchange cleanup, derivatization and quantitation by gas–liquid chromatography, *J. Chromatogr.* **94**, 113–133 (1974).

5. C. W. Gehrke, L. L. Wall, J. S. Absheer, Sr., F. E. Kaiser, and R. W. Zumwalt, Focus: Amino acid analysis. Sample preparation for chromatography of amino acids: Acid hydrolysis of proteins, *J. Assoc. Anal. Chem.* **68**, 811–821 (1985).

6. C. W. Gehrke and D. L. Stalling, Quantitative analysis of the twenty natural protein amino acids by gas–liquid chromatography, *Separations Sci.* **2**, 101–138 (1967).

7. D. Roach and C. W. Gehrke, Direct esterification of the protein amino acids: Gas-liquid chromatography of the *N*-TFA *n*-butyl esters, *J. Chromatogr.* **44**, 269–278 (1969).

8. F. Shahrokhi and C. W. Gehrke, Gas-liquid chromatography of iodine-containing amino acids, *Anal. Biochem.* **24**, 281–291 (1968).

9. S. L. MacKenzie and D. Tenaschuk, Gas–liquid chromatography of *N*-heptafluorobutyryl isobutyl esters of amino acids, *J. Chromatogr.* **97**, 19–24 (1974).

10. S. L. MacKenzie and D. Tenaschuk, Quantitative formation of *N(O, S)*-heptafluorobutyryl isobutyl amino acids for gas chromatographic analysis. I. Esterification, *J. Chromatogr.* **171**, 195–208 (1979).

11. S. L. MacKenzie and D. Tenaschuk, Quantitative formation of *N(O, S)*-heptafluorobutyryl isobutyl amino acids for gas chromatographic analysis. II. Acylation, *J. Chromatogr.* **173**, 53–63 (1979).

12. J. Degrés, D. Boisson, and P. Padieu, Gas liquid chromatography of isobutyl ester, *N(O)*-heptafluorobutyrate derivatives of amino acids on a glass capillary column for quantitative separation in clinical biology, *J. Chromatogr. Biomed. Appl.* **162**, 133–152 (1979).

13. I. M. Moodie, G. S. Collier, J. A. Burger, and B. C. Werb, An improved method based on gas chromatography for determining the amino acid composition of commercially produced fishmeal, *J. Sci. Food Agric.* **33**, 345–354 (1982).

14. I. M. Moodie, G. S. Shephard, and D. Labadarios, Gas–liquid chromatography of amino acids with Supelcoport as solid support, *J. Chromatogr.* **362**, 407–412 (1986).

15. E. Gil-Av, Present status of enantiomeric analysis by gas chromatography, *J. Mol. Evol.* **6**, 131–144 (1975).

16. E. Gil-Av and S. Weinstein, Resolution of α-amino acids and Dns-α-amino acids by high performance liquid chromatography with mobile phases containing a chiral ligand, in W. S. Hanchock (Ed.), *Handbook for the HPLC Separation of Amino Acids, Peptides, and Proteins*, Vol. I, CRC Press, Boca Raton, FL, 1994.

17. H. Frank, G. J. Nicholson, and E. Bayer, Rapid gas chromatographic separation of amino acid enantiomers with a novel chiral stationary phase, *J. Chromatogr. Sci.* **15**, 174–180 (1977).

18. H. Frank, G. J. Nicholson, and E. Bayer, Chiral polysiloxanes for separation of optical antipodes, *Angew. Chem. Int. Ed. Eng.* **17**, 363–365 (1978).

19. H. Frank, A. Rettenmaier, H. Weicker, G. J. Nicholson, and E. Bayer, Determination of enantiomer-labeled amino acids in small volumes of blood by gas chromatography, *Anal. Chem.* **54**, 715–719 (1982).

20. H. Frank, W. Woiode, G. J. Nicholson and E. Bayer, Determination of optical purity of amino acids in proteins, in E. R. Klein and P. D. Klein (Eds.), *Stable Isotopes, Proc. 3rd Int. Conf.*, Academic Press, New York, 1979.

21. H. Frank, G. J. Nicholson, and E. Bayer, Gas chromatographic–mass spectrometric analysis of optically active metabolites and drugs on a novel chiral stationary phase, *J. Chromatogr.* **146**, 197–206 (1978).

22. A. Darbre and K. Blau, Gas chromatography of volatile amino acid derivatives. I. Alanine, glycine, valine, leucine, isoleucine, serine, and threonine, *J. Chromatogr.* **17**, 31–49 (1965).

23. K. Blau and A. Darbre, Gas chromatography of volatile amino acid derivatives. III. Aspartic acid, lysine, ornithine, tryptophan, and tyrosine, *J. Chromatogr.* **26**, 35–40 (1967).

24. G. Gamerith, Gas–liquid chromatographic determination of $N(O,S)$-trifluoroacetyl n-propyl esters of protein and non-protein amino acids, *J. Chromatogr.* **256**, 267–281 (1983).

25. G. Gamerith, C. F. X. Zuder, A. Guiliani, and H. Brantner, Gas–liquid chromatographic determination of free amino acids in fermentation broth during growth of *Clostridium oncolyticum* M 55, *J. Chromatogr.* **328**, 241–252 (1985).

26. C. Ponnamperuna, Organic compounds in the Murchison meteorite, *Ann. NY Acad. Sci.* **194**, 56–70 (1972).

27. C. Ponnamperuma, *Cosmochemistry and the Origin of Life*, D. Reidel, Amsterdam, 1983.

28. W. A. König, Separation of enantiomers by capillary gas chromatography with chiral stationary phases, *J. High Resol. Chromatogr.* **5**, 588–595 (1982).

29. W. A. König, I. Benecke, and S. Sievers, New procedure for gas chromatographic enantiomer separation: Application to chiral amines and hydroxy acids, *J. Chromatogr.* **238**, 427–432 (1981).

30. I. Abe, K. Izumi, S. Kuramoto, and S. Musha, GC resolution of various D,L-amino acid derivatives on a Chirasil-Val capillary column, *J. High Resol. Chromatogr.* **4**, 549–552 (1981).

31. I. Abe, T. Kohno, and S. Musha, Resolution of amino acids on optically active stationary phase by gas chromatography, *Chromatographia* **11**, 393–396 (1978).

32. G. Bengstsson, G. Odham, and G. Westerdahl, Glass capillary gas-chromatographic analysis of free amino acids in biolgoical microenvironments using electron or selected ion-monitoring detection, *Anal. Biochem.* **111**, 163–175 (1981).

33. G. Odham, L. Larsson, and P.-A. Mårdh, Quantitative mass spectrometry and its application in microbiology, in G. Odham, L. Larsson, and P.-A. Mårdh (Eds.), *Gas Chromatography/Mass Spectrometry Applications in Microbiology*, Plenum Press, New York, 1983, Chap. 9.

34. P. Hušek, Gas chromatography of cyclic amino acid derivatives—a useful alternative to esterification procedures, *J. Chromatogr.* **234**, 381–393 (1982).

35. P. Hušek, V. Felt, and M. Matucha, Single-column gas chromatographic analysis of amino acid oxazolidinones, *J. Chromatogr.* **252**, 217–224 (1982).

36. C. W. Gehrke, L. L. Wall, J. S. Absheer, Sr., F. E. Kaiser, and R. W. Zumwalt, Focus: Amino acid analysis. Sample preparation for chromatography of amino acids: Acid hydrolysis of proteins, *J. Assoc. Anal. Chem.* **68**, 811–821 (1985).

37. C. W. Gehrke, R. W. Zumwalt, K. C. Kuo, C. Ponnamperuna, and A. Shimoyama, Search for amino acids in Apollo returned lunar soil, *Origins Life* **6**, 541–550 (1975).

38. C. W. Gehrke and K. C. Kuo, Ribonucleoside analysis by reversed-phase high performance liquid chromatography, *J. Chromatogr.* **471**, 3–36 (1989).

39. C. W. Gehrke, R. A. McCune, M. A. Gama-Sosa, M. Ehrlich, and K. C. Kuo, Quantitative RP-HPLC of major and modified nucleosides in DNA, *J. Chromatogr.* **301**, 199–219 (1984).

40. K. C. Kuo, D. T. Phan, N. Williams, and C. W. Gehrke, Ribonucleosides in biological fluids by a high-resolution quantitative RPLC-UV methods, in C. W. Gehrke and K. C. Kuo (Eds.), *Chromatography and Modification of Nucleosides,* J. Chromatogr. Library Series, Vol. 45C, Amsterdam, 1990, Chap. 2.

41. C. W. Gehrke and K. C. Kuo, High resolution quantitative high-performance liquid chromato-graphy–UV–spectrometry analysis of nucleosides in tRNA, mRNA, DNA, and physiological fluids, *Bull. Mol. Biol. Med.* **10** 119–142 (1985).

42. C. W. Gehrke, J. Desgrés, G. Keith, K. Gerhardt, P. Agris, H. Gracz, M. Tempesta, and K. C. Kuo, Structural elucidation of nucleosides in nucleic acids, in C. W. Gehrke and K. C. Kuo (Eds.), *Chromatography and Modification of Nucleosides, Part A*. J. Chromatogr. Library Series, Vol. 45A, Elsevier, Amsterdam, 1990, Chap. 1, pp. 3–71; Chap. 5, pp. 159–223.

43. K. C. Kuo, C. E. Smith, Z. Shi, P. F. Agris, and C. W. Gehrke, Quantitative measurement of mRNA cap 0 and cap 1 structures by high-performance liquid chromatography, *J. Chromatogr. Biomed. Appl.* **378**, 361–374 (1986).

44. K. C. Kuo, R. A. McCune, C. W. Gehrke, R. Midgett, and M. Ehrlich, Quantitative reversed-phase high performance liquid chromatographic determination of major and modified deoxy-ribonucleosides in DNA, *Nucl. Acids Res.* **8**, 4763–4776 (1980).

45. E. Borek, The morass of tumor markers, *Bull. Mol. Biol. Med.* **10**, 103–117 (1985).

46. E. Borek, Introductions to symposium; tRNA and rRNA modification, in differentiation and neoplasia, *Cancer Res.* **31**, 596–597 (1971).

47. E. Borek and S. J. Kerr, Atypical tRNAs and their origin in neoplastic cells, *Adv. Cancer Res.* **15**, 163–190 (1972).

48. E. Borek, B. S. Baliga, C. W. Gehrke, K. C. Kuo, S. Belman, W. Trollan, and T. P. Waalkes, High turnover rate of transfer RNA in tumor tissue, *Cancer Res.* **37**, 3362–3366 (1977).

49. G. Dirheimer and J. P. Ebel, Fractionnement des rRNA de Levure de biere par distribution en countre-courant, *Bull. Soc. Chim. Biol.* **49**, 1679–1687 (1967).

50. Y. Kuchino, E. Borek, D. Grunberger, J. F. Mushinski, and S. Nishimura, Changes of post-transcriptional modification of wye base in tumor-specific tRNA^{Phe}, *Nucl. Acids Res.* **10**, 6421–6432 (1982).

51. R. W. Zumwalt, T. P. Waalkes, K. C. Kuo, and C. W. Gehrke, Progress and future prospects of modified nucleosides as biological markers of cancer, in C. W. Gehrke and K. C. Kuo (Eds.), *Chromatography and Modification of Nucleosides*, J. Chromatogr. Library Series, Vol. 45C, Elsevier, Amsterdam, 1990, Chap. 1, pp. 15–40.

52. R. W. Trewyn and M. R. Grever, Urinary nucleosides in leukemia: Laboratory and clinical applications, *CRC Crit. Rev. Clin. Lab. Sci.* **24**, 71–93 (1986).

53. K. C. Kuo, F. Esposito, J. E. McEntire, and C. W. Gehrke, Nucleoside profiles by HPLC-UV in serum and urine of controls and cancer patients, in F. Cimino, G. D. Birkmayer, J. V. Klavins, E. Pimentel, and F. Salvatore (Eds.), *Human Tumor Markers*, Walter de Gruyter, Berlin/New York, 1987, pp. 519–544.

54. J. Desgrés, G. Keith, K. C. Kuo, and C. W. Gehrke, Presence of phosphorylated O-ribosylade-nosine in the T–Ψ stem of yeast methionine initiator tRNA, *Nucl. Acids Res.* **17**, 865–882 (1989).

55. R. P. Martin, A. Sibler, C. W. Gehrke, K. C. Kuo, J. A. McCloskey, and G. Dirheimer, 5-Carboxymethylaminomethyluridine is found in the anticodon of yeast mitochondrial tRNAs recognizing two-codon families ending in a purine, *Biochemistry* **29**, 956–959 (1990).

56. G. Keith, J. Desgrés, A. L. Glasser, K. C. Kuo, and C. W. Gehrke, Final structure characteriz-ation of phosphorylated O-ribosyladenosine located at position 64 in T–Ψ stem of yeast meth-ionine initiator tRNA, Paper presented at Int. Conf. tRNA, Vancouver, British Columbia, Canada, June 1989.

57. J. Desgrés, T. Heyman, K. C. Kuo, C. W. Gehrke, and G. Keith, Primary structure of yeast, chicken and bovine tRNA^{pro} presence of 5-carbamoylmethyluridine as the first nucleotide of anticodons, Paper presented at Int. Conf. tRNA, Vancouver, British Columbia, Canada, June 1989.

58. R. W. Holley and S. H. Merrill, Counter-current distribution of an active ribonucleic acid, *J. Am. Chem. Soc.* 5, 735 (1959).

7.F. CHROMATOGRAPHY IN CLINICAL STUDIES—SELECTED ABSTRACTS PRESENTED AT THE 2008 PITTSBURGH CONFERENCE (Charles W. Gehrke)

Several papers were presented at PITTCON 2008 on the use of chromatography in the medical sciences to better understand the disease process and in biomarkers trial. Hyun Joo An, for example in abstract 1 below, reported on a clinical glycomics approach for the early detection of cancers. Abstracts 1–8 are representative.

1. Hyun Joo An (Univ. California, Davis), Scott Kronewitter, Carlito B. Lebrilla, and Lorna DeLeoz: *Clinical Glycomics Approach for the Early Detection of Diseases,* **(Abstract 2510-2).** Glycosylation is highly sensitive to the biochemical environment and has been implicated in development and disease. For this reason, monitoring changes in glycosylation may be a more specific and sensitive method for biomarker discovery and disease diagnosis than indirect methods using monoclonal antibodies or proteomics analysis. In a number of diseases including cancer, the number of sialic acids, fucoses, biantennary, and triantennary N-linked oligosaccharides has been known to vary. We have developed a glycomic approach to identify oligosaccharide marker for the disease. Oligosaccharides are globally released from human serum and have been profiled as a rapid screen for identifying oligosaccharide biomarkers for the specific disease. Clinical glycomics is in its infancy. Tools are needed for its successful development. Central to this effort is the use of mass spectrometry. In this study, mass spectrometry, specifically Fourier transform ion cyclotron resonance (FT-ICR), is used to analyze complicated serum samples. This approach is being developed to examine serum for cancer, specifically ovarian, breast, and prostate. Direct glycan analysis of human serum without protein analysis is a new and innovative approach to early disease detection.

2. Kathryn R. Rebecchi (University of Kansas, Lawrence), Melinda L. Toumi, Eden P. Go, and Heather Desaire: *Development of a Quantitative, Label-Free Approach for Glycopeptide Analysis: A First Step Towards Glycopeptide-Based Biomarker Discovery* **(Abstract 2510-1).** Rebecchi et al. report that glycosylation, an important posttranslational modification, occurs in many proteins. The carbohydrate moiety assists in cellular functions such as protein folding and cell–cell interactions. The type of glycosylation depends, in part, on the local cell environment because specific enzymes and cofactors in the cell regulate glycosylation. Therefore, with the onset of certain diseases, glycosylation profiles of glycoproteins can be changed. Hence, developing methods that quantify changes in glycosylation will be useful in predicting disease development.

Changes in glycosylation can be assessed by observing changes in glycopeptide profiles. Our group has developed a method that can distinguish between glycosylation change and glycoprotein concentration change in a mass spectrum of glycopeptides. Glycopeptides are generated by subjecting glycoproteins to reduction and alkylation followed by proteolysis with trypsin. The resultant glycopeptides are enriched via in-solution purification with Sepharose CL-4B®. The glycopeptide portion is then introduced into an LTQ-FTICR-MS by direct infusion. Data analysis is performed by obtaining a ratio for all glycopeptide peaks present in the mass spectrum, and the ratio is used to rank the individual glycopeptide peaks from smallest to largest.

Multiple samples of RNase B and asialofetuin have been used to demonstrate reproducibility of this method. These glycoproteins were also mixed at various concentrations to confirm whether the method discerns glycoprotein concentration changes without changing the glycosylation profile from a given glycoprotein. The method is being applied to compare RNase B samples with altered glycosylation. We observed the rank order of the glycans to change for samples with altered glycosylation.

3. **Benjamin F. Mann (Indiana University, Bloomington), Yehia Mechref, Milan Madera, Iveta Klouckova, Pilsoo Kang, and Milos V. Novotny:** *Multimethodological Approach to Glycoproteomic and Glycomic Mapping of Healthy and Breast Cancer Blood Serum Samples* (Abstract 2510-8). Proteins endure different posttranslation modifications (PTMs), of which glycosylation is most common and has been frequently implicated in cancer. Specifically, increases in fucosylated and sialylated glycan structures have been repeatedly observed in cancer studies. Because of the extremely wide range of protein concentrations present in serum, the detection of proteins and glycoproteins at low concentrations can be attained only with multidimensional separation capabilities. In this presentation, Mann et al. used a multi-methodological approach for the evaluation of glycoproteomic and glycomic changes as a result of cancer progression. The glycoproteomic approach is based on the use of lectin affinity columns in conjunction with macroporous reversed-phase chromatography to enrich and separate glycoproteins, followed by standard LC-MS/MS experimentation. On the other hand, the glycomic approach is based on our recently developed comparative glycomic mapping (C-GlycoMAP), which involves the analysis of a mixture of methylated and deuteromethylated glycans derived from healthy individuals and breast cancer patients, respectively.

4. **Simon J. North (Imperial College, Division of Molecular Biosciences, Faculty of Natural Sciences, London):** *Mass Spectrometric Strategies for Glycomic and Glycoproteomic Biomarker Discovery* (Abstract 2510-7). Sugar residues on complex glycans expressed at the cell surface have a wide variety of biological and biomedical functions, including the mediation of cell-cell interactions important in cell adhesion and cell migration. North states that Identifying the specific sugars and the glycan structures that carry them is key to defining the roles of sugars

and identifying potential biomarkers. It is also important to establish the range of structures that can be synthesized by given cell types and organisms, which defines their glycome. In order to determine a wide range of glycan structures on small amounts of biological samples, highly sensitive, high-throughput mass spectrometric glycomic screening techniques have been developed and are being applied to cell and tissue samples. Data from such glycomic studies performed by our laboratory on cell types of the immune system, parasitic and free-living nematodes, mice organs, and human tissues were presented. Additionally, we are defining the glycans present on individual glycoproteins and sites of glycan occupancy by application of glycoproteomic strategies.

5. Taufika I. Williams (North Carolina State University—Raleigh, Mass Spectrometry Laboratory, Department of Chemistry) and David C. Muddiman: *Glycan Profiling in Plasma: Fundamentals and Applications to Ovarian Cancer* (Abstract 2510-6).

Epithelial ovarian cancer (EOC), the major classification of ovarian cancers, is the fifth leading cause of cancer deaths among women in the United States and the most fatal of gynecologic malignancies. Because of the lack of effective screening strategies, most patients are diagnosed at a late stage, when prognosis is decidedly poor. Previous studies indicated that glycosylation patterns in plasma glycoproteins in EOC patients differ from that of age-matched controls [*J. Proteome Res.*, 5, 1626 (2006)]. Our studies aim to investigate carbohydrates from glycoproteins in *early-stage EOC plasma* in attempts to elucidate an effective biomarker for early diagnosis.

Important posttranslational modifications such as glycosylation can perform key functions in signaling pathways that enable the metamorphosis of a normal cell into a cancerous one. Proteomic and glycomic profiling has revealed a populated arena for biomarker discovery, due in no small part to the miscellany and sheer number of proteins and carbohydrates in the human body. Recent studies have shown that focus on carbohydrate patterns of glycosylated proteins can better direct biomarker discovery so as to improve the probability of favorable outcomes.

We detail the sample preparation and MALDI-FT-ICR-MS analysis of *O*-linked glycans selectively cleaved from plasma glycoproteins. In this approach, protein information is necessarily lost as the primary focus is differential glycan pattern recognition, a less complicated system than the study of the entire plasma glycoproteome. Separately optimized experimental parameters were integrated in a method that permitted the examination of these carbohydrates in plasma from early- and late-stage EOC patients, as well as age- and menopause-matched controls.

6. David M. Good (University of Wisconsin—Madison, Department of Chemistry), Petra Zürbig, Eric Schiffer, Ronald Krebs, Harald Mischak, and Joshua J. Coon: *Sequence Analysis of Urinary Peptides to Monitor in Vivo Disease Dependent Protease Regulation* (Abstract 2510-3).

The typical "bottom–up" approach employed in most mass spectrometry-based proteomics analyses limits the amount of information that one can obtain on the in vivo state of the analyte. This is due to the requisite in vitro digestion of samples, where enzymes are used to cleave parent peptides and proteins into smaller fragments. Trypsin is predominantly employed in this approach, as it limits the number of internal basic residues a peptide can contain. These residues can block random protonation and thus limit the utility of tandem mass spectrometric analyses, which employ the conventional fragmentation methodology— collision activation dissociation (CAD)—for sequence analysis. Naturally occurring peptides (i.e., biomarkers) do not result from such controlled digestions and can therefore be challenging to sequence. Here we describe LC-MS/MS analysis, using complementary fragmentation techniques [electron transfer dissociation (ETD) and CAD], for characterization of the human urinary peptidome. To date >1200 unique sequences have been characterized, and this information has been coupled to existing diagnostic, disease-specific CE-MS databases through the development of an algorithm specific for this task. Our work has led to the discovery of several biologically interesting, and statistically significant, peptides that were previously uncharacterized. We demonstrate the utility of this method for assessment of disease-driven in vivo protease activity by relating peptides characterized from clinical samples to the preexisting CE-MS databases.

7. James A. Atwood (University of Georgia, Complex Carbohydrate Research Center, Atlanta), Zuzheng Luo, and Ron Orlando: *A Novel Glycoproteomics Approach for the Characterization of Glycopeptides from Complex Biological Mixtures* **(Abstract 2510-4).** Glycoprotein characterization for biomarker discovery is typically divided into two categories: glycomics or proteomics. In glycomics, glycans are released from glycoproteins then analyzed by MS either in their native state or following chemical modification. However, this procedure results in a complete loss of information regarding the glycoproteins from which the glycans originated. To analyze glycoprotein expression, lectin affinity chromatography is routinely employed to separate the glycopeptides from the nonglycosylated species, and the glycans are released prior to peptide identification. Unfortunately, the specificity of the lectin, does not facilitate the global isolation of glycopeptides with a diverse population of glycan structures, and this method prevents the assignment of glycans to glycosylation sites.

The goal of our work was to develop methodologies to characterize intact glycopeptides including, glycan compositions, and glycosylation sites on glycopeptides from complex mixtures. The first step in this method was the development of a normal-phase chromatography (NPLC)-based glycopeptide enrichment procedure. Here we show that glycopeptides can be effectively separated from their nonglycosylated counterparts using NPLC. The captured glycopeptides are then analyzed by MS/MS using an LTQ-FTICR and Q TRAP mass spectrometers. This glycoproteomics approach was validated using standard glycoproteins and complex glycoprotein mixtures from human serum. Our initial studies successfully characterized glycosylation sites and glycan compositions on 102 unique glycopeptides from human serum with over four orders of magnitude range in concentrations. This approach is currently being employed to study changes in glycopeptide expression in the cerebrospinal fluid associated with Alzheimer's disease.

8. Mahalakshmi Rudrabhatla (Varian Inc., Walnut Creek, CA) and John E. George: *Identification of Beta Amyloid and Nitrated and Oxidized Ubiquitins on an Ion Trap Mass Spectrometer* **(Abstract 2510-5).** Alzheimer's is a degenerative neural disease of aging characterized by dementia. Oxidative stress has been proposed as a chief pathological mechanism in Alzheimer's. In vivo, the free-radical superoxide reacts with nitric oxide to form peroxyhitrite. Peroxynitrite, in turn, reacts with cerebrospinal fluid proteins (τ protein, β-amyloid, and ubiquitin), resulting in oxidation and nitration. These oxidized and nitrated protein products in the cerebrospinal fluid serve as important biomarkers for Alzheimer's. Although the free-radical-induced oxidation of β-amyloid has been extensively studied, ubiquitin's role under oxidative stress in Alzheimer's is still under investigation.

In the present study, peroxynitrite induced oxidation products of ubiquitin were characterized in the cerebrospinal fluid on an ion trap mass spectrometer. Peroxynitrite treatment of ubiquitin was done in vitro. The treated mixture was initially analyzed by LC/MS in the full-scan MS mode, and the charge states obtained were deconvoluted to their main mass. Two main products with a mass difference of $+16$ amu and $+44.5$ amu corresponding to oxidation and nitration were observed in the deconvoluted spectra. In the next step, tryptic digest of the above mixture was analyzed using the data-dependent software and the MS/MS spectra exported to the MASCOT server for a protein identification search against the MSDB database displayed excellent sequence coverage (92%). Exact locations of modifications, namely, nitration at tyrosine and oxidation at methionine, were identified at subfemtomole levels. In addition, β-amyloid peptide was also identified at subfemtomole levels in plasma and cerebrospinal fluid. Results indicate that excellent sensitivity, dynamic range, and throughput of an ion trap mass spectrometer is ideally suited for identifying 3-nitrotyrosine and oxidized methionine on ubiquitin and also β amyloid peptides, which are important biomarkers for Alzheimer's disease.

7.G. AFFINITY MONOLITHIC CHEMISTRY

DAVID S. HAGE

Figure 7.26. David S. Hage.

David S. Hage (Fig. 7.26) is a Professor of Analytical and Bioanalytical Chemistry in the Chemistry Department at the University of Nebraska-Lincoln (UNL). He received his B.S. in Chemistry and Biology from the University of Wisconsin-La Crosse (1983), and his Ph.D. in Analytical Chemistry from Iowa State University (1987), and he was a Postdoctoral Fellow in Clinical Chemistry at the Mayo Clinic. He joined the faculty in the Chemistry Department at UNL in 1989 and was named a Charles Bessey Professor at this institution in 2006. Dr. Hage is the author of approximately 150 research publications and reviews in the fields of chromatography, electrophoresis, and related separation methods. He recently edited a handbook on the topic of affinity chromatography (2005) and has several patents related to his work in this field. He did a Postdoctoral on Clinical Chemistry at Mayo Clinic, Rochester, MN in 1989. He has received many awards for his work, including the 1995 Young Investigator Award from the American Association for Clinical Chemistry, the 1995 Young Investigator Award from the Society of Molecular Recognition, and the 2005 Excellence in Graduate Education Award from UNL. He has also received two "Top Cited Article" awards from the *Journal of Chromatography* for the time periods covering 2001–2006 and 2002–2007. Five more recent publications are listed below.

REFERENCES

J. Chen and D. S. Hage, Quantitative analysis of allosteric drug-protein binding by biointeraction chromatography, *Nature Biotechnol.* **22**, 1445–1448 (2004).

T. Jiang, R. Mallik, and D. S. Hage, Affinity monoliths for ultrafast immunoextraction, *Anal. Chem.* **77**, 2362–2372 (2005).

R. Mallik and D. S. Hage, Affinity monolith chromatography, *J. Separations Sci.* **29**, 1686–1704 (2006).

C. M. Ohnmacht, J. E. Schiel, and D. S. Hage, Analysis of free drug fractions using near infrared fluorescent labels and an ultrafast immunoextraction/displacement assay, *Anal. Chem.* **78**, 7547–7556 (2006).

C. Wa, R. L. Cerny, and D. S. Hage, Identification and quantitative studies of protein immobilization sites by stable isotope labeling and mass spectrometry, *Anal. Chem.* **78**, 7967–7977 (2006).

See Chapter 1, Table 1.3, a, h, i, q.

The general area of Dr. Hage's research is in the theory, design, and application of rapid analytical systems that are based on affinity ligands. Much of his past work has involved an exploration of novel assay formats using immobilized antibodies and related binding agents in the creation of chromatography-based immunoassay. These assays have been examined for use with a variety of analytes, such as drugs, hormones, proteins, and environmental contaminants. An example of his more recent work in this area is illustrated in Figure 7.27, in which his group reported a new class of methods for measuring the free or nonbound fraction of a drug or hormone in serum by using ultrafast extraction and an HPLC-based immunoassay. This work involved the creation of a miniaturized antibody-based extraction column that could isolate a small target from a sample in the subsecond time domain (typically 60–200 ms). This approach makes it possible to selectively measure the free fraction of a drug or hormone within 20–30 s of sample injection.

Along with the displacement format shown in Figure 7.27, Dr. Hage's group has studied and characterized the use of sandwich immunoassays, competitive binding immunoassays, and one-site immunometric assays in HPLC systems. His group was also one of the first to use miniaturized antibody columns for online immunoextraction in HPLC and to combine such columns with reversed-phase liquid chromatography to produce rapid multidimensional separations for specific groups of analytes. In addition, he has been active in creating devices that make use of this approach for the

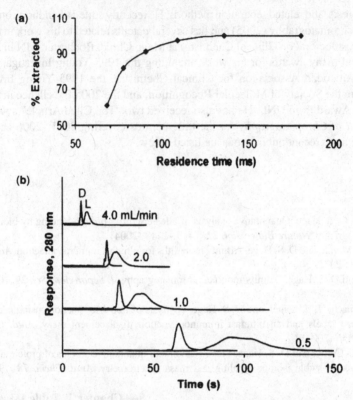

Figure 7.27. (a) Use of ultrafast immunoextraction to isolate fluorescein by a 4.5-mm-i.d. × 1-mm monolithic disk containing immobilized antifluorescein antibodies and (b) chiral separation of D- and L-tryptophan on 4.6-mm-i.d. × 5-cm affinity monolith containing immobilized human serum albumin (HSA). [These figures are reproduced with permission from T. Jiang, R. Malik, and D. S. Hage, *Anal. Chem.* **77**, 2362–2372 (2005) and R. Mallik, T. Jiang, and D. S. Hage, *Anal. Chem.* **76**, 7013–7022 (2004) respectively.]

group-selective analysis of biological compounds and environmental contaminants in clinical samples or during field studies. Dr. Hage's group has explored many detection schemes for use with these methods, including absorbance, fluorescence, mass spectrometry, chemiluminescence, and near-infrared fluorescence detection.

A related area of work by Dr. Hage's group is in the use of affinity ligands in HPLC and capillary electrophoresis to study important binding processes within the body. His particular interest has been in the creation of HPLC devices that can be used to effectively model and study the binding of drugs and small solutes with transport proteins and receptors in the human body. These studies make use of separation theory and careful measurements of retention and peak profiles to provide thermodynamic or kinetic information on the interactions of drugs and other targets with binding agents such as proteins. Procedures that have been refined and developed by Dr. Hage's group now make it possible to quickly obtain high-quality estimates of equilibrium constants and rate constants for these interactions through the use of HPLC. Experimental methods have also been created by his group to study and characterize drug–drug competition and allosteric effects. This research has resulted in improved approaches for the creation and use of columns that contain such proteins as human serum albumin and α_1-acid glycoprotein for use in pharmaceutical studies. These same columns have been examined and evaluated for use in HPLC as the basis for chiral separations.

As part of his work with both immunoassays and studies of biological processes, Dr. Hage has been investigating and developing methods for the use of antibodies, transport proteins, and other binding agents with monolithic supports. The resulting affinity monolith columns have been employed by his group for both ultrafast immunoextraction and for the study of drug–protein binding and chiral separations. Other areas of interest to Dr. Hage's group are the use of aptamers and molecularly imprinted polymers as alternatives to antibodies for the selective detection and isolation of chemicals of biological or environmental interest and the creation of new immobilization methods for biological ligands. In addition, his group is interested in the combined use of affinity methods and mass spectrometry to characterize the binding properties and structure of complex protein populations, and in the theory of how affinity ligands behave when used as selective binding agents in either HPLC or CE systems.

Affinity monolith chromatography (AMC) is a separation method that has only recently been developed that combines the use of monolith support with the use of biologically based binding agent as a stationary phase. This combination offers both increased selectivity over other types of liquid chromatography (e.g., reversed-phase or ion-exchange methods) along with the increased performance, improve flow properties, and ease of structural modification that is made available through the use of monoliths over particle-based support materials. The binding agent, or affinity ligand, that is used in AMC is generally chosen to be an agent capable of specific and reversible interactions with the desired target. The applications for the resulting column will depend on the affinity ligand that is being used and the mobile-phase conditions that are selected for application and elution of the target.

In a typical operating scheme for AMC, a sample containing the target is applied to the affinity column in the presence of an application buffer that allows strong binding between the target and affinity ligand while allowing other sample components to wash from the column. A strongly retained target is eluted by changing the pH, polarity, or ionic strength of the mobile phase or by switching to a mobile phase that contains a competing agent capable of interfering with the analyte–ligand interaction. These elution methods are commonly used with affinity ligands such as antibodies and enzymes. If a target has weak-to-moderate binding to the affinity ligand (i.e., having an association equilibrium constant of $10^6\ M^{-1}$ or less), it is often possible to elute this target under isocratic conditions.

Affinity-based separations are commonly characterized according to the type of ligand being utilized. In bioaffinity chromatography the immobilized ligand is a biological compound. For example, protein A and protein G have both been immobilized with GMA/EDMA monoliths for use in antibody purification in less than 30 s. An important subset of bioaffinity chromatography is immunoaffinity chromatography, in which the binding agent is an antibody or antibody-related ligand. The resulting support can be used in AMC for the rapid purification or analysis of targets that bind to the immobilized antibodies. This method has been shown be useful for the ultrafast immunoextraction of specific drugs or solutes from samples in as little as 100 ms (see Fig. 7.27a).

Chiral separations are another potentially important application of AMC. Serum proteins such as human serum albumin (HSA) and α_1-acid glycoprotein (AGP) have been used in AMC for this type this work (see Fig. 7.27b), along with enzymes such as penicillin G acylase. Related studies have been reported in which AMC columns containing these and other ligands have been used to study drug–protein interactions.

Other ligands that have been used in AMC include immobilized metal ion chelates and synthetic dyes or combinatorial ligands. Metal ion chelates have been placed in monoliths to conduct immobilized metal ion affinity chromatography (IMAC) for the isolation of His-tagged proteins or other proteins that interact with metal ions. It has been shown that the dye Cibacron Blue can be placed into monoliths for use in dye–ligand affinity chromatography and the isolation of proteins such as lysozyme and t-plasminogen activator. Selected peptides from combinatorial libraries have also been used in monoliths for the isolation of specific proteins.

A number of materials have been used to prepare monolithic supports for AMC. Copolymers of glycidyl methacrylate and ethylene dimethacrylate (GMA/EDMA) have been most frequently used in AMC (see Fig. 7.28). In this case, azoisobutyronitrile (AIBN) is used as a thermal initiator to begin the polymerization, and cyclohexanol and dodecanol are used as porogens around which the polymer forms the porous monolith network. These monoliths are relatively easy to prepare, have low nonspecific binding, contain groups that are readily available for modification or immobilization, and can be prepared with a variety of surface areas and pore sizes.

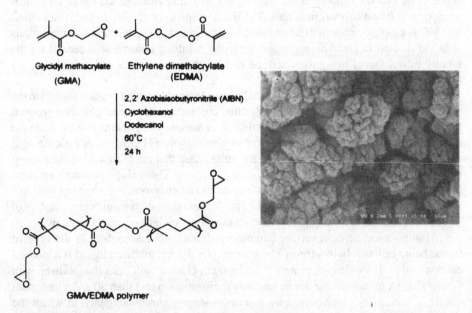

Glycidyl methacrylate
(GMA)

Ethylene dimethacrylate
(EDMA)

2,2′ Azobisisobutyronitrile (AIBN)
Cyclohexanol
Dodecanol
60°C
24 h

GMA/EDMA polymer

Figure 7.28. General scheme for the production of GMA/EDMA affinity monolith and an example of such a monolith. [These figures are reproduced with permission from R. Mallik, T. Jiang, and D. S. Hage, *Anal. Chem.* **76**, 7013–7022 (2004).]

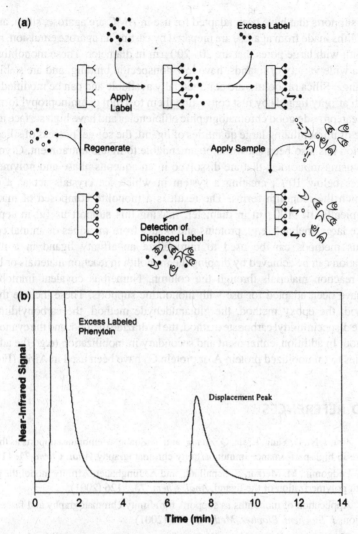

Figure 7.29. (a) General scheme for an ultrafast immunoextraction/displacement assay and (b) a typical chromatogram for such an assay when used to measure the drug phenytoin in serum. This assay method is designed for the detection of the free from of a drug (o) in a complex sample by utilizing a small column containing immobilized antibodies previously loaded with a labeled analog (or "label") of the analyte (•). As the sample passes through the column, the nonbound from the analyte in the sample will be extracted and compete with the label for antibody binding sites, resulting in a displacement peak for the label that is proportional in size to the concentration of the free analyte in the sample. The sample in (b) was injected at 6 min following application of the label to a small column containing immobilized antibodies that can bind to phenytoin. [This figure is reproduced with permission from C. M. Ohnmact, J. E., Schiel, and D. S. Hage, *Anal. Chem.* **78**, 7547–7556 (2006).]

Other supports that have been adapted for use in AMC are agarose, silica, and cryogels. Monoliths made from agarose are prepared by casting an agarose emulsion, resulting in a monolith with large pores that are $20-200$ μm in diameter. These monoliths can be used with a wide variety of ligands, have low nonspecific binding, and are stable over a wide pH range. Silica monoliths are commercially available and can be modified directly for use with affinity ligands by first converting them to a diol or aminopropyl form. Silica monoliths can provide good chromatographic efficiencies and have high surface areas that are capable of immobilizing large quantities of ligand; the sol-gel process of silica monolith formation has also been shown to be amendable to ligand entrapment. Cryogels are formed by using monomers that are dissolved in an aqueous phase and polymerized at temperatures below $10°C$, creating a system in which ice crystals act as a porogen around which the monolith forms. The result is a monolith composed of macropores that are typically $10-100$ μm in diameter, making this support useful in separations that involve large analytes (e.g., proteins and matter from microbes or animal cells).

Various methods can be used to immobilize an affinity ligand in a monolith. Immobilization can be achieved by dipping the monolith in reaction materials or by circulating the reaction materials through the column. Numerous covalent immobilization methods have been adapted for use with monolithic supports. These include the Schiff base method, the epoxy method, the glutaraldehyde method, the carbonyldiimidazole method, the disuccinimidyl carbonate method, the hydrazide method, and the cyanogen bromide method. In addition, entrapment and secondary immobilization (e.g., the adsorption of antibodies to immobilized protein A or protein G) have been used in AMC (Fig. 7.29).

RELATED REFERENCES

D. Zhou, H. Zou, J. Ni, L. Yang, L. Jia, Q. Zhang, and Y. Zhang, Membrane supports as the stationary phase in high-performance immunoaffinity chromatography, *Anal. Chem.* **71**, 115 (1999).

R. Hahn, A. Podgomik, M. Merhar, E. Schallaun, and A. Jungbauer, Affinity monoliths generated by in situ polymerization of the ligand, *Anal. Chem.* **73**, 5126 (2001).

A. B. Josic, Application of monoliths as supports for affinity chromatography and fast enzymatic conversion, *J. Biochem. Biophys. Meth.* **49**, 153 (2001).

R. Mallik, T. Jiang, and D. Hage, High-performance affinity monolith chromatography: Development and evaluation of human serum albumin columns, *Anal. Chem.* **76**, 7013 (2004).

T. Jiang, R. Mallik, and D. Hage, Affinity monoliths for ultrafast immunoextraction, *Anal. Chem.* **77**, 2362 (2005).

G. A. Platonova and T. B. Tennikova, Affinity processes realized on high-flow-through methacrylate-based monoliths, *J. Chromatogr. A* **1065**, 19 (2005).

D. S. Hage, *Handbook of Affinity Chromatography*, 2nd ed., CRC Press/Taylor & Francis, Boca Raton, FL, 2005.

R. Mallik and D. S. Hage, Affinity monolith chromatography, *J. Separations Sci.* **29**, 1686 (2006).

F. Svec, Monolithic materials: Promises, challenges, achievements, *Anal. Chem.* **78**, 2100 (2006).

R. Mallik, H. Xuan, and D. Hage, Development of an affinity silica monolith containing α_1-acid glycoprotein for chiral separations, *J. Chromatogr. A* **1149**, 294 (2007).

8

CHROMATOGRAPHY— ADVANCES AND APPLICATIONS IN PHARMACEUTICAL ANALYSIS IN THE CORPORATE SECTOR

Pat Noland and Terry N. Hopper

ABC Laboratories, Columbia, Missouri

Michael W. Dong

Genetech, South San Francisco, California

Yong Guo

Johnson & Johnson Pharmaceutical & Research Development LLC, Rariton, New Jersey

Todd D. Maloney

Lilly Research Laboratories, Indianapolis, Indiana

Raymond N. Xu

Abbott Laboratories, Abbott Park, Illinois

CHAPTER OUTLINE

8.A. ANALYSIS OF PHARMACEUTICALS FOR NEW DRUG REGISTRATION (Pat Noland and Terry N. Hopper)

PATRICK A. NOLAND

Patrick Noland (Fig. 8.1a) is currently Vice President of Pharmaceutical Program Development at Analytical Bio-Chemistry Laboratories in Columbia Missouri. He has 25 years' experience in regulated sciences working under guidelines from FDA, EPA, NRC, and MHRA (MCA). He has been a primary author for the method validation, specifications, and container closure sections of four New Drug Applications. He has served as Director of Regulatory Strategy at Analytical BioChemistry Laboratories in Columbia, Missouri and Senior Director of Commercial Quality Assurance and Director of Quality Control at Sepracor, Inc. in Marlboro, Massachusetts.

Figure 8.1a. Patrick A. Noland.

Mr. Noland received his B.S and M.S. in Chemistry (1975, 1976) from the University of Missouri—Rolla. He is currently focused on business process, lifecycle management, quality by design, and statistical process control. Mr. Noland is a member of various professional organizations, including the American Association of Pharmaceutical Scientists, American Chemical Society, Parental Drug Association, American Society for Quality, and Society of Quality Assurance.

TERRY N. HOPPER

Terry N. Hopper (Fig. 8.1b) is the Director of GMP Analytical Services at ABC Laboratories in Columbia, Missouri. Prior to joining ABC Laboratories, Mr. Hopper held various technical and management positions in the areas of analytical drug development and quality. He has been employed in the pharmaceutical, biopharmaceutical, and medical device industries for over 31 years and has made contributions to numerous drug applications during this time. Mr. Hopper has contributed to the drug development and quality processes for small, medium, and large companies and has served as the chair for various analytical project teams. Mr. Hopper's accomplishments include building all aspects of an R&D analytical, stability, and validation department, including the development and implementation of comprehensive laboratory CGMP systems, analyst training programs, method/equipment/laboratory/computer validation, laboratory information management systems (LIMS), document control, and metrology programs.

Figure 8.1b. Terry N. Hopper.

Mr. Hopper received his M.S. degree, B.S. degree (mathematics major; physics minor), and B.S. in Chemistry, all from the University of Memphis (formerly Memphis State University). His interests include designing and implementing pharmaceutical systems for drug development, developing learning programs for analytical chemists, LIMS, employing "design of experiments" in analytical technology, utilizing statistical quality control in the analytical laboratory, and anything that involves validation. He is a member of the American Chemical Society, American Association of Pharmaceutical Scientists, Mathematical Association of America, and American Society for Quality.

Analysis of Pharmaceuticals for New Drug Registration. Drugs are generally classified as large or small molecules. Traditional analytical methodologies are utilized primarily with the small molecules, although they can be applied to some biologically produced peptides with low molecular weights. Most drugs are neutral or basic entities. In order to be adequately bioavailable in a primarily aqueous system, such as the alimentary tract, they are formulated as salts. For this reason, most methods employ liquid chromatography, although headspace gas chromatography is utilized for residual solvent in drug substance. Some of the chromatographic challenges for all phases of drug development include stationary-phase selection, pH of mobile phase, solubility in mobile phases, selectivity from matrices, and spectroscopic profile. In line with recent Food and Drug Administration (FDA) initiatives, "design of experiments" (DOE) is now frequently employed to optimize these vectors during the development phase. Column-switching instruments with advanced software can conduct multivector experimental designs automatically, leaving only interpretation to the operator.

During the course of development of a new drug, three basic types of analysis are required. Earlier in the development cycle, methods are required to separate the drug from biological matrices and quantitate it at nanogram per milliliter (ng/mL) or even picogram per milliliter (pg/mL) levels. As the drug development cycle continues, methodology for analyzing the drug and related synthetic impurities or degradation products is required. In the final stages of drug development, methodology that can be used to assess levels of the drug, drug impurities, and drug degradation products, in the presence of excipients and possibly other drugs, is required. Each of these phases requires a slightly different paradigm for analytical development.

In the preclinical and early clinical phases, the focus of the analyst is on selectivity in a biological matrix. Separation of a biologically active compound from a biological matrix, when the biologically active compound is present at sub-ppm levels, has always been challenging. Historically multistep, wet chemical isolations were employed. As solid-phase materials became more readily available and more consistent, rough chromatographic separations were utilized before analytical chromatography. Mixed-mode solid-phase extraction containing both ion-exchange and reversed-phase materials are particularly effective at separating a drug from the matrix [1]. The required selectivity and sensitivity were achieved by using gradient programs and long columns coupled with fluorescence, electrochemical, or mass spectral instruments, when they could be employed. Advances in mass spectrometry and the widespread availability of stable high performance tandem mass spectrometers (HPLC-MS-MS) transformed the chromatographic requirements for the analysis of biological fluids. The introduction of small-particle, high-efficiency-column ultra performance liquid chromatography (UPLC) further increased chromatographic selectivity [2]. Most methods now consist of a few simple steps, often automated, to remove proteins and solids. The resultant solution is subjected to a very short chromatographic separation on a high-efficiency column. The analyte and internal standard selectivity from the matrix is further enhanced in the detector by utilizing analysis of an appropriate parent–daughter pair of ions.

Before the drug is introduced to humans, a complete characterization of drug properties and impurities is required. These characterization studies for the drug substance involve identifying and often isolating drug impurities that are present in the drug substance. Impurities in the drug substance are those compounds that are present at low levels. There are essentially three types of impurities that are characterized for drug substances: (1) impurities that are inherent to the route of synthesis used to produce the drug substance, including residual solvents that remain in the drug matrix after the final purification step in the production process; (2) degradation products that are typically related substances to the drug that are formed from thermal exposure, hydrolysis, oxidation, reduction, or photolysis reactions; and (3) impurities that are introduced into the drug matrix from contact with its container–closure system during storage. While impurities formed during synthesis and degradation products are generally closely related to the drug substance, container–closure impurities generally are not. Container closure impurities are typically neutral CHO compounds and include antioxidants, monomers, oligomers, initiators, adhesive volatiles, and colorants. Gas chromatography with mass spectral detection, HPLC-MS, inductively coupled plasma with mass spectrometry (ICP-MS), and HPLC with corona charged aerosol detectors (CAD) [3] are utilized to characterize the drug substance and impurities. Advances in hyphenated techniques exist with time-of-flight and ion trap instruments that provide higher resolution and increased molecular weight determination capabilities, facilitating the identification of impurities. New column technologies and techniques such as the HILIC (Section 8.C, below) and small particle technologies (Section 8.D, below) with gradient are employed to cover the wide polarity range necessary to separate drugs from packaging impurities.

Drug formulation work involves the combination of the drug substance with excipients that are characteristic of the type of dosage form being developed to administer the drug to the patient. This is the last phase of method development. Some of the developmental methods used for impurity identification may be employed prior to the design of the final drug product control methods. The drug product contains a controlled amount of drug substance with excipients and any impurities already present in the drug substance at low levels; it may also contain new impurities that may be formed from interactions of the drug substance with the excipients and interactions of the drug product with its container–closure system during storage. Excipients are often lubricants surfactants, and inorganic fillers.

The methodology developed for the final drug product must be stable, appropriately selective, and reproducible, so that the drug can be released for use and the stability of the drug over its shelf life can be assessed. Analytical methods employing chromatographic techniques (or any other principle of analysis) must be validated as part of the drug development process. Validation of an analytical method involves the conduct of studies to demonstrate the accuracy, precision, linearity, range of use, sensitivity, specificity, ruggedness, and robustness of the method. Further, the analytical methods developed and validated in one laboratory are typically transferred to another laboratory (e.g., quality control laboratory). Often, the brand and model of chromatographic instrumentation may not be the same. Equivalency of performance must be demonstrated between transferring and receiving laboratories.

Not only is the quantitative performance of instrumental techniques and analytical methods critical to the drug development process; the speed of analysis is also a significant factor. The emergence of short HPLC columns in the 1980s gave promise to reduced chromatographic run times with increased sample testing capability. However, the precision of the results was greatly compromised by the length and bore size of the fluidic tubing employed in HPLC instruments of that era. By the mid-to-late 1990s, HPLC technology had evolved so that short tubing lengths and microbore tubing was integral to modern instrumentation. This led to the reemergence of short and microbore HPLC columns with acceptable levels of precision. Further evolution of HPLC instrumentation has produced UPLC and rapid resolution systems that have further reduced chromatographic run times. It is now possible to reduce a 45-min chromatographic run time to less than 5 min while maintaining the ability to resolve and quantitate analytes to the same or a higher degree of precision.

Once they are validated, these methods are filed in a New Drug Application with the FDA and are often submitted to the United States Pharmacopeia (USP). The FDA will use these methods to check drugs on the market to ensure potency and impurity profile throughout the drugs shelf life. In order to generate sufficient data to ensure the shelf life, stability studies are conducted, wherein samples are stored in environmental chambers under various conditions of temperature and humidity. Typical storage conditions and more extreme conditions ("accelerated aging") are utilized, so that real-time data as well as predictive data are available. For instance, the accelerated aging of a drug generates data in 6 months to provide an idea of impurity profile at 24 months. This helps bring a drug to market in a shorter period of time.

REFERENCES

1. J. X. Shen, H. Wang, S. Tadros, and R. N. Hayes, Orthogonal extraction/chromatography and UPLC, two powerful new techniques for bioanalytical quantitation of desloratadine and 3-hydroxydesloratadine at 25 pg/mL, *J. Pharm. Biomed. Anal.* **40**(3), 689–706 (2006).
2. N. Ellor, F. Gorycki, and C.-P. Yu, *Increasing Sensitivity and Throughput for LC/MS/MS Based Bioanalytical Assays Using UPLC* (2005). Application Note 720001122en, Waters Corp., Beverly, MA.
3. P. H. Gamache, R. S. McCarthy, S. M. Freeto, D. J. Asa, M. J. Woodcock, K. Laws, and R. O. Cole, *HPLC Analysis of Nonvolatile Analytes Using Charged Aerosol Detection,* Chromatography Online by LC-GC North America, Feb. 1, 2005.

8.B. ULTRA-HIGH-PRESSURE LC IN PHARMACEUTICAL ANALYSIS: BENEFITS, IMPACTS, AND ISSUES (Michael W. Dong)

MICHAEL W. DONG

Michael W. Dong (Fig. 8.2) is Senior Scientist in Small Molecule Analytical Chemistry at Genentech, South San Francisco. He was formerly Research Director at Synomics Pharmaceutical Services, Research Fellow/Group Leader at Purdue Pharma, Senior Staff Scientist at Perkin-Elmer, and Section Head in Hoechst Celanese. He holds a B.Sc. in Chemistry from Brooklyn College and a Ph.D. in Analytical Chemistry from the City University of New York Graduate Center. He has conducted numerous HPLC training courses at national meetings on advanced HPLC, fast LC/HTS, stability testing, and HPLC method development at EAS, PITTCON, HPLC, ACS and AAPS. He pioneered fast LC and has over 80 publications in chromatography and analytical chemistry. He authored *Modern HPLC for Practicing Scientists*, Wiley-Interscience, 2006, a bestseller in chromatography and co-edited *Handbook of Pharmaceutical Analysis by HPLC*, published by Elsevier/Academic Press, 2005.

Figure 8.2. Michael W. Dong.

See Chapter 1, Table 1.3, h, r, v.

Since the 1960s, HPLC resolution has been bounded by a system pressure of 6000 psi, limiting practical performance to column efficiency (N) of $\sim$20,000 plates or peak capacity (n) of $\sim$200 peaks [1,3]. The pioneering research in ultra-high-pressure LC (UHPLC) of Jorgenson [4] and Lee [5] and the subsequent commercialization of low dispersion 15,000-psi systems [6] used with sub-2 μm particle columns have allowed the setting of new performance benchmarks. In this article, the benefits, impacts, and practical issues of UHPLC in pharmaceutical analysis are reviewed and discussed.

Benefits—Increased Speed and Resolution. The fundamentals and benefits of UHPLC in pharmaceutical analysis are well documented [2,7–13]. Often touted is the dramatic improvement in analysis speed. For instance, a short 50×2.1-mm column packed with 1.7-μm particles ($N = 12,000$) can separate many mixtures isocratically in 1–2 min or 2–5 min under ballistic gradient conditions. Compared with conventional HPLC or fast LC with short 3-μm columns [14], UHPLC can increase speed by two- fivefold with similar resolution performance. This enhancement is important for high-volume analyses such as potency assays (content uniformity, toxicology dose checks, dissolution testing), bioanalytical testing, and high-throughput screening. The use of low-dispersion systems is mandatory [1–3,12], since these short columns produce small peak volumes (e.g., 13 μL at a retention factor of 3), which are readily affected by extracolumn band-broadening effects. A low-dispersion system (e.g., 4 μL system bandwidth of $\sim$10 μL) means low-volume fluidics in the sample flow path consisting of a micro injector, small i.d. connection tubing (0.002–0.005 in.), and a small detector flow cell (e.g., 0.5–2 μL) [2]. Secondary system requirements are lower system dwell volumes (e.g., 100–500 μL), to reduce gradient delay times and detectors (UV or MS) with high sampling rates. For assays using short columns (e.g., 50 mm), very high pressures might not be as critical; thus modified HPLC systems (6000 psi) or enhanced systems ($\sim$9,000–10,000 psi) can be effectively used.

Higher system pressure, however, is required for more difficult pharmaceutical applications such as stability-indicating assays or impurity testing methods. Currently, 100– 150-mm-long 3-μm columns (or 150–250-mm 5-μm columns) with $\sim$20,000 plates are used for these applications with run times of 20–40 min. UHPLC using 100-mm sub-2 μm columns ($N = 24,000$) can yield equivalent or better resolution in 10–20 min. Higher column temperatures can be used to reduce operating backpressure, although temperatures $>60°$C are rarely used because of the possibility of on-column degradation of the sample components. Longer sub-2 μm (e.g., 150-mm) columns are uniquely suited for very complex samples that are problematic for conventional HPLC [2,12]. Examples are impurity methods for combinational drug products, drug substances with multiple chiral centers (diastereomers), formulations with natural products, and analysis of proteomics or physiological fluids [2,11,12]. A comparative case study of an over-the-counter (OTC) laxative product shown in Figures 8.3 and 8.4 clearly illustrates the performance advantage of UHPLC.

Another advantage of UHPLC lies in the area of rapid method development, particularly ICH-compliant impurity methods for drug substances and products [2]. UHPLC offers quick feedback of the iterative process [2], often reducing time for column screening and mobile-phase optimization from days to hours.

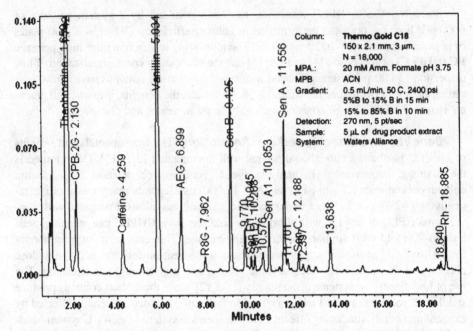

Figure 8.3. HPLC chromatogram of an OTC drug product extract from natural materials using a 150 × 2.1-mm-i.d. 3-μm column ($N = 18{,}000$) on conventional HPLC. Run time is 38 min. ΔP is 2400 psi. Assay potency is calculated by the summation of 8 active ingredient peaks. This chromatogram represents an optimum separation by conventional HPLC. (Reprinted with permission from Ref. 2.)

Practical Issues and Operating Nuances. As a practical technique, higher speed and higher-resolution performance should be achieved without sacrificing precision, sensitivity, and overall system reliability. Potential issues and operating nuances of UHPLC have been described in earlier publications [2,8,9]. These are reviewed and summarized here:

1. *Safety Concerns and Viscous Heating.* Safety concerns of UHPLC have been addressed by Lee et al. [15]. Due to low flow rates and compressibility of liquids, UHPLC poses little inherent dangers to users. Viscous heating of sub-2 μm-particle columns generates a thermal gradient effect, resulting in column efficiency loss [3,7,12]. This loss appears to be acceptable for columns <2 mm i.d., although can be quite significant in larger-inner-diameter columns [16].

2. *Column Selection and Lifetime.* Selection of columns and bonded phases was limited at first, but has improved considerably since 2004. Over 10 manufacturers now offer sub-2 μm-particle columns [12]. Newly introduced fingertight fittings (15,000 psi) allow easy connection to columns from all manufacturers. Column lifetime seems to be quite acceptable (e.g., >4000 injections) as reported in several studies [2,9,12].

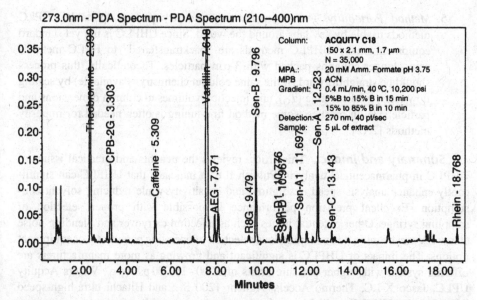

Figure 8.4. UHPLC chromatogram on the analysis of the same OTC drug product extract using a 150 × 2.1-mm 1.7-μm column (N = ~35,000) on UHPLC equipment, demonstrating improved resolution over that in Figure 8.3. Run time is 32 min, ΔP is 10,200 psi (lb/in.²). (Reprinted with permission from Ref. 2.)

3. *Injection Precision and Carryover.* The autosampler is perhaps the weakest link in UHPLC. Poor precision performance under partial loop injection has been reported in earlier studies [2,10,12]. Our data indicated dramatic improvements in relative standard deviation (~0.2–0.3% RSD for 1–7 μL injection volumes) were achieved by replacing the standard 100-μL sampling syringe with a 50-μL syringe [2]. Injector carryover is a potential problem for UHPLC that typically uses x–y–z push-loop-type autosamplers [1]. The higher operating pressures may also accentuate carryover problems of basic analytes. Judicious selection of the strong wash solution in addition to a blank injection at the end of each sequence are recommended precautions.

4. *Detector Sensitivity.* UHPLC uses columns and flow rates ideally suited for LC/MS analysis with electrospray ionization. However, for ICH-compliant impurity analysis, UV is the primary detector. Here small flow cells to reduce system dispersion must be balanced with long path lengths for signal enhancement. A newer generation of flow cells using fiberoptic technology appears to meet this challenge [6]. Another caveat arises from the use of low-volume mixers (e.g., 50–100 μL) to reduce system dwell volumes. Wavy detection baselines are often observed at low UV caused by inadequate blending of the mobile phases by high-pressure mixing pumps [2]. These issues can be reduced by using larger (e.g., 500-μL) or dynamic mixers, smaller piston strokes, and balanced absorbance mobile phase.

5. *Method Portability.* The pharmaceutical industry prefers "portable" HPLC methods usable by any labs around the world. Since UHPLC is not yet standard equipment, many UHPLC methods are "backtransferred" to HPLC methods using longer columns packed with 3-μm particles. Theoretically, this process should be straightforward (if the same column chemistry is available) by scaling flow rates, gradient time (T_G), and injection volumes to column dimensions and particle size. In practice, some method fine-tuning is often needed for impurity methods [2,12].

Summary and Impact. This article reviews the benefits and practical issues of UHPLC in pharmaceutical analysis. Published data indicated that UHPLC can significantly enhance analysis speed, resolution, and sensitivity while reducing solvent consumption. Excellent precision performance is possible with proper selection of sampling syringe. Other potential issues such as injection carryover and blending noise at low UV can be mitigated with judicious selection of operating parameters or equipment options. The impact of UHPLC is significant and growing as more manufacturers are offering system with higher pressure limits of 9000–15,000 psi (e.g., Waters Acquity UPLC, Jasco X-LC, Thermo Accela, Agilent 1200 SL, and Hitachi ultra-high-speed LC). Clearly, UHPLC is highly desirable. A system capable of pressure exceeding 15,000 psi might even be more marketable if other performance characteristics can be maintained. The issues and operating nuances should be viewed as technical challenges by manufacturers as well as learning curves for new users in transition to higher liquid chromatographic performance.

REFERENCES

1. M. W. Dong, *Modern HPLC for Practicing Scientists*, Wiley-Interscience, Hoboken, NJ, 2006.

2. M. W. Dong, Ultrahigh-pressure LC in pharmaceutical analysis: Performance and practical issues, *LC-GC* **25**(7), 656 (2007).

3. U. D. Neue, *HPLC Columns: Theory, Technology, and Practice*, Wiley-VCH, New York, 1997.

4. J. E. MacNair, K. E. Lewis, and J. W. Jorgenson, Ultrahigh-pressure reversed-phase liquid chromatography in packed capillary columns, *Anal. Chem.* **69**, 983 (1997).

5. N. Wu, J. A. Lippert, and M.L. Lee, Practical aspects of ultrahigh pressure capillary liquid chromatography, *J. Chromatogr. A* **911**, 1 (2001).

6. M. E. Swartz, UPLC (TM): An introduction and review, *J. Liquid Chromatogr.* **28**(7–8), 1253 (2005).

7. A. de. Villers, F. Lestremau, R. Szucs, S. Gelebart, F. David, and P. Sandra, Evaluation of ultra performance liquid chromatography: Part I. Possibilities and limitations, *J. Chromatogr. A* **1127**(1–2), 60 (2006).

8. I. D. Wilson, J. Plumb, J. Granger, H. Major, R. Williams, and E. M. Lenz, HPLC-MS-based methods for the study of metabonomics, *J. Chromatogr. B* **817**(1), 67 (2005).

9. S. A. Wren and P. Tchelitcheff, Use of ultra-performance liquid chromatography in pharmaceutical development, *J. Chromatogr. A* **1119**(1–2), 140 (2006).

10. A. D. Jerkovich, R. LoBrutto, and R. Vivilecchia, The use of ACQUITY UPLC in pharmaceutical development, *LC-GC* (Suppl.) **23**, 15 (May 2005).

11. H. Wang, R. W. Edom, S. Kumar, S. Vincent, and Z. Shen, Separation and quantification of two diastereomers of a drug candidate in rat plasma by ultra-high pressure liquid chromatography/mass spectrometry, *J. Chromatogr. B* **854**, 26 (2007).

12. N. Wu and A. M. Clausen, Fundamental and practical aspects of ultrahigh pressure liquid chromatography for fast separations, *J. Separation Sci.* **30**, 1167 (2007).

13. M. Petrovic, M. Gros, and D. Barcelo, Multi-residue analysis of pharmaceuticals in wastewater by ultra-performance liquid chromatography-quadrupole-time-of-flight mass spectrometry, *J. Chromatogr. A* **1124**, 68 (2006).

14. M. W. Dong, How hot is that pepper, *Today's Chemist at Work* **9**(2), 46 (2000).

15. Y. Xiang, D. Maynes, and M. L. Lee, Safety concerns in ultrahigh pressure capillary liquid chromatography using air-driven pumps, *J. Chromatogr. A* **991**, 189 (2003).

16. U. Neue, Waters Corp., private communication.

8.C. HYDROPHILIC INTERACTION CHROMATOGRAPHY (HILIC) IN PHARMACEUTICAL ANALYSIS (Yong Guo)

YONG GUO

Figure 8.5. Yong Guo.

Dr. Yong Guo (Fig. 8.5) is currently a Principal Scientist in Analytical Development, Johnson & Johnson Pharmaceutical Research and Development, LLC. He is a team leader managing a group of scientists performing method validation, clinical release, and stability testing. His responsibility also includes leading a cross-functional team managing late-development projects in a matrix environment. Prior to his employment at Johnson & Johnson, Dr. Guo worked as a Senior Analytical Chemist at Apotex, a generic company in Toronto, Canada. Dr. Guo is also an Adjunct Professor of Pharmaceutical Chemistry at Fairleigh Dickinson University, New Jersey. Dr. Guo received his Ph.D. degree in Analytical Chemistry from the State University of New York at Buffalo in 1996, followed by a year of postdoctoral training at the Biotechnology Research Institute, National Research Council, Montreal, Canada. Dr. Guo's research interest includes drug stability, HILIC, and ultra-high-performance liquid chromatography.

See Chapter 1, Table 1.3, h, r, v.

Introduction. The term *hydrophilic interaction chromatography* (HILIC) was first coined by Andrew Alpert in 1990 to describe a mode of liquid chromatography that can separate polar compounds on polar stationary phases using an aqueous organic mobile phase [1]. The practice of HILIC, however, may be traced to earlier years in the literature [2]. HILIC provides much stronger retention for polar compounds than does the reversed-phase mode, thus making it a viable alternative to the reversed-phase methods widely used in pharmaceutical analysis. HILIC has experienced a tremendous growth in the pharmaceutical industry as evidenced by the number of applications. The popularity of HILIC is also reflected by the number of columns that became commercially available in recent years.

Polar Stationary Phases for HILIC. HILIC separation of polar compounds relies on the help of a polar stationary phase. The simplest polar phase is underivatized silica, which is popular in HILIC, especially for bioanalytical applications. Irreversible adsorption is not as problematic in HILIC with a high level of water in the mobile phase ($>10\%$), as in normal-phase LC. Silica columns have shown greater resistance to overloading for ionized basic compounds than have reversed-phase columns. Bonded polar phases (e.g., amino, cyano, and diol phase) conventionally used in normal-phase LC have also been applied to HILIC. However, it is important to use the columns packed and stored in aqueous organic solvents instead of normal-phase solvents (e.g., hexane) to ensure reliable performance. In addition to these familiar phases, other polar phases, although not well known, have also been used in HILIC for a long time, such as the amide and aspartamide phases. The amide phase with a carbamoyl group attached to the silica surface through proprietary linkers has shown strong retention for polar compounds. The polymeric hydroxyethyl aspartamide phase on a PolyHydroxyethyl-A column was probably the first stationary phase designed for HILIC separation, and has been widely used in HILIC separation of proteins, peptides, and carbohydrates. Zwitterionic phases such as sulfobetaine and phosphocholine are also specifically designed and developed for HILIC. These phases possess both positive and negative charges on the same functional group, but exhibit only weak electrostatic interactions with charged solutes. The polymeric polyhydroxy phase is another newly marketed phase for HILIC application, but its performance needs further evaluation. In addition, many other polar phases traditionally used for different purposes, such as cyclodextrin and cation-exchange phases, have also been attempted for HILIC.

Retention and Selectivity of Polar Stationary Phases. In reversed-phase chromatography, solute retention is directly related to stationary-phase hydrophobicity (e.g., C8 vs. C18). In contrast, retention on polar columns in HILIC varies significantly with the structures of both solutes and functional groups on the polar phases [3]. When a mixture of nucleosides and nucleic acid bases are separated on six different polar phases (i.e., amino, amide, diol, silica, polyhydroxy, and sulfobetaine phases) under the same conditions, the relative retention on various phases is in the order of amide > amino > sulfobetaine > diol > silica > polyhydroxy phase. For a group of acidic compounds, in comparison, the relative retention order becomes amino $\gg$ amide $\sim$ diol > polyhdroxy > silica > sulfobetaine phase. The amino phase shows moderate retention for the nucleosides, but much stronger retention for the acids due to ion-exchange effect between the negative charged acids and positively charged amino groups. In contrast, the sulfobetaine phase has much weaker retention for the acids than for the nucleosides, possibly due to electrostatic repulsion by negative charges on the sulfobetaine phase. In addition, selectivity is also found to vary significantly with different phases and solutes. Similar selectivity is observed for the nucleosides and nucleic acid bases on the amide and amino phases as well as on the diol and polyhydroxy phases, but the silica phase and sulfobetaine phases show completely different selectivity despite similar retention on the two phases. For the acids, all the polar phases exhibit different selectivities. The amino and amide phases have similar electivities for four out of the six acids, but the silica and sulfobetaine phases have completely different selectivities for all the acids, even though both phases have significant amounts of negative charges. The large difference in retention and selectivity observed on various polar phases cannot be explained simply by different partitioning coefficients of the solutes between the mobile phase and the liquid layer on the surface of the stationary phase based on the hydrophilic partitioning model [1]. There must be strong secondary interactions between the solutes and functional groups on the stationary phases, resulting in the changes in retention and selectivity for different solutes and polar phases.

Effect of Chromatographic Parameters on Retention. In addition to the stationary phases, other chromatographic variables (e.g., organic solvent, mobile-phase pH, buffer concentration, and column temperature) are also critical to achieving desired separation. Acetonitrile is the most commonly used organic solvent in HILIC and is typically above 60% (v/v) in the mobile phase. A minimum of 3% water in the mobile phase is also necessary to maintain the stagnant water layer on the surface of the stationary phase. Many studies have shown nonlinear relationships between log k' and volume fraction of acetonitrile in the mobile phase, and the retention is very sensitive to small changes in organic content, particularly at higher levels. Other organic solvents, such as methanol, isopropanol, and tetrahydrofuran, have also been used for HILIC, and usually provide less retention than acetonitrile, but can introduce different selectivity. In HILIC, ionizable solutes are preferred to be charged to enhance retention. Therefore, the mobile phase pH is very important not only to the ionizable solutes but also to the polar phases with ionizable functional groups (e.g., the amino and silica phases). Electrostatic interactions between the charged solutes and stationary phases can lead to variation in retention and selectivity, but can be attenuated by the salt concentration in the mobile phase. Our study on the acidic compounds indicates that an increase in salt concentration results in a large decrease in retention on the amino phase due to reduced electrostatic attraction. In contrast, an increase in salt concentration leads to a small but significant increase in retention not only on the silica and sulfobetaine phases but also on the amide and aspartamide phases. This is attributed to the minimized electrostatic repulsion between the negatively charged acids and negative charges present even on the neutral phases (i.e., amide and aspartamide phases). Statistical evaluation reveals that the salt concentration is more critical than the acetonitrile content for the retention of the acids on the amino phase, and the effect of the salt concentration is more significant for the sulfobetaine and silica phases than for the amide and aspartamide phases [4]. Column temperature, on the other hand, is shown to have only minor effect on the retention of the acids.

Pharmaceutical Application. HILIC has found a wide range of applications in the pharmaceutical industry ranging from analysis of parent drugs and metabolites in biological samples to potency and purity analysis in drug substance and formulated dosage forms. Coupled with MS, HILIC/MS becomes a very powerful technique for bioanalysis with simplified sample preparation and enhanced sensitivity [5]. HILIC also offers different selectivity than the commonly used reversed-phase methods. The separation of an active drug compound (Comp 2) and three potential degradation products (Comp 1, 3

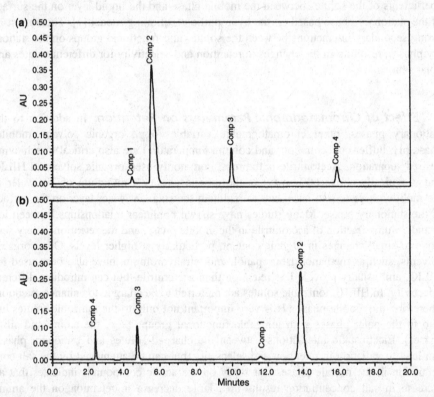

Figure 8.6. Chromatograms of the specificity solution on (a) RP-HPLC and (b) HILIC. (a) RP-HPLC conditions: column 100 × 4.6 mm 5 μm Xterra MS C18; gradient elution with mobile phase A, 0.09% phosphoric acid; and mobile phase B, acetonitrile 2–25% in 20 min; column temperature 35°C; flow rate 1 mL/min; UV detection 215 nm. (b) HILIC conditions: column 250 × 4.6 mm 5 μm YMC-pack Diol-120 NP; mobile phase 10 mM NH_4Cl in acetonitrile–water (95:5, v/v); column temperature 30°C; flow rate 1.5 mL/min; UV detection 215 nm. (Reprinted with permission from Ref 6.)

and 4) in both the reversed-phase and HILIC modes, as shown in Figure 8.6 [6], clearly demonstrates the potential for the use of HILIC as an orthogonal method for impurity or degradation product analysis. In addition, HILIC has also been successfully applied to counterion analysis for drug substance [7]. The HILIC method for counterion analysis does not require special instrument (e.g., ion chromatography) and can analyze either cations or ions on the same column.

Conclusion and Perspective. HILIC has shown great potential as an analytical method for polar compounds. A wide variety of polar stationary phases for HILIC application offer different retention and selectivity to achieve the desired separation. However, more research is needed to further understand the relationship between the functional groups and retention and selectivity. Despite its increasing popularity in the pharmaceutical industry, it should be noted that HILIC is not as versatile as reversed-phase chromatography. Few drug compounds are sufficiently hydrophilic to ascertain successful use of the HILIC method, and only a small portion of highly polar compounds may be amenable to HILIC. It is important to carefully evaluate the structure of the compounds before devoting significant time and effort to developing HILIC methods. Nevertheless, HILIC application to bioanalysis will continue to grow with widespread use of MS and its potential to perform fast analysis at relatively low backpressure, due to the relatively low viscosity of the mobile phase containing high levels of organic solvent.

REFERENCES

1. A. J. Alpert, Hydrophilic-interaction chromatography for the separation of peptides, nucleic acids and other polar compounds, *J. Chromatogr.* **499**, 177 (1990).

2. P. Hemstrom and K. Irgum, Hydrophilic interaction chromatography, *J. Separations Sci.* **29**, 1784 (2006).

3. Y. Guo and S. Gaiki, Retention behavior of small polar compounds on polar stationary phases in hydrophilic interaction chromatography, *J. Chromatogr. A* **1074**, 71 (2005).

4. Y. Guo, S. Srinivasan, and S. Gaiki, Investigating the effect of chromatographic conditions on retention of organic acids in hydrophilic interaction chromatography using a design of experiment, *Chromatographia* **66**, 223 (2007).

5. W. Niadong, Bioanalytical liquid chromatography tandem mass spectrometry methods on underivatized silica columns with aqueous/organic mobile phases, *J. Chromatogr. B* **796**, 209 (2003).

6. X. Wang, W. Li, and H. T. Rasmussen, Orthogonal method development using hydrophilic interaction chromatography and reversed-phase high-performance liquid chromatography for the determination of pharmaceuticals and impurities, *J. Chromatogr. A* **1083**, 58 (2005).

7. D. S. Risley and B. W. Pack, Simultaneous determination of positive and negative counterions using a hydrophilic interaction chromatography method, *LC-GC North Am.* **24**, 776 (2006).

8.D. SMALLER PARTICLES, HIGHER PRESSURES, AND FASTER SEPARATIONS: THE CURRENT STATUS OF PHARMACEUTICAL METHOD DEVELOPMENT 5 YEARS LATER (Todd D. Maloney)

TODD D. MALONEY

Figure 8.7. Todd D. Maloney.

Todd D. Maloney (Fig. 8.7), Ph.D., is a Senior Research Scientist with Eli Lilly and Company, Indianapolis, IN. He received his B.S. in Chemistry from the State University of New York at Oswego in 1996, and his Ph.D. in Analytical Chemistry in 2002 from the University at Buffalo, The State University of New York under the direction of Dr. Luis. A Colón. Maloney is currently investigating the utility of UHPLC for pharmaceutical analysis, automated method of development platforms, and application of chromatographic tools to real-time reaction monitoring/process analytical technologies. His other research interests are stationary-phase characterization, predictive software tools, and fundamentals of column performance in chromatography.

See Chapter 1, Table 1.3, e, h, r, v.

There has been a significant increase in the use of sub-2 μm particle columns with ultra-high-pressure liquid chromatography (UHPLC) in the pharmaceutical industry [1–3]. The increased use of this technology can be directly linked to the demand to reduce cost, improve productivity, and reduce the time required to bring new therapies to market. The theoretical advantages of sub-2 μm particles combined with UHPLC enabled faster method development, increased sample throughput, and increased mass sensitivity, which are well suited to these demands. But as with any emerging technology, widespread implementation and use in a regulated industry can be challenging.

While UHPLC has made numerous advances in instrument capabilities, sub-2 μm column technology has encountered a more difficult path toward widespread acceptance. Practitioners have had success achieving fast separations on the order of minutes or even seconds with 2-μm columns, but achieving speed with concomitant increase in separation efficiency has been more difficult to achieve in practice. Many of the shortcomings of sub-2 μm column performance are attributed to frictional heating and/or variability in flow due to solvent compressibility within the columns, both of which have been shown to have a detrimental effect on separation efficiency and reproducibility [4,5]. The uncertainty of sub-2 μm column performance is of considerable concern in the pharmaceutical industry, where column ruggedness and lot-to-lot variability are critical aspects of method development.

As an alternative to sub-2 μm columns, we have investigated columns packed with 2.5-μm porous particles in a 4.6 × 75-mm geometry. This column geometry achieves superior separation efficiency when compared with 2.1 × 50-mm sub-2 μm columns while exhibiting backpressures amenable to many HPLC instruments (<400 bar). Although the 4.6 × 75-mm 2.5-μm column geometry may result in up to a doubling of analysis time relative to the 2.1 × 50-mm sub-2μm column, the lower backpressure of the 2.5-μm column is amenable to flow rates >2 mL/min in conventional HPLC with no decrease in column performance. It should be noted that while increasing flow rate reduces analysis time, it is not conducive to direct coupling with MS without the use of a splitter. Another advantage of the 4.6 × 75-mm 2.5-μm column geometry is the increased loading capacity of the column relative to a 2.1 × 50 mm, which can offset the loss of sample from splitting when operating at high flow rates.

Another approach to the fast, highly efficient separations that we have investigated is the use of semiporous particles. Semiporous particles consist of a 1.7-μm solid core, surrounded by a 0.5-μm porous silica shell ($d_p = 2.7$ μm). This porous shell reduces axial diffusion while enabling increased mass transfer and separation efficiency. Semiporous particles also exhibit lower backpressures relative to sub-2 μm particles (~30% lower), enabling coupling of columns together to achieve a longer effective length column and hence higher separation efficiency. We have successfully coupled four 4.6 × 150-mm 2.7-μm semiporous columns operated in isocratic and gradient modes. Efficiencies of 130,000 plates are achieved in as little as 5 min (peak capacity of 200 when operated in gradient mode) for a 60-cm-long column.

These two examples of achieving increased separation efficiency with 2.5- and 2.7-μm particles are important for pharmaceutical method development where stationary-phase selectivity is seldom sufficient to separate critical pairs of impurities and degradation products that are structurally very similar (Fig. 8.8). Peaks 6 and 7 in Figure 8.8 were not resolved on a variety of different stationary phases in a 4.6 × 150 mm geometry, but are easily separated on a 4.6 × 150 mm (600 mm effective length), 2.7 μm column where column efficiencies exceed 125,000 theoretical plates.

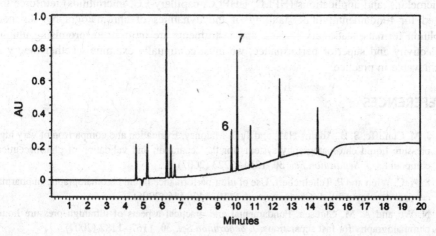

Figure 8.8. Separation of a 10-component pharmaceutical mixture on four 4.6 × 150 mm (600 mm effective length) 2.7-μm Halo C18 columns. Gradient conditions: mobile phase A, 0.1% TFA in water; mobile phase B, 0.1% TFA in ACN; 60/40 A/B, linear gradient to 100% B over 10 min, hold at 100% B for 10 min (20 min total run time). Injection, 2 μL; column temperature, 70°C; detection, PDA at 220 nm.

In this new era of narrow-bore columns and sub-2 μm particle sizes, the often overlooked value of the 4.6-mm i.d. column format is the ability to span both HPLC and UHPLC instrument platforms. As methods are developed in R&D laboratories, they will ultimately need to be transferred to a quality control or manufacturing lab where HPLC instruments are more prevalent. While scaling of column geometries and particle sizes is an alternative, increasing the diameter of the particle or column will require revalidation of the method, which is time-consuming and undesirable. Standardizing method development and column screening on a 4.6 × 75-mm 2.5-μm geometry not only provides fast, efficient separations but also enables greater flexibility in method transfer.

Our work has demonstrated that there are alternatives to performing fast, high-efficiency chromatography without the use of high pressures. The focus of our work is not intended to ignore the increased speed of analysis and capabilities offered by sub-2 μm columns; rather, we have carefully balanced the value of speed of analysis, quality of information, and flexibility of the analytical method for performing fast, efficient chromatography while enabling transfer of the method between HPLC and UHPLC platforms without changing the column. As UHPLC instrumentation continues to evolve, and as sub-2 μm column technology matures, it will continue to find increased acceptance and utility in pharmaceutical method development.

Perspective. More recent advances in column technology and instrumentation have stimulated interest in ultra-high-performance separation techniques. Whether one uses high pressures, small particles, high temperatures, or high flow rates, the goal is to reduce analysis time with a concomitant increase in quality of information. With these goals in mind separations scientists in the twenty-first century should revisit fundamentals of column performance and stationary-phase characterization. The abundance of new materials (semiporous silica, polymers), column formats (packed column, monolith, microchip), and applications (HPLC, UHPLC, capillary-LC, microfluids) reinforce the need for fundamental understanding of the dynamics of chromatography. As new column formats, stationary phases, and instruments are introduced, promising unique selectivity and superior performance, we must continually examine whether theory is achievable in practice.

REFERENCES

1. J. M. Cunliffe, S. B. Adams-Hall, and T. D. Maloney, Evaluation and comparison of very high pressure liquid chromatography systems for the separation and validation of pharmaceutical compounds, *J. Separation Sci.* **30**, 1214–1223 (2007).
2. S. A. C. Wren and P. Tchelitcheff, Use of ultra-performance liquid chromatography in pharmaceutical development, *J. Chromatogr. A* **1119**, 140–146 (2006).
3. N. Wu and A. M. Clausen, Fundamental and practical aspects of ultrahigh pressure liquid chromatography for fast separations, *J. Separation Sci.* **30**, 1167–1182 (2007).
4. F. Gritti and G. Guiochon, Consequences of the radial heterogeneity of the column temperature at high mobile phase velocity, *J. Chromatogr. A* **1166**, 47–50 (2007).
5. K. Kaczmarksi, F. Gritti, and G. Guiochon, Prediction of the influence of the heat generated by viscous friction on the efficiency of chromatography columns, *J. Chromatogr. A* **1177**, 92–104 (2008).

Key Publications

1. J. M. Cunliffe, and T. D. Maloney, Fused-core particle technology as an alternative to sub-2-micron particles to achieve high separation efficiency with low backpressure, *J. Separation Sci.* **30**, 3104–3109 (2007).

2. J. M. Cunliffe, S. B. Adams-Hall, and T. D. Maloney, Evaluation and comparison of very high pressure liquid chromatography systems for the separation and validation of pharmaceutical compounds, *J. Separation Sci.* **30**, 1214–1223 (2007).

3. J. A. Anspach, T. D. Maloney, and L. A. Colón, Ultrahigh-pressure liquid chromatography using a 1-mm ID column packed with 1.5-micron porous particles, *J. Separation Sci.* **30**, 1207–1213 (2007).

4. J. A. Anspach, T. D. Maloney, R. W. Brice, and L. A. Colón, Injection valve for ultrahigh-pressure liquid chromatography, *Anal. Chem.* **77**, 7489–7497 (2005).

8.E. THE USE OF MONOLITHIC MATERIAL AS ONLINE EXTRACTION SUPPORT FOR HIGH-THROUGHPUT BIOANALYTICAL APPLICATIONS (Raymond Naxing Xu[1], Matthew J. Rieser, and Tawakol A. El-Shourbagy)

RAYMOND N. XU ET AL.

Raymond Naxing Xu (Fig. 8.9), Ph.D., is an Associate Research Investigator at Global Pharmaceutical Research and Development division of Abbott Laboratories. He received his B.S. in Chemistry from University of Science and Technology of China in 1991. He graduated from Michigan State University in 1997 with a Ph.D. in Analytical Chemistry. He was a postdoctoral fellow in Pacific Northwest National Laboratory from 1997 to 1998. He has been with Abbott Laboratories since 1998.

His current research interests are in bioanalytical applications for pharmaceutical compound development and laboratory automation. He has served as a bioanalytical principal investigator on a number of toxicology, pharmacology, and clinical studies for drug candidates under development by Abbott Laboratories. He has authored or co-authored 17 peer-reviewed research and review articles in the areas of analytical chemistry and biochemistry.

Figure 8.9. Raymond Naxing Xu.

See Chapter 1, Table 1.3, h, p, v, z.

Chromatography has long been considered as the premier separation technique in drug discovery and development and has played an important role in numerous pharmacokinetics and metabolism studies. In recent years, hyphenated chromatographic techniques have been used more extensively for both quantitative and qualitative analyses in the pharmaceutical industry as a growing trend. Our research interests focus on developing fast and cost-effective bioanalytical methods using hyphenated chromatographic techniques for quantitative measurement of drug and metabolites in biological matrices (plasma, blood, serum, urine, and tissue).

One of the most prevalent hyphenated techniques, liquid chromatography tandem mass spectrometry (LC-MS/MS), has led to major breakthroughs in the field of quantitative bioanalysis since the 1990s because of its inherent specificity, sensitivity, and speed. It is now generally accepted as the preferred technique for quantitating small-molecule drugs, metabolites, and other xenobiotic biomolecules in biological samples. Samples from biological matrices are rarely directly compatible with LC-MS/MS analyses. Sample preparation has traditionally been done using protein precipitation (PPT), liquid–liquid extraction (LLE), or solid-phase extraction (SPE). Manual operations associated with these processes are very labor-intensive and time-consuming. A sample extraction method that has generated a lot of interest is the direct injection of plasma using an online extraction method. A major advantage of online SPE over offline extraction techniques is that the sample preparation step is embedded into the chromatographic separation and thus eliminates most of the sample preparation time traditionally performed at the bench. Different extraction supports have been explored allowing direct injection of biological fluids or extracts. These extraction supports or sorbents include restricted-access media (RAM), large-size particle, and disposable cartridges. Most online SPE approaches use column-switching to couple with the analytical columns. Various column dimensions can be configured for the fast analysis of a drug and its metabolites in biological matrix at the ng/mL level or lower.

We have investigated the use of monolithic material as extraction support for online solid phase extraction (SPE) in quantitation LC-MS/MS applications. Monolithic columns have attracted considerable interests lately for fast chromatographic separation. Because of their high permeability, separations on monolithic columns can be performed at a flow rate 5–10 times higher than that generally used with conventional supports. We have developed an automated procedure using online extraction with monolithic sorbent for determination of two HIV protease inhibitors, amprenavir (APV, Agenerase) and atazanavir (AZV, Reyataz), in human plasma. A short monolithic C18 cartridge was used for high flow extraction at 4 mL/min. Plasma samples were subjected to protein precipitation first with acetonitrile, and the supernatant was diluted and loaded onto the monolithic cartridge. Sample elution was accomplished with narrow-bore LC-MS/MS system with a total analysis time of 4 min. Figure 8.10 shows representative ion chromatograms of a lower-limit-of-quantitation (LLOQ) sample and a middle concentration quality control sample (mid-QC) by the monolithic-phase-based online extraction method. Very low carryover, on the order of 0.006%, was demonstrated using the monolithic-phase based method. The method has high recovery and good tolerance to the matrix effect, which was demonstrated in 12 lots of plasma. The backpressure of the monolithic extraction cartridge remained unchanged after 450 samples were injected. The performance of the monolithic-phase online extraction method was compared with that performed by an automated 96-well liquid–liquid extraction procedure, carried out using hexane and

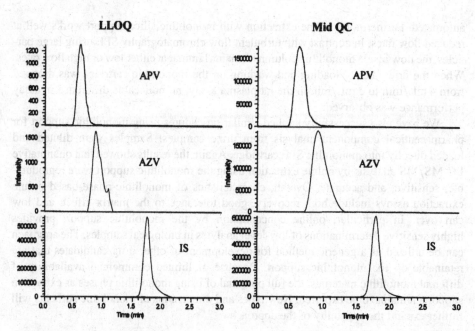

Figure 8.10. Representative ion chromatograms of a lower-limit-of-quantitation (LLOQ) sample and a mid-QC sample by monolithic-phase-based online extraction method (APV—Amprenavir; AZV—Atazanavir; IS—internal standard).

ethyl acetate as the extraction solvent. The results from both methods produced similar precision and accuracy.

We have applied the online extraction approach using a monolithic support to support the development process of a drug candidate from animal studies to clinical studies. Assays in rat, monkey, and human plasma have been developed and validated using a generic approach to sample processing procedures. Online concentration provided for highly sensitive determinations of the analyte in human plasma with a lower limit of quantitation of 0.077 ng/mL. All three assays exhibited high recovery and good tolerance to the matrix effect. The results show that quantitative LC-MS/MS analysis by online extraction using the monolithic support is reproducible, sensitive, and accurate enough to satisfy the requirements of both preclinical and clinical drug development. The online extraction approach using the monolithic support stems from a "dilute-and-shoot" scheme. Sample preparation with protein precipitation (PPT) is a commonly used technique in bioanalysis for plasma samples because of its simplicity. However, when analyzing supernatant from a plasma sample using PPT, salts and endogenous materials that are still present can cause ion suppression or enhancement leading to higher sample-to-sample variation in results. Our approach simply combines protein precipitation and solid-phase extraction. Online extraction efficiently adds another dimension of sample purification to protein precipitation by taking advantage of high-speed loading, extraction, and washing on the monolithic support. The extra dimension is added without additional sample preparation time or procedures because it is highly automated. Furthermore, online extraction with monolithic silica support works well at reduced flow rates. In contrast with turbulent flow chromatography SPE using large particles, the flow inside monolithic columns remains laminar at either low or high flow rates. When the flow rate for loading and washing on the monolithic cartridge was dropped from 4 mL/min to 2 mL/min in the rat plasma assay, no noticeable difference in assay performance was observed.

We have also successfully applied similar procedures using monolithic support for pharmaceutical component analysis from urine samples. Samples were diluted and loaded directly into monolithic SPE cartridge. Again, the results showed that quantitative LC-MS/MS analysis by online extraction using the monolithic support were reproducible, sensitive, and accurate. Overall, characteristics of monolithic-phase-based online extraction assays include high recovery, good tolerance to the matrix effect, and low carryover. In particular, online concentration by the monolithic support provides highly sensitive determinations of low-level analytes in biological samples. The approach can be utilized as a generic method for development of other drug candidates that are retainable on the monolithic support. Because of limited commercial availability of different monolithic materials, the full potential of using monolithic phases as extraction supports remains to be explored. New types and dimensions of monolithic sorbents will further expand the capability of the approach.

Perspective on Chromatography for the Twenty-First Century. Chromatographic technique will continue to evolve for pharmaceutical research and development in the twenty-first century. Speed of separation will be further advanced to satisfy the needs of shortened pharmaceutical development cycle time and rising sample volume. Resolution of chromatography is another key issue to be addressed to face the challenges of complex problems such as proteomic analysis and metabolomic analysis. Among the techniques used for pharmacokinetic and drug metabolism studies, we see that monolithic chromatography and ultra-performance liquid chromatography with sub-2 μm particles will make a significant impact on analytical scientists in either developing strategies for a new method or modernizing an analytical laboratory.

REFERENCES

1. R. N. Xu, L. Fan, M. Rieser, and T. A. El-Shourbagy, Recent advances in high-throughput quantitative bioanalysis using LC-MS/MS (review), *J. Pharm. Biomed. Anal.* **44**, 342–355 (2007).

2. R. N. Xu, B. Boyd, M. Rieser, and T. A. El-Shourbagy, Simultaneous LC-MS/MS determination of a highly hydrophobic compound and Its metabolite in human urine by on-line solid phase extraction, *J. Separation Sci.* **30**, 2943–2949 (2007).

3. R. N. Xu, G. Kim, B. Boyd, J. Polzin, M. Rieser, and T. A. El-Shourbagy, High-throughput quantitative LC-MS/MS assays by on-line extraction using monolithic support, *LC-GC* **25**, 396–405 (2007).

4. R. N. Xu, L. Fan, G. Kim, and T. A. El-Shourbagy, A monolithic-phase based on-line extraction approach for determination of pharmaceutical components in human plasma by PLC–MS/MS and a comparison with liquid–liquid extraction, *J. Pharm. Biomed. Anal.* **40**, 728–736 (2006).

9

CHROMATOGRAPHY— ADVANCES IN ENVIRONMENTAL AND NATURAL PRODUCTS, CHEMICAL ANALYSIS AND SYNTHESIS

Del Koch and Lyle Johnson

Analytical Biochemistry Laboratories (ABC), Columbia, Missouri

Charles W. Gehrke

Department of Biochemistry and the Agricultural Experiment Station Chemical Laboratories, College of Agriculture, Food and Natural Resources, University of Missouri, Columbia

Pat Sandra

Laboratory of Organic Chemistry, University of Gent, Ghent, Belgium

Richard B. Van Breemen

Department of Medicinal Chemistry and Pharmacognosy, UIC/NIH Center for Botanical Dietary Supplements Research, University of Illinois, College of Pharmacy, Chicago, Illinois

CHAPTER OUTLINE

Chromatography: A Science of Discovery. Edited by Robert L. Wixom and Charles W. Gehrke
Copyright © 2010 John Wiley & Sons, Inc.

9.A. CHROMATOGRAPHY—ADVANCES IN TECHNOLOGY FOR AGRICULTURAL, ENVIRONMENTAL AND NATURAL PRODUCTS, CHEMICAL ANALYSIS AND SYNTHESIS (Del Koch and Charles W. Gehrke)

DEL A. KOCH AND CHARLES W. GEHRKE

Figure 9.1. Del A. Koch.

Del A. Koch (Fig. 9.1) is currently a Principal Scientist in the Chemical Services Group Analytical Bio-chemistry Laboratories (ABC) Laboratories, Inc., Columbia, Missouri. His current responsibilities include program management for one of ABC's strategic alliance partners and providing regulatory and analytical consultation (within ABC and to external clients as well). His activities at ABC have included trace metal analysis (via ICP), performing or supervising pesticide residue chemistry studies (using GC and HPLC techniques), and bioanalysis of human blood samples. Prior to joining ABC Laboratories in 1980, Mr. Koch worked for the Washington University Medical School, St. Louis, Missouri and for the Grand Forks Energy Technology Center in Grand Forks, North Dakota. Mr. Koch received a B.S. in Chemistry from Centre College, Danville, Kentucky in 1976 and an M.S. in Analytical Chemistry from the University of North Dakota, Grand Forks, in 1979.

See Chapter 1, Table 1.3 d, h, w.

Figure 9.2. Charles W. Gehrke.

Charles William Gehrke (Fig. 9.2) was born on July 18, 1917 in New York City. He studied at The Ohio State University, receiving a B.A. in 1939, a B.Sc. in Education (1941), and an M.S. in Bacteriology in (1941). From 1941 to 1945, he was Professor and Chairman of the Department of Chemistry at Missouri Valley College, Marshall, Missouri teaching chemistry and physics to World War II Navy midshipmen (from destroyers, battleships, and aircraft carriers in the South Pacific) for officer training. These young men returned to the war as deck and flight officers. In 1946, he returned as instructor in agricultural biochemistry to The Ohio State

University, receiving his Ph.D. in 1947. In 1949, he joined the College of Agriculture at the University of Missouri—Columbia (UMC), retiring in Fall 1987 from positions as Professor of Biochemistry, Manager of the Experiment Station Chemical Laboratories, and Director of the University Interdisciplinary Chromatography Mass-Spectrometry facility. His duties also included those of State Chemist for the Missouri Fertilizer and Limestone Control laws. He was Scientific Coordinator at the Cancer Research Center in Columbia until 1997. See Editors section for additional information.

See Chapter 1, Table 1.3, a, c, d, e, h, p, t, x, z.

In 1968 with private funds. I founded the Analytical Bio-Chemistry (ABC) Laboratories along with Jim Ussary and David Stalling (see also Chap. 7.B of this book about ABC Labs). The University of Missouri was only mildly receptive. Technicon Corporation of Tarrytown, NY wanted me to set up an automated laboratory for testing of agricultural feeds, fertilizers, and amino acids. This was a natural, as my job in the Experiment Station Chemical Laboratories was the same for the College of Agriculture and its staff. The Graduate School offered help and a site in the University Research Park near the Reactor facility. After due consideration I declined, as I could not accept the work and avoid a conflict of interest. We then raised $200,000 of private money, bought 70 acres of land on the outskirts of Columbia, and started ABC Laboratories. Jim Ussary was ABC's first CEO and served from 1968 to 1980.

Introduction. As noted earlier in Chap. 7 of this book, since the 1960s, developments in analytical chemistry have markedly increased our understanding of life and environmental effects on the basis of molecular chemistry. We have been able to detect and quantify trace amounts of compounds affecting the very basics of life processes. The existence and persistence of some industrial and agricultural chemicals since 1960 has caused concern in the reproduction and longevity of higher-foodchain animals. The discovery and innovation of progressively more sensitive and accurate analytical procedures allowed scientists and consequently government regulators to develop chemistries and to implement enforcement guidance, harmonious with nature and beneficial to humankind and wildlife environment.

Detector and Column Developments. Two significant developments in gas chromatographic analytical separation techniques had a large impact on the environmental regulatory process. J. E. Lovelock, in 1958, observed a physical characterization of noble gases ionized by radioactive beta emitters. This characteristic was quenched by electrophilic compounds passing through the gaseous plasma, which led to the development of the argon-ionization detector, followed by the electron capture detector (ECD). The latter is highly sensitive to halogen-containing compounds and many other organic species (see also Lovelock presentation in Chap. 5 of our 2001 book [1]).

Although chromatographic separation of chlorophyll pigments was reported by Tswett decades earlier, gas–solid chromatography was reported using a common detergent as the solid phase, with an inert gas passed through the mobile phase, was used for the separation of pesticides [1]. Early detectors of gas–solid chromatography used biological detectors and other awkward devices; the science was desperately in need of an electronic means of monitoring the elution of the analyte. The development of the Lovelock ECD was such a significant event, which, through its capacity to detect pesticides in just about everything, in all kinds of matrices. This set the scene for Rachel Carson in 1962 and the environmental movement. It detected picomolar quantities of halogenated compounds, such as for industrial control releases in the 1940s, and a couple of decades following. These chlorinated bullets were knocking holes in the reproductive function of higher-foodchain birds, most notable our national avian hero, the American Bald Eagle. The process of eggshell thinning, and hatchling mortality caused a significant decline in the population of the eagle. The GC analyses showed that the use of poly(chlorinated biphenyls) (PCBs), an excellent electrical insulator in transformers (though resistant to biodegradation), found its way into the *environmental* foodchain. Similar chemicals used for agricultural use, such as DDT, chlordane, heptachlor, aldrin, dieldrin, endrin, BHC, lindane, methoxychlor, and toxaphene, were suspected of also contributing to adverse environmental effects, which were complicated by their long biological half-lives and their tendency to accumulate in body tissues. The availability of the GC-ECD to selectively detect and quantify these compounds in various matrices facilitated the development of screening and monitoring programs (sponsored by governmental agencies, academic research groups, and industrial organizations) for wildlife, soil, water, air, and foodstuffs, yielding a total biosystem perspective of chemical exposure.

Other selective detectors that were developed for the GC included the nitrogen–phosphorus detector (NPD), which utilized a glass or ceramic bead coated with alkali, which responds with increased thermionic emission response in the presence of nitrogen or phosphorus.

The flame photometric detector (FPD), which utilized a photomultiplier tube to enhance the signal from flame emission, could be operated to selectively detect phosphorus, sulfur, or even tin, by use of the appropriate filter between the flame and the PM tube. The electrolytic conductivity detector (ELCD) could be operated very selectively for either halogens (in the ELCD-X mode) or nitrogen (in the ELCD-N mode). While this detector was highly specific for these targeted heteroatoms, the earlier models were more expensive and harder to maintain than many of the other detector options, and hence it was seldom the detector of choice in the analytical laboratory.

This array of detectors was sufficient for the analysis of many classes of pesticides, as the majority of pesticides contained either one or more halogens (as discussed previously), phosphorus (such as the widely used organophosphorus insecticides) and/or nitrogen (such as atrazine and other triazine herbicides). Indeed, the multiresidue screening procedures developed by the US FDA as part of their pesticide residue food surveillance programs are based on analysis of food extracts, using standardized GC columns and temperature parameters, in conjunction with all of the detectors described previously. Later, the coupling of GC with mass spectral detection allowed for, in general, better sensitivity and increased selectivity. While this technique has broad applicability, it is limited to compounds that can be successfully gas-chromatographed (or be derivatized to produce a chromatographable species).

The maturation of high-performance liquid chromatography (HPLC) in the 1970s as a separation tool offered a new analytical approach for some of the newer agricultural chemicals that were too polar (or lacked the requisite thermal stability) to be analyzed by GC methods. HPLC offered a more nearly universal analytical capability, with its ability to analyze for a specific molecule primarily limited only by the compound's ability to be dissolved into some combination of HPLC-compatible solvents. The earliest available HPLC detectors, such as refractive index (RI) or ultraviolet (UV) absorbance, were somewhat lacking in selectivity. Fluorescence detection was more specific, but was limited to those compounds with fluorescent properties, or that could be derivatized to add a fluorescent "tag" to the molecule, either prior to analysis or after elution from the HPLC column prior to entering the flowthrough detector (the latter technique is commonly described as *postcolumn derivatization*). This technique was particularly useful for small, polar compounds with poor UV absorbance, such as the widely used glyphosate herbicide.

As was the case with GC, once the technology was developed to successfully interface the HPLC column with a mass spectral detector, (LC/MS), this became a means of detecting (typically with improved sensitivity and selectivity) a wide variety of compounds, with the primary limitation of LC/MS being the ability to ionize the targeted analytes in the detection system. The development of triple quadrupole (LC-MS/MS) and ion trap mass spectra detectors introduced a new degree of selectivity. With these detectors, response at a given retention time can be screened not only on the basis of characteristic ions related to the parent molecule; these ions may be further fragmented to characteristic product ions, thus accomplishing a second level of screening. LC-MS/MS analysis has matured to the degree that it is now the analytical technique of choice in the analytical residue laboratory, despite the relatively high cost of the instrumentation.

Chromatography in Synthesis Work. In the past 10 or more years prior to 2010, ABC Laboratories has pursued the custom small-scale synthesis market as a complement to the core business in toxicology and analytical chemistry. One specialized service area in which ABC has been particularly successful is the synthesis and/or purification of radiolabeled chemicals, such as ^{14}C-labeled compounds for use as radiotracers in metabolism investigations. Chromatography plays an important role in the production of the finished synthesis products. Purification of the synthetic product is commonly required to achieve the desired purity. Thin-layer chromatography (TLC) may be used to purify the product by separating impurities on the TLC plate and recovering the compound by scraping the appropriate area of the plate and resolubilizing in solvent. Another technique commonly used for purification is HPLC with a semipreparative column; by collecting and combining the product-containing fractions from the column output, impurities may be eliminated and the product reconstituted by evaporating the HPLC solvent. Following the isolation of the product, either HPLC (with an analytical column in place) or GC may be used to measure the purity of the finished product. When coupled with mass spectrometric detection, confirmation of identity may be accomplished as well.

Significance. In the years since the 1960's "environmental movement", there have been many monitoring programs for food by USDA and FDA to determine the safety of food items. Superfund cleanup sites have required a tremendous amount of effort and accountability. More recently, in August 1996, Congress entered into legislation for the "Food Quality and Protection Act," which required a more holistic approach to agricultural chemical exposure. It required evaluation of the effects of chemicals with "common mechanistic" modes, a 10-fold safety factor for children, screening of compounds for possible endocrine disruption, and a total perspective evaluation for all exposures, whether the source is by food, home, lawn, workplace, water, and/or air. In order to measure the actual levels of exposure, instead of perceived or estimated exposure, refinements in chromatographic techniques for quantification and confirmation of compounds were developed and effectively implemented. In particular, the successful interfacing of mass spectrometric detectors with GC and HPLC separation techniques has allowed for improved accuracy at much lower levels of quantification than were achievable when environmental exposures first became a major concern. Many of the newer agricultural chemicals are efficacious at much lower rates than those from years past; correspondingly, the developments in chromatographic techniques have proved capable of measuring much lower concentrations in foods, soil, and water.

Conclusions. The ever-increasing desire for accurate, low-level determination of environmental contaminants has been satisfied by remarkable advances in chromatographic separation techniques and the associated detection and data-handling systems. Although it may sometimes be difficult to ascertain which is driving the other, the end result has been the availability of more definitive data with which safety assessments may be made. ABC Laboratories, with its core competency in the field of analytical chemistry, has taken advantage of the many developments in the field of chromatography to offer state-of the-art testing services to satisfy the most up-to-date data requirements. The development of ABC's synthesis service line complements the analytical team by being able to supply both radiolabeled and unlabeled standard materials for use in a variety of analytical studies.

LC-MS/MS also has applications in the area of protein sequencing, which provides for better understanding of metabolic systems. Similarly, understanding and identifying gene traits via analytical protein mapping have been, and will continue to be, important techniques for developing and producing transgenic plants capable of resisting disease and insects. Herbicidal resistance allowing selective plant control, and engineering of plants for selected food or energy traits are common goals of many agricultural industries. Continued development of advanced analytical technology will allow our twenty-first century civilization to enjoy higher standards of living, with curtailed adverse impact on our environment.

Summary and Perspective. Advances in gas chromatographic (GC) and high-performance liquid chromatographic (HPLC) separation techniques, coupled with the development of more selective and sensitive detectors, have facilitated the measurement of environmental contaminants in complex matrices at increasingly lower levels. By allowing more accurate measurements at extremely low concentrations, these advances in the field of chromatography have to a great extent satisfied the public demand for data that are meaningful to the assessment of the safe usage of agricultural chemicals. Since its founding in 1968 by Professor of Biochemistry, Charles W. Gehrke and two of his staff and former graduate students, Dr. David Stalling and James Ussary, ABC Laboratories has grown along with the increasing need to generate comprehensive data for use by the regulatory community in their decisionmaking activities. ABC, from its inception, has been involved in analytical chemistry and its application to environmental evaluation of agricultural (as well as other) chemicals. ABC has more recently developed expertise in custom synthesis, with a successful specialization in the preparation of radiolabeled chemicals for use in metabolism investigations. Chromatography plays an important role in this service line as well, in both the purification of synthetic products (thin-layer chromatography and semipreparative HPLC) and the purity assay of the finished product (GC and HPLC).

REFERENCE

1. C. W. Gehrke, R. L. Wixom, and E. Bayer (Eds.), *Chromatography: A Century of Discovery (1900–2000)—the Bridge to the Sciences/Technology*, Vol. 64, Elsevier, Armsterdam, 2001.

9.B. SOLVING THE BELGIAN DIOXIN CRISIS BY ANALYZING PCBs IN FATTY MATRICES (P. Sandra, F. David, and T. Sandra)

P. SANDRA, F. DAVID, and T. SANDRA

Figure 9.3. Pat J. Sandra.

Pat J. Sandra (Fig. 9.3) (born Oct. 20, 1946 in Kortrijk, Belgium) received his M.S. in Chemistry in 1969 and his Ph.D. in Sciences in 1975 from Ghent University, Belgium. Since then, he has been on the Faculty of Sciences at the same university, where he is currently Professor in Separation Sciences in the Department of Organic Chemistry. In 1986 he founded the Research Institute for Chromatography (RIC) in Kortrijk, Belgium, a center of excellence for research and education in chromatography, mass spectrometry and capillary electrophoresis. He is currently also extraordinary professor at the Department of Chemistry at the University of Stellenbosch, South Africa, at the Department of Analytical Chemistry, Evora, Portugal and Director of the Pfizer Analytical Research Center at Ghent University.

Pat Sandra is active in all fields of separation sciences (GC, LC, SFC, and CE) and major areas of his research are high-throughput, high-resolution, miniaturization, hyphenation, and automation to study chemicals, pharmaceuticals, natural products, food products, and pollution.

See Chapter 1, Table 1.3, d, e, h, k, p, w.

Introduction. In Belgium, 1999 will be remembered as the year of the dioxin crisis. Early in the year (January–February), chicken farmers observed premature death (up to 25%) and nervous disorders among chicks, combined with a high ration of eggs failing to hatch. Initially, different hypotheses were forwarded, including the possibility of mortal doses of an antibiotic (salinomycin) used as growth promoter in animal feed. One month after the beginning of these problems, a producer of animal feed took the initiative to send a sample of suspected animal feed and of chicken fat to a laboratory specialized in the analysis of poly(chlorinated dibenzodioxins) (PCDDs) and poly(chlorinated dibenzofurans) (PCDFs). These compounds are often referred to as "dioxins", but in fact they consist of two classes of chlorinated aromatic compounds. It has been shown in the past that chickens are very sensitive to the toxicity of chlorinated aromatic compounds. The response time for the dioxin analysis, however, was 4 weeks. At the end of April, the Belgian authorities were informed that very high concentrations of dioxins were detected in animal feed and chicken fat, but hardly any measurements were taken at that time. Another set of samples was taken and sent to the same laboratory for confirmation, which took another 4 weeks. Only at the end of May, 4 months after the problems at the farms started, the public was informed about a food contamination and strong measurements were taken, including the destruction of several lots of eggs, chicken meat, and related food products. At that stage the source of the contamination was still unknown. The analyses, however, showed a 1000-fold higher concentration of PCDDs and PCDFs in the animal feed fat and in the chicken fat $(1-2\,\text{ng/g}$ fat) versus normal background values $(1-5\,\text{pg/g}$ fat). It was also observed that the PCDFs were present in much higher concentrations than the PCDDs.

PCDDs and PCDFs are not technical chemicals, but are byproducts in the synthesis of several chlorinated compounds such as pentachlorophenol (PCP, an insecticide, used mainly in wood preservation), poly(chlorinated biphenyls) (PCBs), or the chlorophenoxy acid herbicides 2,4-D and 2,4,5-T. Dioxins are also formed during poorly performed combustion of chlorinated material and can therefore also be present in the emissions of waste incinerators.

At the Research Institute for Chromatography (RIC), we did not believe that dioxins as such could contaminate food products at the ppb level without the presence of other chlorine-containing contaminants in much higher levels. Moreover, the high PCDF/PCDD ratio supported a hypothesis of PCB contamination. Poly(chlorinated biphenyls) are, in contrast to PCDDs and PCDFs, technical products that have been used intensively in electric capacitators, transformers, vacuum pumps, and even as adhesive fire retardant, in inks, etc. Although their production has been banned for many years, they are still widely present in old electric installations. Since the disposal costs of PCB-containing materials is very high, mixing PCB oils with mineral or edible oils becomes an attractive and very profitable alternative. For many years we have noticed on several occasions that fats used in the production of animal feed were of low quality, not only containing polymerized lipids but also contaminants such as mineral oil. The link between dioxins, PCBs, contaminated oils, fat, animal feed, and finally chicken and eggs was very logical to us. Consequently we analyzed some samples for PCBs present in animal feed, chicken fat, and eggs. The link was made, and it was very clear that, in this case, used (spent) PCB oil was the source that later indeed could be traced and confirmed.

Poly(chlorinated biphenyls) consist of a group of 209 possible congeners ranging from mono- to decachlorobiphenyls. Although their acute toxicity is lower than that of dioxins, several studies have shown that PCBs are linked to several negative health effects, including endocrine-disrupting activity, reproductive function disruption, developmental deficits in newborn and decreased intelligence in school-aged children who had in uteri exposure. Moreover, the toxicity of dioxins and organochloro pesticides increases in the presence of PCBs, PCBs accumulate in fat and build up in the foodchain. Levels of 50–500 ppb (ng/g fat) have also been detected in human fat samples [1] and in breast milk.

Because PCBs are present at higher levels than dioxin, PCB analysis is much faster and cheaper (by a factor of 10) than dioxin analysis. The analysis of PCBs was finally accepted by Belgian and European authorities. The norm for PCBs in fatty food (more than 2% fat content) was set at 200 ppb (200 ng/g fat) for the sum of seven PCB congeners (PCBs 28, 53, 101, 118, 138, 153, and 180). Using the much faster PCB-monitoring analysis, some 50,000 analyses were performed in various laboratories. This enabled us to deliver certificates for noncontaminated food and to localize all contaminated farms.

An overview is given of the sample preparation and analytical methods that can be used to perform PCB analysis in fatty matrices. A short description of the rapid screening method developed at RIC is presented here; more details may be found in Chap. 5 of Ref. 2. During the crisis, the GC analysis was further improved by using the retention-time-locking concept, and the time was reduced by using narrow-bore capillary columns and optimization via the method translation software [3].

Methodologies for the Analysis of PCBs in Fatty Matrices. The analytical scheme for the analysis of PCBs in fatty matrices consists of three steps: extraction, cleanup, and analysis. In the first step, the lipophilic contaminants are extracted from the matrix using apolar solvent. The extract contains the lipids, PCBs, PCDDs, PCDFs, and other apolar solutes, such as pesticides, polycyclic aromatic hydrocarbons, and mineral oil. For the extraction of fat and apolar solutes from meat, eggs, and the fatty food matrices in general, different extraction techniques are available. These include Soxhlet extraction (or the automated versions Soxtec or Soxtherm), solvent extraction using ultrasonic agitation, accelerated-solvent extraction (ASE), microwave-assisted solvent extractions (MASE), and supercritical-fluid extraction (SFE). All these techniques perform similarly and are efficient in extracting fat and contaminants such as PCBs. In general, classical sample extraction techniques (Soxhlet) can handle larger samples than the more recent techniques such as ASE or MASE, but the latter have the advantages of automation, speed, and low solvent consumption. Selection of the best technique is therefore based on factors such as sample throughput, laboratory history, and experience with PCBs in other (environmental) samples.

In the second step, the PCBs are fractionated from the (coextracted) fat matrix. For this fractionation, several techniques can be applied, including treatment with sulfuric acid (fat destruction), gel permeation chromatography (GPC), column chromatography on a polar adsorbent (silica, aluminum oxide), and solid-phase extraction (SPE) [4]. A critical parameter in the selection of the sample cleanup method is sample capacity. Column chromatography can handle more fat than GPC or SPE, but the latter techniques can be fully automated. In combination with state-of-the-art PCB analyzers, sufficient sensitivity can be reached using miniaturized sample preparation techniques.

Finally, the cleaned extract can be analyzed using capillary gas chromatography with electron capture detection (CGC-ECD) or capillary gas chromatography with mass selective detection (CGC-MS). CGC-ECD is extremely sensitive and in most cases sufficiently selective for the detection of PCBs extracted from fat. CGC-MS in the selected-ion monitoring mode is somewhat less sensitive, but more specific since the presence of PCBs is confirmed by the detection of several ions per congener in a well-defined relative ratio.

Conclusion. The contamination of Belgian food products was initially referred to as a "dioxin crisis." The contamination source, however, was poly(chlorinated biphenyls). PCBs can be analyzed much more rapidly and economically than PCDDs and PCDFs, and therefore the analysis of PCBs was the only possible solution to the crisis. A new method, using ultrasonic extraction, followed by matrix solid-phase dispersion cleanup and CGC-μECD or CGC-MS, was used for rapid screening of the samples. Retention time locking and fast capillary GC completed the introduction of new methodologies in the PCB screening.

REFERENCES

1. A. Pauwels, F. David, P. Schepens, and P. Sandra, Automated gel permeation chromatographic clean-up of human adipose tissue for analysis of organichlorine compounds, *Int. J. Environ. Anal. Chem.* **73**(3), 171–178 (1999).

2. C. W. Gehrke, R. L. Wixom, and E. Bayer (Eds.), *Chromatography: A Century of Discovery (1900–2000)—the Bridge to the Sciences/Technology*, Vol. 64, Elsevier Science, Amsterdam, 2001.

3. F. David, D. R. Gere, F. Scalan, and P. Sandra, Instrumentation and application of fast high-resolution capillary gas chromatography, *J. Chromatogr. A*, **842**, 309–319 (1999).

4. S. A. Barker, Matrix solid-phase dispersion, *Curr. Trends Devel. Sample Prepar. Suppl. to LC · GC* 37 (May 1998).

9.C. DEVELOPMENT OF LIQUID CHROMATOGRAPHY–MASS SPECTROMETRY AND APPLICATIONS TO NATURAL PRODUCTS RESEARCH (Richard van Breeman)

RICHARD VAN BREEMEN

Professor Richard van Breemen (Fig. 9.4) received his Ph.D. in Pharmacology from the Johns Hopkins University in Baltimore, Maryland, in 1985, and a B.A. in Chemistry from Oberlin College in Oberlin, Ohio, in 1980. During 1985–1986, he carried out postdoctoral research in the Middle Atlantic Mass Spectrometry Facility at the Johns Hopkins University School of Medicine. After establishing a mass spectrometry facility for biotechnology research as Assistant Professor of Chemistry at North Carolina State University, Prof. van

Figure 9.4. Richard van Breemen.

Breemen joined the University of Illinois College of Pharmacy in 1994 as Associate Professor of Medicinal Chemistry. Since 2000, he has been Professor of Medicinal Chemistry at the University of Illinois. He is Co-Director of the UIC/NIH Center for Botanical Dietary Supplements Research, which was founded in 1999, and is also the Scientific Director of the Mass Spectrometry Laboratory of the UIC Research Resources Center.

Professor van Breemen is author or co-author of more than 200 scientific papers and two patents. Since 1997, he has been Editor-in-Chief of the journal *Combinatorial Chemistry & High Throughput Screening*. In addition, he is on the Editorial Board of *Biomedical Chromatography*. He has been recognized with awards for outstanding papers by the International Society for the Study of Xenobiotics and the American Oil Chemists Society. In 1995, he was named by the Editors of *Spectroscopy* as one of 19 "Bright Young Stars" in the field of analytical spectroscopy. Finally, he is the recipient of the University Scholar faculty award from the University of Illinois as well as the 2008 Harvey W. Wiley Award from the AOAC International.

See Chapter 1, Table 1.3, h, p, r, t, z.

Natural products have been the subject of mass spectrometric analysis since the beginning of organic mass spectrometry more than 50 years ago. However, until the development of desorption ionization techniques and liquid chromatography–mass spectrometry (LC-MS), natural product studies using chromatography and mass spectrometry were limited to volatile and thermally stable compounds that could be analyzed in the gas phase using electron impact or chemical ionization and then later using gas chromatography–mass spectrometry (GC-MS). As ionization technology advanced, isolated nonvolatile compounds could be analyzed using mass spectrometers equipped with desorption ionization techniques such as fast-atom bombardment (FAB), laser desorption, or field desorption. When direct liquid introduction methods, such as continuous-flow FAB mass spectrometry were introduced, mixtures of natural products, such as botanical extracts containing carotenoids or chlorophylls, could be analyzed for the first time using LC-MS and then LC-MS/MS. The mass chromatogram and mass spectra shown in Figure 9.5 represent early (ca. 1990) examples of LC-MS and MS/MS of carotenoids using continuous-flow FAB mass spectrometry.

Since desorption ionization techniques tended to produce abundant molecular ions, protonated molecules or deprotonated molecules with minimal fragmentation, little structural information was usually obtained using LC-MS (see Fig. 9.5b). Therefore, the use of

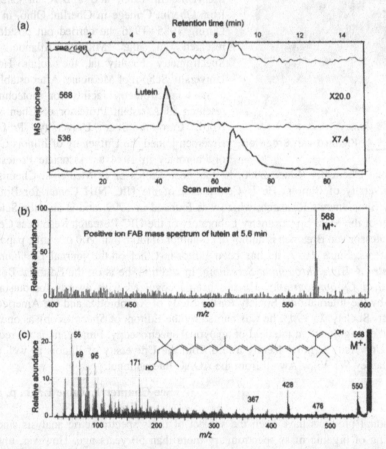

Figure 9.5. Continuous-flow fast atom bombardment LC-MS and MS/MS analysis of carotenoids. (a) Lutein was separated from α-carotene and β-carotene using a C30 reversed-phase column. The fast-atom bombardment (FAB) matrix, 3-nitrobenzyl alcohol, was added postcolumn, resulting in some band broadening. (b) Positive-ion FAB mass spectrum of lutein eluting at a retention time of 5.6 min. (c) product ion tandem mass spectrum of lutein obtained using collision-induced dissociation. Structurally significant fragment ions were detected such as the ion of m/z 428, which corresponds to loss of the terminal ring not in conjugation with the polyene chain.

collision-induced dissociation of selected-ion precursors followed by tandem mass spectrometry is usually required to obtain structurally significant fragment ions of natural products as shown in Figure 9.5c. As tandem mass spectrometers became more widely available in the 1980s, the use of MS/MS for natural products analysis resulted in increased structural information (due to the use of collision-induced dissociation), greater selectivity (MS1 and MS2 could be used to select particular molecular ions and structurally significant product ions, respectively) and enhanced SNR (noise could be reduced at each stage of tandem mass spectrometry).

During the 1990s, LC-MS using electrospray and atmospheric pressure, with chemical ionization, eclipsed all previous LC-MS techniques. These innovations have enabled biomedical mass spectrometrists to characterize and measure natural products belonging to virtually any chemical class—even those present in complex mixtures and matrices. An example of the use of electrospray LC-MS to analyze an extract of the mushroom *Inonotus obliquus* for betulinic acid is shown in Figure 9.6. Betulinic acid is under investigation as a cancer chemoprevention agent and occurs in the bark of the birch tree. Since *Inonotus obliquus* infects birch trees, the source of betulinic is probably in this the host fungus.

During LC-MS, accurate mass measurement and high-resolution mass spectrometry may be used to enable the determination of the elemental compositions of natural products in mixtures such as complex extracts. In Figure 9.6, LC-MS with accurate mass measurement was used to detect two isomeric compounds in an extract of *I. obliquus* that corresponded to betulinic acid. Comparison with a betulinic acid standard facilitated the identification of the peak eluting at 9.6 min as betulinic acid. The other peak was probably isomeric trametanolic acid.

In addition to compound identification, LC-MS and LC-MS/MS may be used for the quantitative analysis of natural products in complex mixtures. Quantitative analyses are useful for the standardization of botanical dietary supplements and in support of clinical trials of the safety and efficacy of botanical dietary supplements and isolated natural products. As an example, LC-MS-MS assays were developed for the quantitative analysis of isoflavones and triterpene glycosides in extracts of red clover (*Trifolium pratense* L.) and black cohosh (*Cimicifuga racemosa* L. Nutt.) at the UIC/NIH Botanical Center for Dietary Supplements Research [1]. Then, these standardized extracts were used to measure isoflavones and triterpene glycosides in blood and urine samples obtained during phase I pharmacokinetics studies and phase II clinical trials of the safety and efficacy of these botanical dietary supplements in the relief of hot flashes in menopausal women [1].

For the discovery of pharmacologically active compounds, LC-MS screening assays have been developed that facilitate the rapid isolation and identification of the active natural products in complex extracts. As an example, ultrafiltration LC-MS was developed and is being used at the UIC/NIH Botanical Center for Dietary Supplements Research for the discovery of estrogenic natural products in botanical dietary supplements used by menopausal women [2]. Additional LC-MS screening assays based on targets such as RXR, cyclooxygenase-2, and Keap1 are being used to screen extracts of marine bacteria and botanicals for cancer chemoprevention agents [3]. After pharmacologically active compounds are identified, LC-MS and LC-MS-MS may be used to facilitate in

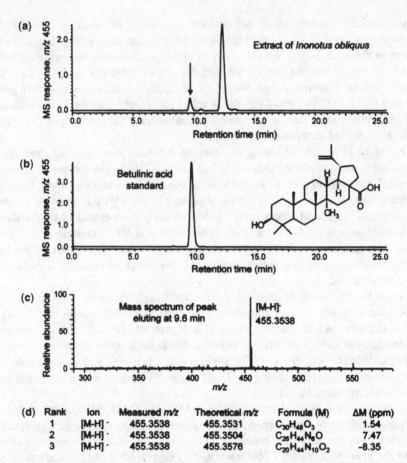

Figure 9.6. Electrospray LC-MS analysis of an extract of the mushroom *Inonotus obliquus* using a high-resolution ion trap time-of-flight mass spectrometer. (a) Betulinic acid was separated and detected in the extract using a C18 reversed-phase column. (b) LC-MS analysis of a standard facilitated the identification of the peak eluting at 9.6 min in the extract as betulinic acid. (c) Negative-ion electrospray mass spectrum of the peak eluting at 9.6 min in the mushroom extract showing a deprotonated molecule of *m/z* 455.3538 consistent with betulinic acid. (d) Comparison of the accurate mass measurement of *m/z* 455.3538 with theoretical exact masses indicates an elemental composition of $C_{30}H_{48}O_3$ (within 2 ppm of the theoretical mass) confirming the identification of the peak at 9.6 min as betulinic acid.

vitro, in vivo, and clinical trials of pharmacokinetics, bioavailability, metabolism, and toxicity. As an example of safety testing, an LC-MS-MS assay has been developed to screen natural products for metabolic activation to electrophilic intermediates that can alkylate proteins and DNA and potentially cause toxicity [4].

As a result of technological developments in mass spectrometry, highly selective and sensitive assays based on LC-MS and LC-MS-MS now enable the determination of mass, elemental composition, and structural features that facilitate the rapid identification of natural products in complex biological matrices such as plant or bacterial extracts. Screening assays based on LC-MS accelerate the discovery of pharmacologically active compounds from natural product sources. Quantitative assays based on LC-MS-MS are being used to support the chemical standardization of botanical dietary supplements. Finally, LC-MS-MS assays are being used to expedite studies of the safety and efficacy of pharmacologically active compounds in both preclinical and clinical studies.

Mass spectrometry is an extremely fast and sensitive analytical technique. However, LC-MS is relatively slow and requires minutes per analysis, due to the long chromatographic separation step. Currently, chromatographic techniques using higher-pressure pumps (e.g., UPLC) are being introduced that reduce the time required for separations. Therefore, LC-MS separations that currently require 30 min might be reduced to a few seconds in the future. Another slow step in carrying out LC-MS analyses is sample preparation. To address this issue, high-throughput screening methods have been developed that enable massively parallel sample extraction and filtration. Therefore, LC-MS analyses of large sample sets can be carried out faster than ever before with the use of robotics and multiwell plate technology for sample preparation. In the future, sample preparation and nanoscale chromatography for natural products analysis might be carried out on a single microchip similar to the early chip-based electrospray LC-MS systems now being introduced for proteomics research. Undoubtedly, the next 20 years of natural products LC-MS will be at least as exciting as the last two decades.

REFERENCES

1. C. E. Piersen, N. L. Booth, Y. Sun, W. Liang, J. E. Burdette, R. B. van Breemen, S. E. Geller, C. Gu, S. Banuvar, L. P. Shulman, J. L. Bolton, and N. R. Farnsworth, Chemical and biological characterization and clinical evaluation of botanical dietary supplements: A phase I red clover extract as a model, *Curr. Med. Chem.* **11**, 1361–1374 (2004).

2. Y. Sun, C. Gu, X. Liu, W. Liang, P. Yao, J. L. Bolton, and R. B. van Breemen, Ultrafiltration tandem mass spectrometry of estrogens for characterization of structure and affinity for human estrogen receptors, *J. Am. Soc. Mass Spectrom.* **16**, 271–279 (2005).

3. B. M. Johnson, D. Nikolic, and R. B. van Breemen, Applications of pulsed ultrafiltration-mass spectrometry, *Mass Spectrom. Rev.* **21**, 76–86 (2002).

4. B. M. Johnson, J. L. Bolton, and R. B. van Breemen, Screening botanical extracts for quinoid metabolites, *Chem. Res. Toxicol.* **14**, 1546–1551 (2001).

10

THE CHROMATOGRAPHY STORY UNFOLDS

Charles W. Gehrke

Department of Biochemistry and Agricultural Experiment Station Chemical Laboratories, College of Agriculture, Food and Natural Resources, University of Missouri, Columbia

Robert L. Wixom

Department of Biochemistry, University of Missouri, Columbia

Many theses, research papers, review articles, research-oriented books, and serial reviews on chromatography have been written since the 1903 research of M. S. Tswett. To minimize undue repetition, some references are now cited. Our 2001 book [1] has references at the end of each of seven chapters. Further documentation may be found in our online Supplement [2] containing nine chapters and seven appendixes (with 48 pages of references). Leslie S. Ettre has been known since 1979 as one of the key historians of chromatography [3,4]. Ettre has prepared an excellent anthology [5] of his earlier articles "that explore the evolution of chromatography, the background of key developments in their historical context, plus the life and work of the pioneers."

Chromatography: A Science of Discovery. Edited by Robert L. Wixom and Charles W. Gehrke
Copyright © 2010 John Wiley & Sons, Inc.

REFERENCES

1. C. W. Gehrke, R. L. Wixom, and E. Bayer (Eds.), *Chromatography: A Century of Discovery (1900–2000)—the Bridge to the Science/Technology*, Vol. 64, Elsevier, Amsterdam, 2001.

2. C. W. Gehrke, R. L. Wixom, and E. Bayer (Eds.), *Chromatography: A Century of Discovery (1900–2000)*, Vol. 64, Elsevier, Amsterdam; see Supplement Chapters S7–S15 on the Internet at Chem. Web Preprint Server (http://www.chemweb.com/preprint).

3. L. S. Ettre and A. Zlatkis (Eds.), *75 Years of Chromatography—a Historical Dialogue*, Vol. 17, Elsevier, Amsterdam, 1979.

4. L. S. Ettre, *A Comprehensive Bibliography of Publications with Commentary on the History and Evolution of Chromatography*, taken from Ref. 2, Supplement S8 (2001) on the Internet and milestone articles to 2008 by Leslie S. Ettre.

5. L. S. Ettre and J. B. Hinshaw (Eds.), *Chapters in the Evolution of Chromatography*, Imperial College Press, London, 2008.

11

CHROMATOGRAPHY IN THE MILLENNIUM—PERSPECTIVES

Robert L. Wixom

Department of Biochemistry, University of Missouri, Columbia

Charles W. Gehrke

Department of Biochemistry and the Agricultural Experiment Station Chemical Laboratories, College of Agriculture, Food and Natural Resources, University of Missouri, Columbia

The urge to discover, to invent, to know the unknown, seems so deeply human that we cannot imagine our history without it. Eventually that passionate urge conquers the fear of the foreign, the fear of the gods, even the fear of personal danger and death. What remains is the pure exhilaration of discovery.

—Alan Lightman, *The Discoveries—Great Breakthroughs in Twentieth Century Science*, (2005).

CHAPTER OUTLINE

Chromatography: A Science of Discovery. Edited by Robert L. Wixom and Charles W. Gehrke
Copyright © 2010 John Wiley & Sons, Inc.

11.A. FIFTY YEARS OF PITTCON—CHROMATOGRAPHY INNOVATION (John A. Varine)

JOHN A. VARINE

John Varine, 2008 President of PITTCON, teaches chemistry laboratory classes at the California University of Pennsylvania. Prior to his current teaching position, from 1964 to 2002, with a 2-year hiatus for military service, Varine taught laboratory classes at the University of Pittsburgh (Greensburg), chemistry at the secondary level, and was the science department head for kindergarten through grade 12. John received his B.S. degree from Pennsylvania State University and his M.S. degree from Clarkson College of Technology, and took continuing education courses at City College of New York, Temple University, University of Pittsburgh, Indiana University of Pennsylvania, Eastern Michigan University, and the University of Wisconsin. He has served as Workshop Leader for Project SERAPHIM and for the Institute for Chemical Education (ICE), which are organizations funded by the National Science Foundation and headquartered at the University of Wisconsin at Madison. Over the years, John has conducted more than 30 Project SERAPHIM and ICE workshops for science teachers throughout the eastern United States. John is a member of the American Chemical Society, the Spectroscopy Society of Pittsburgh (SSP), and the Society for Analytical Chemists of Pittsburgh (SACP). He served as Chairman of the SACP during 2002–2003. Varine is a member of the organizing committee of the Pittsburgh Conference on Analytical Chemistry and Applied Spectroscopy (PITTCON), which is jointly owned and operated by the SSP and SACP. As a member of the PITTCON organizing committee, he has chaired the Science Week, Transportation, Exposition, Registration, and Treasury committees, as well as serving on the Board of Directors.

A Rich History in Chromatography Innovation. The first Pittsburgh Conference on Analytical Chemistry and Applied Spectroscopy, with its Exposition of Modern Laboratory Equipment, was held on February 15, 1950 at the William Penn Hotel in downtown Pittsburgh. The birth of the conference was driven by its two sponsoring societies, the Spectroscopy Society of Pittsburgh (SSP) and the Society for Analytical Chemists of Pittsburgh (SACP), whose mission was to explore new analytical techniques and how they would enhance the future.

Dedication to the Advancement of Scientific Education. The SACP was founded in 1942 as a nonprofit organization dedicated to the advancement of analytical chemistry. The Society fosters programs for the education of its membership, the community, and the future scientists in the nation's schools. The SACP held its first Analytical Symposium in February 1946. The Symposium attracted keen interest, and in 1949 the first exposition of Modern Laboratory Equipment was held in conjunction with the symposium. The SSP was founded in 1946 to advance the science and practical application of spectroscopy with a strong emphasis on education in high schools and colleges. Like the SACP, the SSP held annual conferences and was closely allied with the SACP. The two societies recognized the value in providing a symposium to address the needs of the fast-growing analytical community and decided to cosponsor a new event they called the "Pittsburgh Conference and Exposition on Analytical Chemistry and Applied Spectroscopy," which debuted in 1950.

The first combined Pittsburgh Conference hosted 56 technical papers, 14 exhibiting companies, and 800 attendees. Today, the annual gathering of scientists from industry, academia, and government offers more than 2000 scientific presentations, 1100 exhibiting companies, and more than 20,000 attendees.

A Devoted Group of PITTCON Volunteers.

Since its inception, PITTCON has been run by a volunteer organization consumed by a passion for the advancement of science and education [1]. The organization's underlying philanthropic mission is to provide funding to advance education in the sciences in schools at all levels. In recent years, nearly $1M in proceeds from PITTCON have been donated annually in the form of teacher awards, equipment grants, and scholarships. The noble educational mission of PITTCON remains as the driving force behind the efforts of the volunteers, who work tirelessly to organize the annual conference and exhibition. Although it has been transformed in size and scope, PITTCON remains true to its tradition of volunteerism.

The Rise of Modern Chromatography Instrumentation.

PITTCON has evolved to become the meeting ground for discussion and demonstration of the latest scientific technologies and instrumental techniques used in today's laboratories around the world. A number of young entrepreneurs in the 1950s and 1960s, including Maurice Hasler, Walter Baird, Dick Jarrell, Jim Waters, Arnold Beckman, and Dave Nelson, to name just a few, exhibited their inventions at PITTCON, which later gave birth to the sophisticated instrumental techniques used today. One can easily track the early beginnings of instrument-based chromatography by taking a historical look at PITTCON [1–4].

Chromatography filter papers and filter media for column chromatography were first exhibited at PITTCON in 1952. The following year, Keene P. Dimick, a chemist with the US Department of Agriculture, presented his research on the isolation of the essence of strawberries. This early research contributed to the development of a gas chromatograph (GC), and Dimick later presented the "magical" results at PITTCON. The first commercial gas chromatograph was demonstrated by Perkin-Elmer 2 years later, in 1955.

In 1959, Waters Associates introduced a refractometer that was used by early researchers in liquid chromatography to measure changes in the refractive index of sample constituents and to correlate them with concentration.

In 1963, Waters introduced an instrumental technique for characterizing polymers using gel chromatography columns, followed by the first commercial automated liquid chromatograph (LC) 2 years later. At the time, most chemists who used liquid chromatography were performing research within university laboratories. Jim Waters believed that the technology would be useful far beyond the research laboratory and would find its way into a mass market that included production, quality control, and clinical testing.

Of course, Water's vision was accurate, and today, chromatography instrumentation and applications have advanced to nearly every area of research. As more and more scientific discoveries were made through the use of chromatography, instrument companies developed new highly sophisticated instruments that were demonstrated on the PITTCON exposition floor. Early chromatographs were replaced by modern high-performance liquid chromatography (HPLC), ultra-performance liquid chromatography (UPLC), and most recently hyphenated techniques where LC and GC instruments are integrated with other sophisticated instruments, such as mass spectrometers, advancing science in various areas such a genomics and proteomics, pharmaceutical discovery, and environmental analysis.

Chromatography and Scientific Innovation at PITTCON. A long list of scientists, including several Nobel Laureates, from industry, government, and academia have used PITTCON as a platform to present their most recent breakthroughs to the scientific community that were, in part, facilitated by the advances in modern instrumentation. These important collaborations resulted in many groundbreaking discoveries involving a wide array of chromatographic-based techniques.

In recent years, scientists engaged in some mechanism of chromatography contribute to more than 25% of the PITTCON Technical Program. A quick scan of the PITTCON technical presentations, scientific posters, workshops, and educational short courses include hydrophilic interaction liquid chromatography, ion chromatography, advanced liquid chromatography–tandem mass spectrometry, ion-exclusion chromatography, pulsed amperometric detection, two-dimensional gas chromatography with mass spectrometric detection, inductively coupled plasma–mass spectrometry and chromatography, and many other exotic chromatography-based techniques.

Many of the scientific discoveries enhanced by chromatography techniques that have been presented at PITTCON have important sociological implications. Scientists regularly present results from their work using chromatography techniques in cancer research, HIV studies, pharmacokinetics, drug discovery and pharmaceutical production, environmental pollutant identification and analysis, forensics, and even Homeland Security.

The PITTCON Tradition Continues. The PITTCON committee and its numerous volunteers are well organized to continue to organize and host PITTCON. They continue to recruit volunteers through the SACP and SSP organizations, and each year new individuals are elected to chair the various committees required to orchestrate this important annual gathering of scientists. The stability of the conference is driven by the unwavering passion of its volunteers and their commitment to the advancement of science and education.

REFERENCES

1. J. Wright, Vision, venture, and volunteers, paper presented at Pittsburgh Conf. and Chemical Heritage Society, 1999.

2. B. Howard, The birthplace of the Pittsburgh Conference: Why Pittsburgh? *Am. Lab. Int. Sci. Commun.* (Feb. 2008).

3. T. Ricci, PITTCON: A platform for scientific innovation. Part I: Years 1950 to 1980, LC · GC the PEAK, *Advanstar* (Nov. 2006).

4. T. Ricci, PITTCON: A platform for scientific innovation. Part II: Years 1981 to 2007, LC · GC the PEAK, *Advanstar* (Jan. 2007).

11.B. TWENTY-FIVE YEARS OF CHEMICAL HERITAGE FOUNDATION—VOYAGES OF DISCOVERY (Thomas R. Tritton and Robert L. Wixom)

THOMAS R. TRITTON AND ROBERT L. WIXOM

Chemists and their many scientific allies are well served by the Chemical Heritage Foundation, which has a broad scientific perspective and provides an incomparable focus on chemistry's history, educational programs, and industrial connections. The Chemical Heritage Foundation (CHF) has over 25 years of service for the chemical and molecular sciences and to the wider public.

In 1982, a pilot project, the Center for History of Chemistry (CHOC) was launched by the University of Pennsylvania, the American Chemical Society (ACS), and later (1984), the American Institute of Chemical Engineers (AIChE). After several major additions, the Center assumed its present name in 1992 and moved to its present location in Philadelphia, PA in the historic district of downtown Philadelphia. It has grants to staff over 55 persons, plus the endorsement and support of more than 40 affiliated organizations in its goal "to treasure the past, educate the present and inspire the future." Dr. Arnold Thackeray served as the original visionary President from CHF's inception to 2007.

The new leader of CHF, Dr. Thomas R. Tritton, became President of CHF after many years as a researcher in molecular cancer chemotherapy and later serving as President of Haverford College, near Philadelphia (1997–2007). Building on earlier developments, the CHF is a library, museum, and center for scholars. CHF has several traveling exhibits plus online exhibits that bring to light unique histories, personalities, and developments in chemistry and related sciences, technologies, and industries. Among these prized exhibits is the research notebook of Paul Lauterbur [Magnetic Resonance Imaging (MRI), Nobel Prize, 2003] recording his idea for non-invasive medical imaging.

CHF has a broad-based collection of over 400 oral history interviews that includes well-known scientists, such as David Baltimore (Nobel Prize, 1975), Arnold O. Beckman, Konrad E. Bloch (Nobel Prize, 1964), Mildred Cohn, Donald J. Cram, Carl Djerassi, Karl A. Folkers, Joshua Lederberg (Nobel Prize, 1955), Boris Magasanik, R. B. Merrifield (Nobel Prize, 1984; see Section 3.C, above), Alfred O. E. Nier, Linus C. Pauling (Nobel Prize, 1952, 1962), Vladimir Prelog (Nobel Prize, 1975), Tadeus Reichstein (Nobel Prize, 1950), Harold A. Scheraga, Phillip A. Sharp (Nobel Prize, 1993), R. B. Woodward (Nobel Prize, 1965; see Section 3.C, above), Kurt Wüthrich (Nobel Prize, 2002), and others (see Chapter 3). The oral histories are edited, indexed, available through the CHF Website, and provide autobiographic accounts that supplement other material in the written historical records.

In addition, CHF has many rare books, fine art, photographs, and personal papers of prominent scientists, all related to the chemical and molecular sciences. CHF also hosts conferences and lectures, supports research, offers fellowships and produces educational materials. The organization's programs and publications provide insight on subjects from alchemy's influence on modern science to the social science impact of nanotechnology.

CHF publishes an excellent quarterly magazine, *Chemical Heritage*, which started in 2007 and features the ideas, the chemical leaders, and the academic/industrial institutions that have shaped our present lives and society. The reader should note the sections called "Inspire," "Driving Forces," "Exploring," and "Treasures." This CHF magazine has a circulation of 25,000 readers in chemistry, including its professional allies, educators, policymakers, and historians. A sample of recent articles pertinent to chromatography include "the greening of chemistry" and the commercial development of quadrupole GC-MS by Robert E. Finnigan; the GC-MS quickly became the crucial instrument for analyzing environmental pollutants by the EPA. In another article, James Waters and his team built the first HPLC system, which led to the UPLC system with major increases in speed, sensitivity, and resolution. Waters and his team also developed a number of high-purity solvents for GC and LC, and invented the disposable miniature silica-based adsorbent chromatographic columns for sample enrichment and purification by solid-phase extraction (SPE) (see also J. Waters in Section 11.A).

At its site in Philadelphia, CHF houses the Othmer Library of Chemical History. The print collection has grown to over 100,000 resources for the history of chemical and molecular sciences, technologies, and industries. In addition to books, manuscripts, and archives, the CHF collects works of art related to the history of alchemy and the emergence of academic and industrial chemistry that helps tell the story of the human effort in these areas.

Science would be remiss without instruments. CHF Museum has more than 2000 items and includes several early chromatographic instruments. Much of the collection is on view in the newly opened museum and has both permanent and changing exhibits on the history of chemistry and its role in the modern world. A few specifics show the depth and breadth: early dyes, Bunsen burners, thermometers, Geiger counters, fuel cells, buckyballs (fullerene), Bakelite, nylon, Gore-Tex, a number of different models of early gas chromatographs, and others. Time is marching on; the past tends to disappear, but CHF works with similar national and overseas organizations, senior executives, and experienced individuals in the chemical industries to keep earlier science clear and relevant.

At its recent 25th Anniversary, CHF announced plans to find new ways to support our unique mission, "to treasure the past, educate the present and inspire the future." CHF achieved its initiative goal of $25 million of gifts, one of which was $7.5 million.

The reader is invited to visit and explore in the Chemical Heritage Foundation, 315 Chestnut Street, Pennsylvania, PA 19106-2702 (USA); phone 215-925-2222, or to discover the richness of the organization through its website (www.chemheritage.org).

11.C. CHROMATOGRAPHY IN THE EXTREMES (Mitch Jacoby)

MITCH JACOBY

Mitch Jacoby (Fig. 11.1) was an undergraduate at Cleveland State University and a graduate student at Northwestern University. In 1993, he completed his Ph.D. studies in physical chemistry under the direction of Peter C. Stair. His thesis work focused on electron–surface interactions and the application of surface analytical techniques to probe those phenomena. On graduating, Jacoby managed a Chicago-area environmental laboratory specializing in analysis of hazardous gases. Since 1997, Jacoby has served as a science writer for *Chemical & Engineering News*, where he covers developments in catalysis, materials, alternative energy, and other topics. To date, he has written two cover stories for C&EN on advances in chromatography.

Figure 11.1. Mitch Jacoby.

Chromatography in the Extreme—a Review. The following article was adapted from a cover story published in *Chemical & Engineering News*, April 28, 2008.

Ask a dozen chemists about the forces that drive innovation in high-performance liquid chromatography (HPLC), and nearly all of them will agree that when it comes to chemical separations, the name of the game is speed.

"Nowadays, it comes down to being able to run as many samples as possible in a given amount of time," says Dennis D. Blevins, a senior manufacturing chemist at Agilent Technologies, an analytical instrumentmaker based in Santa Clara, California. Today's chemists are called on to analyze a larger number of samples, often mixtures of active drug compounds, impurities, and synthesis intermediates, more quickly than ever before, he says.

Drug makers concur. "The most critical success factor is speed to market," says Jeremy B. Desai, executive vice president for R&D at Toronto-based Apotex, Canada's largest generic drug manufacturer. Desai goes on to say that reducing analysis times can mean the difference between a product's success or failure.

Industry's pressing need for faster separations coupled with other researchers' calls for greater separating power have prompted chromatography experts to push the limits of HPLC. Although every area of chemistry could benefit from more powerful analysis methods, the enormous popularity of HPLC has kept that technique in the crosshairs of instrument developers for years. HPLC is so ubiquitous in today's laboratories, according to John G. Dorsey, that "liquid chromatography instrumentation now ranks third behind analytical balances and pH meters in number." Dorsey, a chemistry professor at Florida State University, adds that "every major pharma company has an LC instrument on virtually every bench."

It's no surprise, then, that major analytical chemistry conferences, such as PITTCON (see Section 11.A), feature multiple symposia and workshops on "extreme" and "ultra" methods of HPLC. Some of the most talked about changes nowadays are related to the sizes of the tiny particles that fill the inside of a chromatography column and the pressures needed to push liquid through such a column.

"In terms of capabilities and performance, liquid chromatography remained largely unchanged since the early 1970s," says Douglas R. McCabe, a marketing manager at Waters Corp. In those days, HPLC columns were packed with irregularly shaped particles of roughly 10 μm diameters and solvents were forced through the column at pressures of up to 6000 psi.

Incremental improvements were made over the years, McCabe says, but with the introduction in 2004 of Waters' Acquity ultra performance LC (UPLC) system, the field was hit with what he terms "disruptive technology." Soon other manufacturers introduced UPLC systems.

The UPLC instruments, which were designed to provide faster and higher-quality separations than conventional HPLC units, feature columns packed with highly spherical uniform-sized silica particles measuring less than 2 μm in diameter. In addition, because smaller particles led to more tightly packed columns with greater resistance to the flow of liquid, the mobile phase is driven through the columns at pressures of up to 15,000 psi.

In one demonstration of enhanced performance, Waters chemists showed that a mixture of acetaminophen, caffeine, and a few other compounds, was separated into its components by a conventional HPLC system with 5-μm column-packing particles in just under 7 mins. With the UPLC system, the same mixture was separated with greater resolution and increased analyte sensitivity in just over 1 min.

Other researchers report similar improvements in separations. At Eli Lilly, for example, senior research chemist, Todd D. Maloney, says that a typical HPLC separation of an active drug compound and a total of six known impurities and synthesis intermediates via conventional HPLC methods can be completed in about 45 mins. Now, using today's smaller particles and newer instruments, he can complete the analysis and achieve an equivalent separation in just under 10 mins.

Maloney finds that the souped-up LC instruments that he's worked with, which includes units made by five manufacturers, are "well designed and very easy to use." The new advanced LC systems, however, are also more expensive than conventional ones.

At Apotex, Desai finds that standard analyses of impurity levels in drug compounds are reduced on average to one-eighth the conventional HPLC run time—and assays of the active component of drugs are completed two-and-a-half times faster by using UPLC than by HPLC.

Apotex has a strong incentive to invest in the fast analytical technology. As Desai explains, in the United States, the first company to receive approval from the Federal Drug Administration to market a generic pharmaceutical compound is often granted a 180-day exclusive marketing license. To help secure that advantage, Apotex plans to acquire a total of 50 Acquity systems, Desai notes.

Although the pharmaceutical industry is the heaviest user of the new technology, it's not the only user. The new LC systems are being put to use in environmental studies to measure pesticides and other contaminants in water; in food safety tests to quantify drug residues and other impurities in, meat, poultry, and seafood; and to study amino acids, peptides, and other bioanalytes.

Explanations for why reducing the size of column-packing particles leads to enhanced separations come in a few forms. According to classic chromatography theory, increasing the number of theoretical plates improves separations. That number can be increased by lengthening the column or by reducing the plate height, which corresponds to a particle's diameter or thickness.

Another explanation relates the size of the porous particles to the time it takes analyte molecules to diffuse in and out of those particles. In general, small differences in the extent to which the components of a mixture interact with or are retained on the surfaces of the packing materials result in chemical separation. As a mixture flows through a column, those differences ultimately lead to a measurable separation because the analyte molecules repeatedly shuttle back and forth between the solvent and the particles and thereby interact over and over again with the particles. Because molecules can diffuse in and out of small porous particles faster than large ones, smaller particles facilitate faster shuttling and hence improved separations.

Far and away, silica is the most common column-packing material for conventional and ultra-HPLC instruments. The material, which is generally prepared from tetraethoxysilane (TEOS) or related silanes, is polymerized to form high-purity, monodisperse (same-sized) silica spheres that are free from metals that interfere with chromatographic separations. The particles are mechanically strong, have high surface area, and feature pore sizes that can be easily tailored. In addition, depending on the intended application, the properties of the material are customized by manufacturers by treating the surfaces with C8, C18, phenyl, and other functional groups.

Commercial examples of modern column packing materials include Waters' high-strength silica (HSS) and ethylene bridged hybrid (BEH) materials. Particles of both materials, which were designed for the UPLC systems, have diameters in the 1.7–1.8 μm range and feature pores with diameters of either 100 Å (HSS) or 130 Å (BEH).

Unlike the all-silica HSS product, BEH particles contain C—C bridges between pairs of silicon atoms. The hybrid organic–inorganic particles, which are prepared via copolymerization of TEOS and bis(triethoxysilyl)ethane, were designed to maintain silica's mechanical strength while overcoming its tendency to undergo column-damaging hydrolysis in alkaline environments (pH > 8). BEH's covalently bonded Si—C—C—Si units render the hybrid material chemically stable up to a pH level of 12, which is an ideal condition for analyzing some pharmaceutical agents and other types of compounds. "Plain silica" and strengthened ethylene-bridged silica particles are depicted in Figures 11.2 and 11.3, respectively.

Making new types of separation media isn't the only challenge in developing advanced HPLC columns. Packing the particles into a column such that the end result works well chromatographically can also be difficult. That's one of the areas that

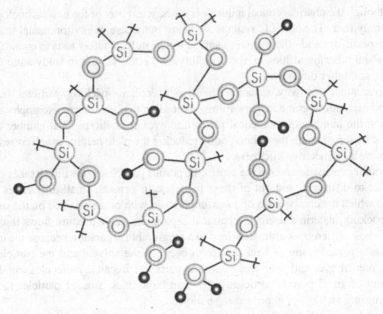

Figure 11.2. Diagram of "plain" silica HPLC column-packing material.

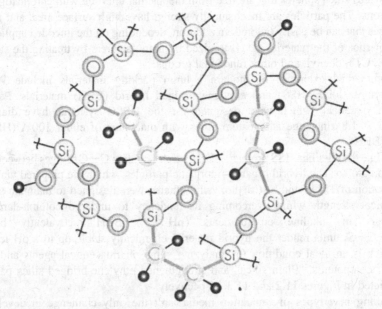

Figure 11.3. Diagram of modified silica particles for HPLC column-packing, which include ethylene bridges.

James W. Jorgenson studies with his research group at the University of North Carolina, Chapel Hill.

"Chromatography theory tells us that we could be doing faster and more efficient separations by using even smaller particles than the ones we use now," Jorgenson says. So, rather than drawing the line at 1.7 μm, the Chapel Hill team works with silica particles measuring 1.0 to 1.5 μm in diameter and makes use of 100,000-psi pumps to do their work.

"As the particles get smaller, they become immensely harder to pack into columns that provide efficient separations," Jorgenson observes. The key problem, he says, is the tendency of tiny particles (∼1.0 μm) to agglomerate. The group uses various solvents and ultrasonication to prepare well-dispersed slurries, then packs capillary columns at high pressure. Then they test the columns chromatographically and examine the microscopic packing structure and other properties with the aim of discovering the most effective column-packing methods.

In addition to sub-2 μm commercial separation materials and the even smaller particles studied by researchers such as Jorgenson, other types of "ultra" or "extreme" packing materials are being studied by the LC community. For example, fused-core

Figure 11.4. Artist rendition of 2.7 μm diameter "HALO" particles for HPLC column packing, with solid silica core and this porous silica coating. The color version of this image appeared as the cover of the April 28, 2008 issue of *Chemical & Engineering News*, courtesy of Advanced Materials Technology.

particles, are unusual in that they are composed of a solid (nonporous) silica core (1.7 μm in diameter) surrounded by 0.5-μm-thick porous silica shell giving an overall size of 2.7 μm (see Fig. 11.4).

Recently developed by Advanced Materials Technology and marketed by Mac-Mod Analytical, the short-diffusion-path core–shell materials were designed to provide high-speed LC separations in a very durable column. Joseph (Jack) Kirkland, AMT's vice president for R&D, says that the 2.7-μm fused-core materials are competitive with sub-2-μm particles in terms of separation efficiencies, but don't require the special high-pressure HPLC equipment.

Head-to-head comparisons by users are consistent with Kirkland's claim. Maloney reported recently that fused-core columns offered a much lower pressure alternative to sub-2μm columns with only "a slight sacrifice" in separation efficiency [*J. Separation Sci.* **30**, 3104 (2007)].

Still other types of materials fit the "extreme" LC bill. According to University of Minnesota chemistry professor, Peter W. Carr, zirconia makes an incredibly stable column-packing material, much more so than silica. "You can cook zirconia in strong acid and strong base and it won't dissolve," Carr says. In fact, his desire to do just that as a means of cleaning HPLC columns led him to help found ZirChrom, a zirconia-based HPLC supplies vendor, with which he is no longer affiliated.

At Kyoto Institute of Technology in Japan, Nobuo Tanaka makes extremely long one-piece, porous, monolithic HPLC columns. These columns, which extend to several meters in length, can provide on the order of one million theoretical plates and can be used to separate extremely similar compounds and highly complex mixtures, such as the ones encountered in proteomics and metabolomics studies. Demonstrating just how similar the compounds can be, Tanaka showed that benzene and monodeuterated benzene, which in many ways are nearly identical, can be separated readily on an 8-m silica mono-lith [*Anal. Chem.* **80**, 8741 (2008)]. Superlong monolithic columns aren't yet available commercially, but Tanaka predicts that will change. "I believe 5–10-m columns will be used practically in a few years," he says.

Solvent temperature is another parameter that can be adjusted to significantly boost LC performance. Raising the solvent temperature lowers its viscosity and in turn decreases the pressure needed to force the liquid quickly through a packed column. By using that approach, together with small column particles and short columns, Carr carries out two-dimensional separations on biological mixtures of hundreds of compounds in one-tenth the time required to separate the compounds via conventional 2D LC methods. High-temperature 2D LC methods can generate a peak capacity of 1000 in just 30 mins, Carr says [*J. Chromatogr. A* **1122**, 123 (2006)].

It has been only a couple of years since instrumentmakers launched advanced LC products that significantly outpace conventional instruments. Yet LC aficionados are already wondering what's coming down the pike. Should the next generation of commercial instruments run at 40,000–50,000 psi, Jorgenson wonders? "The jury is still out" on that question, Maloney says, because the "ultra" instruments are still new to the market place. Jorgenson concurs: "Right now," he says, "it's way too early to decide."

11.D. ANALYTICAL SEPARATIONS AS MOLECULAR INTERACTION AMPLIFIERS (Apryll Stalcup)

A. M. STALCUP

Figure 11.5. Apryll Stalcup.

Dr. Apryll Stalcup (Fig. 11.5) is currently a Professor of Chemistry, at the University of Cincinnati, where she investigates the basic mechanisms responsible for various types of separations. Her research for the last 20 years has focused on development of new strategies for chiral separations and includes introduction of sulfated β-cyclodextrin and heparin as chiral selectors for capillary electrophoresis, preparative electrophoretic chiral separations and synthesis and characterization of new chromatographic stationary phases. More recently she has been investigating the retention properties of surface-confined ionic liquids as stationary phases in liquid chromatography. Before accepting the position at the University of Cincinnati, she was a member of the faculty at the University of Hawaii. She obtained her Ph.D. in Chemistry from Georgetown University under the direction of Professor Daniel E. Martire. A significant part of her graduate work was conducted as a Graduate Co-op student at the National Institute of Standards and Technology under the direction of Dr. Lane C. Sander and Dr. Stephen A. Wise. Her postdoctoral work with Professor Daniel W. Armstrong while at the University of Missouri—Rolla involved development of derivatized cyclodextrins for chiral separations. She currently serves on the Greater Cincinnati Water Quality Advisory Board, the Editorial Board of *Trends in Analytical Chemistry*, and the Advisory Board of *Analytical and Bioanalytical Chemistry*.

Recently, in preparing for a presentation in honor of John Dorsey, recipient of the 2008 Dal Nogare Award in Chromatography, I came across a remarkable statement from Fred McLafferty, Peter J. W. Debye Professor Emeritus at Cornell University and one of the pioneers of mass spectrometry. In reflecting on the Gordon Research Conferences, McLafferty wrote, "The talks on GC by Steve Dal Nogare of DuPont and H. N. Wilson of ICI (Imperial Chemical Industries) were so exciting that I could hardly think of anything else" [1]—and yet, in many respects, chromatography and related separation techniques such as electrophoresis are often viewed as trivial, mature techniques, a necessary evil used in the pursuit of "real science." For example, in 2001, Haleem Isaaq, of the National Cancer Institute, noted that "The unfortunate aspect of the Human Genome Project is that the analytical chemist was not only excluded but completely ignored from the recognition bestowed upon biologists by the United States Administration at a White House ceremony last year" [2].

Part of the trivialization of chromatography and related separation techniques may be traced back to the inception of chromatography by Mikhail Tswett, the Russian botanist generally credited with inventing chromatography. At that time, the invention of a technique, no matter how useful, was considered the province of technicians, not real scientists, and his work was largely ignored or belittled [3]. It should be noted that by 1948, when the Nobel Prize was awarded to Arne Tiselius, for his introduction of electrophoresis [4], the scientific culture had changed somewhat. Part of the continued low regard for chromatography and related electrophoresis-based techniques may also be related to the relative simplicity of the data output. In an era of color-enhanced images of tissues, cells, proteins, organized media, etc., chromatograms and electropherograms can seem somewhat dowdy by comparison. Paradoxically, in the midst of this trivialization, separations scientists are being asked to perform faster separations on increasingly complicated samples (e.g., cell lysate) in support of the various "-omic"-related research areas (proteomics, metabolomics, glycomics, etc.) [5]. The complexity of these samples resides not only in the sheer number and concentration range of sample constituents and the array of interaction modalities presented by these analytes but also the potential convergent isoforms (glycoforms, isoenzymes, etc.) [6] of various biopolymers that may be present.

The separations community, which I consider to include all practitioners of the various chromatographic and electrophoresis modes, has become so very good at what it does that what often gets lost is that the range of analytes successfully separated by either chromatographic or electrophoretic techniques ranges from ions [7–9] to small molecules, polymers and biopolymers [10,11], to cells [12] and even cell organelles [13]. Chiral separations, once considered one of the most difficult of all separations, have become routine in the last 20 years [14].

What also often gets lost is that separation techniques are incredibly sensitive tools for investigating intermolecular interactions. The intermolecular interactions exploited by the separations community are the same intermolecular interactions responsible for the phase behavior of pure substances (e.g., dispersion, hydrogen bonding, dipole–dipole, dipole–induced dipole, electrostatic). Hydrophobic, dispersive, and electrostatic interactions are all known to play a role in the retention properties of proteins [15]. While spectroscopic techniques such as IR and NMR are very good at identifying specific interaction modes (e.g., shift of IR bands denoting interactions through hydrogen bonding [16]; NMR peak shifts denoting p–p interactions [17]), molecules are seldom so well behaved with each other in the liquid or gas state so as to interact through only one interaction mode. Indeed, as noted by Boehm, Martire, and Armstrong [18], interactions between a chiral analyte and a chiral stationary phase reflect all the potential interactions, each weighted by the energy of those specific interactions. The same can be said for any kind of chromatography or electrokinetic chromatography.

In some ways, the separation system reflects the global interactions between the solute and the separation system, which acts as a molecular amplifier of these interactions. A good example is provided by determination of the relative affinity of drugs for a specific protein that can be readily and rapidly accessed through their chromatographic retention on an immobilized protein (e.g., human serum albumin) stationary phase [19–21]. This approach can be extended in support of drug discovery to rapidly screen potential drug candidates for a specific receptor site [21–24]. Even ambiguous phenomena such as allosteric interactions, in which nonsubstrate binding events occur that induce conformational changes in the protein tertiary structure far from a substrate binding site on a protein, can be investigated using immobilized protein stationary phases. These allosterically induced conformational changes can have a profound impact on protein–substrate binding [25] which can be manifested as changes in substrate retention on an immobilized protein stationary phase [26,27]. In addition, separations methods not only can be used to measure very small binding constants (e.g., $K = 5$ M^{-1}, which corresponds to a DG ~ 1 kcal/mol [28]) in the presence of auxiliary agents, but also exploit much smaller differences in binding to achieve separations.

As noted by Nasal et al. [29], the interactions responsible for solute lipophilicity or hydrophobicity largely emanate from the solvent environment and not really from the solutes, themselves. Hence, the ability of chromatographic and electrophoretic methods to capture the global interactions between the solute and the separation system provides a convenient tool for rapidly screening solute aqueous solubility [30], or lipophilicity [31] for environmental [32] and pharmacological or toxicological studies [29] as characterized by octanol–water partitioning. This ability of chromatographic systems can also be extended to probe solute interactions with other types of novel materials such as ionic liquids [33–37].

Chromatographic methods can also be used in the characterization of synthetic and natural polymers. The use of size exclusion or gel permeation chromatography for the determination of polymer molecular weights relies on stationary phase transparency to the solutes and is well established [38]. Critical chromatography, in contrast, represents a particularly elegant exploitation of fundamental molecular interactions in the characterization of heterogeneous polymers such as block copolymers. In the simplest case of a diblock copolymer that contains both aromatic and aliphatic subunits, mobile-phase conditions are selected that render one type of subunit "chromatographically invisible" to the stationary phase [39]. At the "critical" mobile-phase condition, the enthalpic contribution to retention for the "invisible" subunit is exactly compensated by the entropic contributions to retention [40]. Hence, retention is based on the distribution of the nonchromatographically invisible subunits along the polymer chain.

Molecularly based models are critically important in accessing the molecular interactions responsible for separations, and there are increasing levels of sophistication in the models used to describe retention. Linear plots of log retention versus number of carbons in homologous series suggest that retention is basically the sum of the retention of the individual "parts" of the molecules. Hence, some of the most successful of these models assume that retention can be described globally as a linear combination of independent molecular interactions. The linear solvation energy relationships (LSER) model assumes that retention can be described by a linear combination of individual intermolecular interactions and relates fundamental solute descriptors to solute retention through [41–43]

$$SP = eE + sS + aA + bB + vV + c$$

where SP is any free-energy-related property of the stationary-phase and mobile-phase system. The capital and italic variables are molecular descriptors for individual solutes, where E represents excess molar refraction, which reflects polarizability contributions from p and n electrons; S is the solute dipolarity/polarizability; A and B are the overall solute hydrogen bond acidity and basicity, respectively, which scale as the hydrogen bonding propensity of a solute to surrounding solvent molecules; and V is McGowan's characteristic volume. The constant c is thought to include the phase ratio of the chromatographic system. Multilinear regression analysis of retention data, using molecular descriptors for each of a set of selected analytes for which the molecular properties are known, yields the coefficients in the LSER equation. Indeed, the molecular descriptors, themselves are often derived from chromatographic measurements [44]. Because the molecular descriptors are all positive, positive system coefficients correspond to stronger interactions with the stationary phase through a particular type of interaction (e.g., v: hydrophobic interactions). While these models can provide much needed insight into the interactions driving the overall separations [45], they also can be extended to include electrostatic interactions [46] or used to characterize molecular properties of novel materials such as ionic liquids [47,48].

The separations community is also reexamining longstanding practices in liquid chromatography based on understanding of the intermolecular interactions. For instance, the elimination of organic modifiers in reversed-phase chromatography is possible as the extended hydrogen-bonding network of water is eliminated at elevated temperatures [49,50].

In summary, the relatively low regard of the separations community is ironic because there is incredible excitement in the field, right now. In many ways, a very sophisticated material science has emerged as the separations community rises to the challenges presented in the analysis of these complicated samples. Further, separation techniques are incredibly sensitive tools for investigating and amplifying subtle differences in intermolecular interactions. The same intermolecular interactions responsible for the phase behavior of pure substances (e.g., dispersion, hydrogen bonding, dipole–dipole, dipole–induced dipole, electrostatic) are being creatively exploited by the separations community. A variety of novel competitive strategies are evolving to address the challenges of complicated samples as the community reevaluates how columns are engineered. Simultaneously, new developments in separations capability are providing new competencies and new information to a wide variety of disciplines.

ACKNOWLEDGMENTS

The author wishes to extend her congratulations to Professor John Dorsey for his well-deserved Dal Nogare Award, gratefully acknowledges support from the Waters Corporation and the National Institutes of Health (R01 GM 067991), and thanks the editors for the opportunity to contribute to this volume.

REFERENCES

1. McLafferty, http://www.frontiersofscience.org/reflections.aspx?category=1&essay=11.
2. H. J. Issaq, The role of separation science in proteomics research, *Electrophoresis* **22**(17), 3629–3638 (2001).

3. L. S. Ettre and K. I. Sakodynskii, M. S. Tswett and the discovery of chromatography I: Early work (1899–1903), *Chromatographia* **35**(3/4) (1993).

4. P. G. Righetti, Electrophoresis: The march of pennies, the march of dimes, *J. Chromatogr. A* **1079**, 24–40 (2005).

5. S. Garbis, G. Lubec, and M. Fountoulakis, Limitations of current proteomics technologies, *J. Chromatogr. A* **1077**, 1–18 (2005).

6. J. Reinders and A. Sickmann, State-of-the-art in phosphoproteomics, *Proteomics* **5**(16), 4052–4061 (2005).

7. J. S. Fritz, Factors affecting selectivity in ion chromatography, *J. Chromatogr. A* **1085**(1), 8–17 (2005).

8. W. Hu, P. R. Haddad, K. Tanaka, S. Sato, M. Mori, Q. Xu, M. Ikedo, and S. Tanaka, Determination of monovalent inorganic anions in high-ionic-strength samples by electrostatic ion chromatography with suppressed conductometric detection, *J. Chromatogr. A* **1039**(1–2), 59–62 (2004).

9. Q. Wang, G. A. Baker, S. N. Baker, and L. A. Colon, Surface confined ionic liquid as a stationary phase for HPLC, *Analyst* **131**(9), 1000–1005 (2006).

10. F. Fang, M.-I. Aguilar, and M. T. W. Hearn, Temperature-induced changes in the bandwidth behavior of proteins separated with cation-exchange adsorbents, *J. Chromatogr. A* **729**(1–2), 67–79 (1996).

11. F. Fang, M. I. Aguilar, and M. T. Hearn, Influence of temperature on the retention behavior of proteins in cation-exchange chromatography, *J. Chromatogr. A* **729**(1–2), 49–66 (1996).

12. M. J. Desai and D. W. Armstrong, Separation, identification, and characterization of microorganisms by capillary electrophoresis, *Microbiol. Mol. Biol. Rev.* **67**(1), 38–51 (2003).

13. Y. Chen, G. Xiong, and E. A. Arriaga, CE analysis of the acidic organelles of a single cell, *Electrophoresis* **28**(14), 2406–2415 (2007).

14. T. J. Ward and B. A. Baker, Chiral separations, *Anal. Chem.* **80**(12), 4363–4372 (2008).

15. J. Stahlberg, B. Jonsson, and C. Horvath, Combined effect of coulombic and van der Waals interactions in the chromatography of proteins, *Anal. Chem.* **64**(24), 3118–3124 (1992).

16. V. W. Jurgensen and K. Jalkanen, The VA, VCD, Raman and ROA spectra of tri-L-serine in aqueous solution, *Phys. Biol.* **3**(1), S63–S79 (2006).

17. B. Cabovska, G. P. Kreishman, D. F. Wassell, and A. M. Stalcup, Capillary electrophoretic and nuclear magnetic resonance studies of interactions between halophenols and ionic liquid or tetraalkylammonium cations, *J. Chromatogr. A* **1007**(1–2), 179–187 (2003).

18. R. E. Boehm, D. E. Martire, and D. W. Armstrong, Theoretical considerations concerning the separation of enantiomeric solutes by liquid chromatography, *Anal. Chem.* **60**(6), 522–528 (1988).

19. H. S. Kim and I. W. Wainer, Rapid analysis of the interactions between drugs and human serum albumin (HSA) using high-performance affinity chromatography (HPAC), *J. Chromatogr. B* **870**(1), 22–26 (2008).

20. F. Beaudry, M. Coutu, and N. K. Brown, Determination of drug ± plasma protein binding using human serum albumin chromatographic column and multiple linear regression model, *Biomed. Chromatogr.* **13**, 401–406 (1999).

21. C. Bertucci, M. Bartolini, R. Gotti, and V. Andrisano, Drug affinity to immobilized target bio-polymers by high-performance liquid chromatography and capillary electrophoresis, *J. Chromatogr. B* **797**(1–2), 111–129 (2003).

22. R. Moaddel and I. W. Wainer, Development of immobilized membrane-based affinity columns for use in the online characterization of membrane bound proteins and for targeted affinity isolations, *Anal. Chim. Acta* **564**(1), 97–105 (2006).

23. D. S. Hage, T. A. G. Noctor, and I. W. Wainer, Characterization of the protein binding of chiral drugs by high-performance affinity chromatography interactions of R- and S-ibuprofen with human serum albumin, *J. Chromatogr. A* **693**(1), 23–32 (1995).

24. H. R. Luckarift, G. R. Johnson, and J. C. Spain, Silica-immobilized enzyme reactors; application to cholinesterase-inhibition studies, *J. Chromatogr. B* **843**(2), 310–316 (2006).

25. P. Ascenzi, A. Bocedi, S. Notari, G. Fanali, R. Fesce, and M. Fasano, Allosteric modulation of drug binding to human serum albumin, *Mini Rev. Med. Chem.* **6**(4), 483–489 (2006).

26. R. Moaddel, K. Jozwiak, and I. W. Wainer, Allosteric modifiers of neuronal nicotinic acetylcholine receptors: New methods, new opportunities, *Med. Res. Rev.* **27**(5), 723–753 (2007).

27. I. F. Jianzhong Chen and D. S. Hage, Chromatographic analysis of allosteric effects between ibuprofen and benzodiazepines on human serum albumin, *Chirality* **18**(1), 24–36 (2006).

28. E. G. Yanes, S. R. Gratz, M. J. Baldwin, S. E. Robison, and A. M. Stalcup, Capillary electrophoretic application of 1-alkyl-3-methylimidazolium-based ionic liquids, *Anal. Chem.* **73**(16), 3838–3844 (2001) (PMCID: PMC11534705).

29. A. Nasal, D. Siluk, and R. Kaliszan, Chromatographic retention parameters in medicinal chemistry and molecular pharmacology, *Curr. Med. Chem.* **10**(5), 381–426 (2003).

30. P. A. Tate and J. G. Dorsey, Column selection for liquid chromatographic estimation of the k_w' hydrophobicity parameter, *J. Chromatogr. A* **1042**(1–2), 37–48 (2004).

31. S. K. Poole and C. F. Poole, Separation methods for estimating octanol-water partition coefficients, *J. Chromatogr. B* **797**(1–2), 3–19 (2003).

32. R. Guo, X. Liang, J. Chen, Z. Qing, W. Wenzhong, and A. Kettrup, Using HPLC retention parameters to estimate fish bioconcentration factors of organic compounds, *J. Liquid Chromatogr.* **27**, 1861–1873 (2004).

33. D. S. Van Meter, Y. Sun, K. M. Parker, and A. M. Stalcup, Retention characteristics of a new butylimidazolium-based stationary phase. Part II: Anion exchange and partitioning, *Anal. Bioanal. Chem.* **390**(3), 897–905 (2008).

34. C. F. Poole, Chromatographic and spectroscopic methods for the determination of solvent properties of room temperature ionic liquids, *J. Chromatogr. A* **1037**(1–2), 49–82 (2004).

35. D. W. Armstrong, L. He, and Y. S. Liu, Examination of ionic liquids and their interaction with molecules, when used as stationary phases in gas chromatography, *Anal. Chem.* **71**(17), 3873–3876 (1999).

36. E. L. Arancibia, R. C. Castells, and A. M. Nardillo, Thermodynamic study of the behavior of two molten organic salts as stationary phases in gas chromatography, *J. Chromatogr.* **398**, 21–29 (1987).

37. M. E. Coddens, K. G. Furton, and C. F. Poole, Synthesis and gas chromatographic stationary phase properties of alkylammonium thiocyanates, *J. Chromatogr.* **356**(1), 59–77 (1986).

38. W. Radke, P. F. W. Simon, and A. H. E. Mueller, Estimation of number-average molecular weights of copolymers by gel permeation chromatography-light scattering, *Macromolecules* **29**(14), 4926–4930 (1996).

39. S. V. Olesik, Liquid chromatography at the critical condition, *Anal. Bioanal. Chem.* **378**, 43–45 (2004).

40. S. Phillips and S. V. Olesik, Fundamental studies of liquid chromatography at the critical condition using enhanced-fluidity liquids, *Anal. Chem.* **74**, 799–808 (2002).

41. M. Reta, P. W. Carr, P. C. Sadek, and S. C. Rutan, Comparative study of hydrocarbon, fluorocarbon, and aromatic bonded RP-HPLC stationary phases by linear solvation energy relationships, *Anal. Chem.* **71**(16), 3484–3496 (1999).

42. M. Vitha and P. W. Carr, The chemical interpretation and practice of linear solvation energy relationships in chromatography, *J. Chromatogr. A* **1126**(1–2), 143–194 (2006).

43. C. West and E. Lesellier, Characterization of stationary phases in subcritical fluid chromatography with the solvation parameter model IV. Aromatic stationary phases, *J. Chromatogr. A* **1115**(1–2), 233–245 (2006).

44. M. H. Abraham, A. Ibrahim, and A. M. Zissimos, Determination of sets of solute descriptors from chromatographic measurements, *J. Chromatogr. A* **1037**, 29–47 (2004).

45. Z. Ali and C. F. Poole, Insights into the retention mechanism of neutral organic compounds on polar chemically bonded stationary phases in reversed-phase liquid chromatography, *J. Chromatogr. A* **1052**, 199–204 (2004).

46. S. Espinosa, E. Bosch, and M. Roses, Retention of ionizable compounds on high-performance liquid chromatography XI. Global linear solvation energy relationships for neutral and ionizable compounds, *J. Chromatogr. A* **945**, 83–96 (2002).

47. J. L. Anderson, J. Ding, T. Welton, and D. W. Armstrong, Characterizing ionic liquids on the basis of multiple solvation interactions, *J. Am. Chem. Soc.* **124**(47), 14247–14254 (2002).

48. Y. Sun, B. Cabovska, C. E. Evans, T. H. Ridgway, and A. M. Stalcup, Retention characteristics of a new butylimidazolium-based stationary phase, *Anal. Bioanal. Chem.* **382**(3), 728–734 (2005).

49. J. W. Coym and J. G. Dorsey, Reversed-phase retention thermodynamics of pure-water mobile phases at ambient and elevated temperature, *J. Chromatogr. A* **1035**(1), 23–29 (2004).

50. B. Yan, J. Zhao, J. S. Brown, J. Blackwell, and P. W. Carr, High-temperature ultrafast liquid chromatography, *Anal. Chem.* **72**, 1253–1262 (2000).

11.E. PERSPECTIVES BY AWARDEES AND CONTRIBUTORS

Now that chromatography has passed the one-century period, the subject has an extensive literature that has an international distribution of scientists. Several areas that were covered in our 2001 book [1] were not reviewed in this book. Hence some major recent references are now presented to provide the needed balance for this book [1–5]. From another viewpoint, these multivolume references are an appropriate introduction for the following *perspectives* by leading chromatographers for the twenty-first century.

Armstrong, Dan

The efficient and effective separation and characterization of particles (including microbes) will continue to grow in importance relative to the traditional focus of chromatography and electrophoresis, which has been the separation method of these molecules. The current research emphasis on colloidal (now nano-) particles will hasten the growth of this area of separations. The expansion of rapid and comprehensive multidimensional techniques will escalate along with the areas that support them. This will include a push to develop new chromatographic stationary phases, mechanical and miniaturization innovations, and the statistical plus information technology means to identify all/most separated components in real time. Sometimes the "chemistry" that allows or improves an analysis lags behind the development of new hardware. It is only when both are optimized and combined that the best results are obtained. We will continue to be indebted to those that couple useful and innovative chemistries to the latest instrumentation.

Berger, Terry

I will limit my comments about the future of chromatography to supercritical-fluid chromatography (SFC). Most developments in HPLC are toward the use of very small particles to increase speed of analysis. SFC achieves the same objectives without the excess pressure drops caused by small particle size, since the solute binary diffusion coefficients are 3–10 times larger in SFC compared to HPLC. Thus, SFC is inherently faster. Further, the use of very small particles in HPLC is not scalable, and semipreparative SFC should be the primary means of delivering gram-scale purifications in the foreseeable future. SFC is far safer, faster, and cheaper for semipreparative separations. Thus, SFC is likely to both replace HPLC for most semipreparative applications and significantly increase the number and economic importance of such applications.

Burger, Ben

To compensate for inadequate training of personnel, and to ensure the integrity of chromatographic data, equipment will in the future have to be practically fully automated. However, this will also make life difficult for researchers who are interested in the development of novel analytical techniques. For research purposes I myself would prefer to use instruments that allow me to override certain settings embedded in the instruments' software.

It can be foreseen that in the future all GCs on the market could have comprehensive two-dimensional capabilities. The required hardware is relatively simple and inexpensive. However, the problem of the prohibitively high cost of the cooling liquids used for cryomodulation still has to be solved.

Davankov, V. A.

In the twenty-first century, I believe, chromatography will strengthen its positions as a preparative and industrial method of obtaining valuable compounds; new continuous procedures with minimized consumption of reagents and solvents will be developed, that meet requirements of "green" technology. Here, size-exclusion chromatography can show its thus far unrecognized potentials. Analytical applications will further develop toward miniaturization and automation.

Haddad, Paul

The field of separation science of inorganic species continues to develop at a rapid pace. Unlike the broader field of HPLC, where the newest developments are based primarily in the use of small-diameter-particle monolithic materials as stationary phases, and miniaturization on microchips, developments in ion chromatography fall more in the areas of capillary separations, software for rapid optimization of complex eluent compositions, and faster separations. Ion chromatography is also expanding into more diverse and challenging applications. Electrodriven separations of inorganic species have hitherto been limited by deficiencies in detection sensitivity, but these deficiencies are now being addressed through new approaches to online sample enrichment and also new modes of detection. Of particular relevance is contactless conductivity detection, which will soon become the dominant detection method for inorganic ions. Finally, both ion chromatography and capillary electrophoresis are filling important roles in counterterrorism research as effective means of preblast and postblast detection of improvised explosive devices, which are used widely in terrorist attacks.

Hancock, William

The area of HPLC is as fresh as in the mid-1970s, when availability of (then) modern instrumentation and packing materials enabled breakthrough studies of peptide and protein separations. At the time, most HPLC applications were an extension of earlier studies from GC and involved small molecule analysis. The study of polar, high-MW separations was, however, an area awaiting a powerful new approach. Furthermore, this area of analysis has continued to grow from these early applications in protein biochemistry, to the emerging biotechnology industry and then to clinical proteomics. In the initial transition of GC into HPLC, one often found that separations theory migrated in a reasonably linear manner from the gas to liquid phases. The challenge was different in the world of biopolymer analysis! For example, the effect of temperature on retention of protein molecules in reversed-phase LC is driven by conformational changes in the 3D structure. In the area of clinical proteomics the analysis of the plasma proteome is a huge challenge because of the complexity of the sample as well as the need to detect the presence of low-level disease biomarkers. Not only is HPLC married to high-information-content detectors such as mass spectrometry; sample preparation is as important as in an earlier era of developing sensitive environmental applications. In conclusion, HPLC of polypeptides continues to be an important field with fresh applications and challenges for those fortunate enough to be working in the area.

Hennion, Marie-Claire

The recent developments in chromatography and its coupling to mass spectrometry were mainly to provide analytical techniques allowing more and more resolution to solve complex samples and lower and lower detection limits—and in that area, progresses will certainly go on. But one should not forget that one driving force of the twenty-first century is sustainable environment. This is particularly important for a field as chromatography which has a key role to play to ensure better healthcare, a better environment, and more safe food and products. When looking to the routine analyses that are made at present, just a simple lifecycle analysis of our routine protocol to monitor our environment is very bad. There is a strong need for methods that consume far less organic solvent and energy and are robust enough for routine analysis. Miniaturization is a part of the answer, but will not be only a simple miniaturization of chromatography. Microchips are a concept allowing a lot of combination of chemical, physical, and biological processes. Chromatography and electrophoresis are part of them, and several examples suggest that their combination with bioassays on the same microchip, for instance, can be a good answer to rapid and efficient analysis for some toxic environmental micropollutants.

Koch, Del

Since the 1960s, there have been many monitoring programs for food by the USDA and FDA to determine the safety of food items. Superfund cleanup sites have required a tremendous amount of effort and accountability. More recently, in August 1996, Congress entered into legislation for the "Food Quality and Protection Act," which required a more holistic approach to agricultural chemical exposure. It required evaluation of the effects of chemicals with "common mechanistic" modes, a 10-fold safety factor for children, screening of compounds for possible endocrine disruption, and a total perspective evaluation for all exposures, whether the source is by food, home, lawn, workplace, water, and/or air. In order to measure the actual levels of exposure, instead of perceived or estimated exposure, refinements in chromatographic techniques for quantification and confirmation of compounds were developed and effectively implemented. In particular, the successful interfacing of mass spectrometric detectors with GC and HPLC separation techniques has allowed for improved accuracy at much lower levels of quantification than were achievable when environmental exposures first became a major concern. Many of the newer agricultural chemicals are efficacious at much lower rates than those from years past; correspondingly, the developments in chromatographic techniques have proven capable of measuring much lower concentrations in foods, soil, and water.

Lowe, Christopher

Four decades after the term "affinity chromatography" was coined, this highly selective mode of chromatography remains an essential tool in the armory of separation techniques being used in research intensive industries. The technique continues to be favored owing to its high selectivity and predictability, speed, ease of use and yield, although widespread application of affinity chromatography in industry is tempered by the current high cost of the adsorbents. So the question is: What is the future of this technique in the twenty-first century? It is clear that the technique is likely to have a continuing impact on the biopharmaceutical industry to improve production processes for the "well-characterized biologic" in order to make them cost-efficient for the more stringent economic demands of the healthcare providers. Development of new highly selective approaches to the resolution of post-translationally modified proteins and their isoforms, elimination of leachates, and the removal of prion or viral contaminants from therapeutic preparations will require further attention. It is conceivable that ultra-high-throughput synthesis, selection, and optimization of affinity adsorbents may be conducted on-chip in the future. Furthermore, to implement the "omics" revolution, avid, high-resolution, and selective affinity techniques will be required to separate and analyze a range of low-abundance proteins found in biological samples.

Maloney, Todd

Recent advances in column technology and instrumentation have stimulated interest in ultra-high-performance separation techniques. Whether one uses high pressures, small particles, high temperatures, or high flow rates, the goal is to reduce analysis time with a concomitant increase in quality of information. With these goals in mind, separations scientists in the twenty-first century should revisit fundamentals of column performance and stationary-phase characterization. The abundance of new materials (semiporous silica, polymers), column formats (packed column, monolith, microchip), and applications (HPLC, UHPLC, capillary-LC, microfluids) reinforce the need for fundamental understanding of the dynamics of chromatography. As new column formats, stationary phases, and instruments are introduced to promise unique selectivity and superior performance, we must continually examine whether theory is achievable in practice.

Marriott, Phillip

Chromatography is entering a period of rapid transformation—which is probably surprising given its maturity. I envisage a future that will encompass considerably more hyphenation both of separation dimensions and expanded (spectroscopic) detection capabilities. The foundations for this have been established in the late twentieth century, and what we have seen is that multidimensional methods have undergone a significant resurgence in both popularity, and performance. This resurgence has been a consequence of much improved technical implementation of the basic instrumental components, development of new separation media, and the realization that discovery in the chemical sciences must rely on the generation of considerable more peak capacity than has been possible in the past. This is common to both gas and liquid chromatography formats. The improved performance must be matched by detection capabilities, suited to these multidimensional methods. As an example, NMR has been a powerful tool for chemical characterization in off-line methods, but it is hoped that continued improvements in detection limits will make this technology more generally applicable. Routine "multi-mode" mass spectrometry approaches through hyphenation (e.g., IMS with TOFMS), in combination with multidimensional separations, will provide a degree of separation and specificity that will make many previously difficult identification problems tractable. Trends towards much greater separation will run parallel with increased miniaturization and field-deployable equipment.

Neville, David C. A. and Butters, Terry

Accurate structural information of protein- and lipid-bound oligosaccharides is required to evaluate the contribution of this modification to biological function. Improved methods for the nondestructive release of the oligosaccharide, fluorescence labeling, and separation using HPLC techniques has made a complex analytical problem easier to solve and provided tools that are accessible to most laboratories. Significant challenges remain, and improvements in the chromatographic resolution of glycan species in a predictive and rapid fashion will be an important goal. The application of advanced techniques will aid the identification of disease-related changes to protein and lipid glycosylation leading to an understanding of the pathological mechanisms and therapeutic opportunities.

Okamoto, Yoshio

As described above, the polysaccharide-based chiral-phase modifiers (CPMs) show very high chiral recognition and today, these are the most popular CPMs. Many chiral compounds can be analyzed with these CPMs. However, these are still not sufficient for the preparative resolution in an industrial scale, which usually asks a high productivity. To attain a high productivity, a CPM must show a high chiral recognition to a target compound. Until now, we have intended to develop the CPMs with a broad applicability. However, for a large-scale separation, we must find out the efficient methods for synthesizing the CSP with a specifically high chiral recognition for a target compound.

Saelzer, Roberto and Vega, Mario

And what of TLC in the future? Keep it as a quantitative tool with all the advantages that it possesses, like visual side-by-side comparison of samples, standards, and unknowns; complemented with the recent advances in connecting HPTLC with mass spectrometry that confirms the flexibility of planar chromatography.

Sandra, Pat

The contamination of Belgian food products was initially referred to as a "dioxin crisis." The contamination source, however, was polychlorinated biphenyls. PCBs can be analyzed much faster and cheaper than PCDDs and PCDFs and therefore the analysis of PCBs was the only possible solution to the crisis. A new method, using ultrasonic extraction, followed by matrix solid-phase dispersion cleanup and CGC-μECD or CGC-MS, was used for fast screening the samples. Retention-time-locking and fast capillary CG completed the introduction of new methodologies in the PCB screening.

Schurig, Volker

The development of enantioselective chromatography continues unabated. In the pharmaceutical environment, the screening of enantiomeric compositions and the detection of high enantiomeric purities is of utmost importance in drug analysis. In the field of analytical enantioselective chromatography, the following perspectives and challenges are noteworthy: (1) very fast enantiomeric analysis; (2) combinatorial chiral selector screening by multiple high-throughput determinations of the enantiomeric excess (ee); (3) enantiomeric purity analysis in the range of 1 : 100,000 (ee = 99.998%); (4) unraveling the mechanistic aspects of enantiorecognition; and (5) further enantiomerization studies of configurationally labile compounds by dynamic GC, LC, and CE.

In the field of preparative enantioselective chromatography, the following perspectives and challenges can be envisioned: (1) the preparative access to single enantiomers on chiral stationary phases (CSPs) by discontinuous (batchwise) or continuous chromatographic processes (simulated moving-bed technology), complementing or superceding synthetic procedures to single enantiomers (crystallization, chirality pool approaches, catalysis, enzymatic reactions, kinetic resolutions), whereby enantioselective chromatography (as a thermodynamically controlled process) furnishes both enantiomers in 100% enantiomeric purity even on enantiomerically impure CSPs, whereas enantioselective catalysis (as a kinetically controlled process) produces only one enantiomer in one run, not always in an enantiomerically pure form and not free of residual heavy-metal catalyst; (2) enantioselective carbon dioxide–based supercritical-fluid chromatography (SFC) will gain momentum as a "green" alternative to liquid chromatography; and (3) refinement of monoclonal antibodies and synthetic plastibodies (molecularly imprinted polymers, MIPs) as CSPs in enantioselective chromatography.

Stalcup, Apryll

Within the separations community there is incredible excitement right now. A very sophisticated material science has emerged as the separations community has risen to the challenges presented in the analysis of complex samples. Separation techniques now serve as incredibly sensitive tools for investigating and amplifying subtle differences in intermolecular interactions. The same intermolecular interactions responsible for the phase behavior of pure substances (e.g., dispersion, hydrogen bonding, dipole–dipole, dipole–induced dipole, electrostatic) are being creatively exploited by the separations community. A variety of novel competitive strategies are evolving to address the challenges of complicated samples as the community reevaluates how columns are engineered. Simultaneously, new developments in separations capability are providing new competencies and new information to a wide variety of disciplines.

Svec, Frantisek

Since it is unlikely that interest in increasingly rapid separations of ever growing numbers of samples may decline, the war with diffusional limitations of mass transfer within stationary phases will continue. Development of various approaches including decreasing size of particulate packings, formation of novel bead formats, and monoliths appears to lead the way. We need to wait to see which of these techniques will prevail. All have their advantages and weaknesses. It is also possible that a completely new technology will emerge. It may be condemned at the beginning but accepted eventually, just as we experienced with monolith. In addition to the use of "brutal force" of increasing inlet pressure and/or flow rate, throughput can also be enhanced via parallelism. An ideal format for this approach would be microfluidic chips that can include several parallel channels in which separations can run simultaneously. This technology may lead to realization of the lab-on-a-chip concept adding more functions to the separations.

Zooming in on monoliths, I trust that they will be finding new applications with many completely different from chromatography. For example, the initial results with immobilized enzyme reactors containing several enzymes in tandem demonstrate that systems mimicking metabolic paths can be formed and used in environmentally friendly production of valuable compounds. Formation of monolithic thin layers appears to attract attention as well. In combination with photopatterning, they will enable two dimensional separations of peptides, preparation of superhydrophobic materials, and diagnostic devices with flow channels confined just by a large difference in surface tension.

Uden, Peter

The essential experimental features of gas and liquid (and supercritical fluid) mobile phase chromatographies are established and will continue. Optimization with respect to speed, resolution, and sample capacity will be a continued focus with viable two-parameter combinations being paramount. Optimization with respect to analyte molecular weight range will be enhanced. Accurate and repeatable high-speed methods will be pursued with goals of reliable computer-controlled methodology. While the separation methodologies are well set, detection techniques will surely be further developed. Comprehensive online identification of separated analytes at nanogram levels or below will be an overarching goal. Simplification of information-rich "instrumentation detectors" such as mass, resonance, and atomic spectroscopies will move toward integrated analyte determinations. Application fields in biochemical, clinical monitoring, and nanomaterials will undoubtedly expand.

Xu, Raymond

Chromatographic techniques will continue to evolve for pharmaceutical research and development in the twenty-first century. Speed of separation will be further advanced to satisfy the needs of shortened pharmaceutical development cycle time and rising sample volume. Resolution of chromatography is another key issue to be addressed to face the challenges of complex problems such as proteomic analysis and metabolomic analysis. Among the techniques used for pharmacokinetic and drug metabolism studies, we see that monolithic chromatography and ultra-high-performance liquid chromatography with sub-2-μm particles will make a significant impact to analytical scientists in either developing strategies for a new method or modernizing an analytical laboratory.

REFERENCES

1. C. W. Gehrke, R. L. Wixom, and E. Bayer (Eds.), *Chromatography: A Century of Discovery*, Elsevier, Amsterdam, 2001.
2. J. Sherma and B. Fried (Eds.), *Handbook of Thin Layer Chromatography*, Marcel Dekker, New York, 2003.
3. C. W. Gehrke, in Ref. 1 (above), Sec. 1.D.
4. J. A. Marisky, Y. Marcus, and A. K. SenGupta (Eds.), *Ion Exchange and Solvent Extraction—A Series of Advances*, Vols. 3–18, CRC Press, Boca Raton, FL, 1969–2007.

5. C. W. Gehrke, in Ref. 1 (above), Sec. 1.E.

6. J. C. Giddings and R. A. Keller (Eds.), *Advances in Chromatography*, Vols. 1–46, Taylor & Francis, Boca Raton, FL, 1965–2008.

11.F. SYSTEMS BIOLOGY, GENOMICS, AND PROTEOMICS
(Robert L. Wixom and Charles W. Gehrke)

Life in animals, plants, microorganisms, and lower life forms can be studied by the nature of their components and their functions, as followed for over 100 years, *or* as integrated living cells, tissues, and intact organisms—a subject of recent search in the life sciences, namely, systems biology, genomics, proteomics and other subareas. Let us briefly examine each of these areas and their relation to chromatography.

Chromatography as either a laboratory procedure or a scientific discipline has always been a contributor to other scientific disciplines, particularly chemistry, biology, environmental science, biochemistry, and molecular biology, and their related applied disciplines of medicine and agriculture. One of these, the concept of molecular biology in 1938, was based on the insights of Warren Weaver, President of the Rockefeller Foundation, and on the research of Francis Crick and James Watson (Nobel Prize, 1962). Molecular biology aimed to decipher down from the level of cells to the molecules therein, and to understand their function in overall biological processes. Now we turn the page of history to introduce new areas of understanding, which also constitutes a "paradigm shift" (see Chapter 3).

Systems biology was recognized initially by Ludwig von Bertlanffly in his 1969 book [1], who applied general systems theory to biology and later to several other sciences [2]. Contemporary science needs to move beyond the interplay of elementary units to recognize the role of "wholeness—the problems of organization, phenomena not resolvable into local events, dynamic interactions manifest in the difference of parts when isolated, or in higher organisms" [2]. Beyond identifying all the genes and proteins in an organism are the gene regulatory network and their biochemical interactions [2], leading to identification of "four key properties: system structure, system organics, the control method, and the design method" [2]. Progress in these areas requires advances in our understanding of computational sciences, engineering design of systems, genomics, proteomics, separation and measurement technologies (such as chromatography and mass spectrometry), and integration of such discoveries with existing knowledge [2].

Biology has grown in knowledge over the years by the reductional approach to the many cell components. However, the "new discipline of systems biology examines how these components are intact and form networks, and how the networks generate whole cell functions corresponding to observable phenotypes" [3]. Other leaders describe "system biology as the coordinated study of biological systems by investigating the components of cellular networks and their interactions, applying experimental high-throughput and whole-genome techniques, and integrating computational methods with experimental efforts" [4].

One of several leaders in systems biology is Dr. Leroy Hood, Professor at the University of Washington and Founder (2000) and President of its Institute for

Systems Biology. Dr. Hood has received many awards and most recently the PITTCON Heritage Award (2008) [5]. Pertinent to the subject, "he and his colleagues pioneered five instruments—the DNA gene sequencer and synthesizer, and the protein sequencer and synthesizer, and ink-jet oligonucleotide synthesizer"—to bring engineering into biology [5]. His >600 peer-reviewed papers are related to application of these technologies to immunology, neurobiology, cancer biology, molecular evolution, and systems medicine. Hood describes that the aim of system biology is to decipher "what cell or organism components do or function, how these components work together and how these circuits are integrated" [5]. His Institute is based on two types of biological information: digital genome and environmental; that biological networks capture, transmit, integrate, dispense, and execute biological information and that biological information is hierarchical and multi-scaler. His agenda has been to use biology to drive technology and computation—"to invent the future" [5]. He envisions the framework for medicine evolving from its current reactive mode to a predictive, preventive, personalized, and participatory mode ("forming a P4 medicine") over the next 5–20 years [5a,5b]. Several of the many recent books on systems biology are cited [3,4,6]. Consequently, systems biology is inclusive for genomics, proteomics, and other "-omics"; each is driven by both experiments and hypotheses.

Over similar recent decades, the term *gene* has led to *genome* and *genomics*, which refer to the complete set of genetic material of an organism [7]. WorldCat gave the first appearance of *genome* in a Ph.D. thesis in 1942 and in a thesis abstract in 1943. Time marches on. Usage of these terms has grown so that the 2007 year of literature contained genome or genetics in 281 titles of books and 11,719 titles of journal articles! Genomics refers to the examination of the full complement of genes in the cell—their size, physical structure, component nucleotides, sequence information, and function and then relates the organism's DNA to biology, health, and society [8]. One key milestone was the 2002 announcement of the complete DNA sequence of the human genome (Drs. Francis Collins and Craig Venter). Dr. Collins has made major contributions in his leadership of the National Center for Human Genome Research [9]. Key reviews and books [10–15] have followed. In earlier decades, practitioners of molecular biology identified specific mutants and/or their related functional proteins (or enzymes), characterizing them and relating them to specific genes. For more recent genomics, the study of the flow of genetic information is reversed—namely, utilizing DNA sequence information with the assistance of bioinformatics to understand protein structures and their subsequent functions. The rapidly growing area of genomics is highlighted by selected references [8–15].

Proteome, the term first used in 1995, refers to "the hierarchal analysis for the mass screening of proteins and the analysis of genomes via their protein complement or proteome" [16]. The "proteome technology" may be used to unlock the genetic code [17]. *Proteome* refers to "the entire complement of proteins that is (or can be) expressed by cell, tissue or organism" [17]. The proteome is "the time- and cell-specific protein complement of the genome, encompassing all proteins expressed in a cell at any given time" [18]. While "the genome of the cell is constant and nearly identical for all cells of an organ or organism, the proteome is extremely complex, dynamic and in far greater numbers" [19]. Proteomics includes ascertaining the facets of protein structure, function, expression, the many forms of post-translational modifications, degradations, and how proteins interact in biological systems. The proteins are considered to be the real effectors of all biological activities from signal transduction to hormone communication between distal sites [19]. Another description of proteomics is "the encompassing of all aspects of systems-oriented, global protein analysis and function, emphasizing the synergy between the physical and life sciences to yield a multi-disciplinary approach to the understanding of biological processing" [18]. Several selections of the many recent books on the proteome and proteomics are cited in Refs. 17 and 19–25.

The preceding paragraphs are necessarily abbreviated, but indicate the contributory role of different chromatographies (usually HPLC) for repository analysis and fragmentation of nucleic acids (mers), nucleotides, nucleosides, purines, pyrimidines, proteins, peptides, and amino acids for the structure and, in many specific examples, the function of genes and proteins. Proteomics has built mainly on the procedures of 2D gel electrophoresis, protein chips [18], several methods of chromatography [19], and particularly, mass spectrometry and the use of other hyphenated methods [20,22,24].

The most recent literature has additional scientific reports on related areas: glycomics, transcriptomics, peptidomics, metabolomics, bioinformatics, and others. The suffix "-omics" directs attention to a holistic abstraction, an eventual goal of which only a few parts may be initially at hand [26]. The reader is referred to these and other similar subjects in both the printed literature and the online sources.

The numeric values in Table 11.1, although approximate, indicate the high potency and fertility of these three areas. The significant numbers of the correlations in the Boolean searches establish the intimacy or interrelatedness of these terms. Without any doubt, chromatography and spectroscopy will contribute to many future scientific developments and interrelationships.

What are the messages derived from the preceding paragraphs in this section of the chapter? How do these messages illuminate the *"science of discovery"*?

- Biological sciences do not stand still, but march on in depth of understanding, breadth of coverage, integration of experimental findings, and, the comprehensive role(s) of cell components and their interactions.
- The knowledge of and the laboratory experience with the many forms of chromatography (see Refs. 27 and 28 plus our Chapters 3, 4, and 5 with the descriptions of the awardees' research) have enriched many subject areas, but particularly molecular biology, genomics, and proteomics as chromatography moves forward to 2010.
- Research in molecular biology, genomics, proteomics, metabolomics, and other similar focuses leads to the integrative role of systems biology and personalized medicine.
- The paragraphs in this section cover only some of the highlights and several selected references. Considerable details are known [5–30], and are reviewed elsewhere, but must be omitted here because of space limitations. The reader is encouraged to visit the scientific library and/or the online search engines (MEDLINE, PubMed, CAS Online, SCOPUS, etc.) for the stated area.

TABLE 11.1. The Growth and Combination of Systems Biology, Genomics, and Proteomics

	Systems Biology	Genome or Genomics	Proteome or Proteomics
Term first used	1966	Genomics: 1942	Proteomics: 1995
Books in 1990[a]	0	0	0
Books in 2007[a]	11	281	86
Articles in 1990[b]	0	2149	0
Articles in 2007[b]	471	11,719	2732
Boolean search with system biology (books in 2007)[b,c]	–	2	2
Boolean search with genome or genomics (books in 2007)[a,c]	2	–	15
Boolean search with proteome or proteomics (books in 2007)[a]	2	15	–
Boolean search with system biology (articles in 2007)[b,c]	–	97	74
Boolean search with genome or genomics (articles in 2007)[b,c]	97	–	418
Boolean search with proteome or proteomics (articles in 2007)[b,c]	74	418	–
Boolean search with chromatography (articles in 2007)[b,c]	14[d]	87[d]	264

[a]*Date*: June 2008. *Source*: WorldCat. *Notes*: Searched for term in title of book; discarded duplicate entries; includes theses and dissertations.

[b]*Date*: June 2008. *Source*: PubMed. *Notes*: Searched for term in title or abstract of articles.

[c]Boolean search data by Ms. Kate Anderson, Reference Librarian, Health Science Libraries, University of Missouri—Columbia.

[d]Lower frequency of combined terms here due to reliance also on 2D gel electrophoresis, mass spectrometry, and other methods.

11.G. OVERALL CONCLUSIONS—THE EDITORS

Chapters 1–3 introduced chromatography as a discipline of science, discussed the definition of chromatography, and presented the early leadership for chromatography that includes five Nobel Prize awardees. Chromatographic procedures were utilized by an additional 37 Nobel awardees. As the contributions of chromatography grew and spread after the 1940s, multiple trails of research developed as described in Chapter 4 (see also Refs. 31 and 32). The search continued and opened new frontiers with the chromatography awards in the recent (2001–2008) period. Chapter 7 covers advances in space, environmental, biological, and medical sciences, including chromatography of amino acids and nucleosides. Chapter 8 presents applications on natural products. Chapter 10 presents a history and evolution of chromatography over the decades.

Sections 11.A, 11.B, and 11.C in this chapter provide evidence of the strong impact of chromatography on the Pittsburgh Conference on Analytical Chemistry and Applied

Spectroscopy (PITTCON), the Chemical Heritage Foundation, and the American Chemical Society. The significance of analytical chromatography as molecular modifiers is emphasized in Section 11.D. The Section 11.E, on perspectives, contains insights on chromatography in the near future by knowledgeable chromatographers/scientists. The last, present section (11.F) introduces the recent past and now current research on systems biology, genomics and proteomics, and related areas. The areas of genomics and proteomics and other related areas have had considerable input from modern chromatography due to its sensitivity, specificity, reliability, and rapidity. Knowledge about these macromolecules—the genes with their molecular information and proteins with their many post-translational modifications—a long list of cellular enzymes, receptors, and structural proteins—leads to modern molecular biology and systems biology.

A review of the 11 chapters in this book, plus our many citations, plus the considerable other related, but not cited, references leads to some general observations. Science has a history of many connections over time, although some are cloudy. Science employs the pursuit of the unknown and an organized endeavor beyond the partially known areas of life, the physical world, the planet, and the solar system. The science of chromatography has contributed to the advancements of physical chemistry, inorganic/organic chemistry, analytical chemistry, biochemistry, biology, and other areas of study. The borders or boundaries of the classical sciences are melting and changing with the more recent insights and research. Separation techniques have become incredibly sensitive tools for investigating and amplifying subtle differences in intermolecular interactions of analytes, as manifested in hydrogen bonding, dispersion, dipole–dipole, electrostatic, and other molecular interactions. New areas of science appeared during 2000–2008 and, indeed, in almost every earlier decade.

Yes, science involves dedicated individuals, separated by country, language, and century. The scientists in our preceding book and this book come from many different countries. Science is a recognized international pursuit of knowledge. The advances in science depend on the human qualities of creativity, integration, perception, and insight along with initiative, intuition, rational thinking, continuous persistence, and critical reasoning. Coherence and communication of very large areas of knowledge have been achieved by the combination of scientific organizations, professional meetings, journals, related scientific books, and the recent computer sources. The many specific details in the preceding chapters are descriptive of and consistent with the overall pattern—"*the science of discovery.*"

REFERENCES

1. L. Von Bertalanffy, *General System Theory—Foundations, Development, Applications*, George Brazillen Publisher, New York, 1969.
2. L. Chong, L. B. Ray, H. Kitano, et al., Systems biology, *Science* **295**(5560), 1661–1682 (2007).
3. B. O. Palsson, *Systems Biology—Properties of Reconstructed Networks*, Cambridge Univ. Press, New York, 2006.
4. E. Klipp et al., *Systems Biology for Practice—Concepts, Implementation and Application*, Wiley-VCH, Weinheim, Germany, 2005.

5a. L. Hood, Systems biology and systems medicine—plenary lecture, paper presented at Pittsburgh Conf. Analytical Chemistry and Applied Spectroscopy (abbreviated PITTCON), Program of Annual Meeting, New Orleans, LA, March 2, 2008, pp. 1, 12.

5b. R. Mullin, Personalized medicine, *Chem. Eng. News (ACS)* **86**(6), 17–27 (Feb. 11, 2008).

6. A. Kriete and R. Ellis (Eds.), *Computational Systems Biology*, Elsevier, Amsterdam, 2006.

7. Genome, definition from *Oxford English Dictionary Online*, 2008; available from http://dictionary.ocd.com/cgi/entry/500937042.

8. F. Collins et al., A vision for the future of genomics research, *Nature* **422**(6934), 835–847 (2003).

9. S. R. Morrissey, F. Collins leaves NIH, *Chem. Eng. News (ACS)* **86**(31), 33–34 (Aug. 4, 2008).

10a. S. Choudhuri, The path from nuclein to human genome. *Bull. Sci. Technol. Soc.* **23**, 360–367 (2003).

10b. D. Cooper (Ed.), *Nature Encyclopedia of the Human Genome*, Nature Publishing Group, London, 2003.

11. J. W. Dale and M. von Schantz, *From Genes to Genomes*, 2nd ed., Wiley, Hoboken, NJ, 2007.

12. V. Bastolla et al. (Eds.), *Structural Approaches to Sequence Evolution: Molecular Networks, Populations*, Springer, Berlin, 2007.

13. M. Bina, *Gene Mapping, Discovery and Expression: Methods and Protocols*, Humana Press, Totowa, NJ, 2006.

14. H. Kitano, Systems biology—a brief overview, *Science* **295** (1661), 1662–1664 (2002), plus three other associated reviews.

15. G. Kahl, *The Dictionary of Genomics, Transcriptomics and Proteomics*, 4th ed., Wiley-VCH, Weinheim, Germany, 2009.

16. Proteome, definition from *Oxford English Dictionary Online*, and other books, 2008; available at http://dictionary.oed.com/cgi/entry00306962?.

17. M. L. Wilkins, Progress with proteome projects: Why all proteins expressed by a genome should be identified and how to do it, *Biotech. Genet. Eng. Rev.* **13**, 19–50 (1995).

18. R. Abersold, *ASBMB* [American Society for Biochemistry and Molecular Biology] *Today—Bringing Proteomics into the 21st Century* Feb. 2008, pp. 31–33.

19. N. Takahashi and T. Isobe, *Proteomic Biology Using LC-MS*, Wiley, Hoboken, NJ, 2007.

20. J. S. Albala and I. Humphrey-Smith, *Protein Arrays, Biochips and Proteomics—the Next Phase of Genomic Discovery*, Marcel Dekker, New York, 2003.

21. D. Nedelkov and R. W. Nelson (Eds.), *New and Emerging Proteomic Techniques*, Methods in Molecular Biology Series, Vol. 328, Humana Press, Totowa, NJ, 2006.

22. A. G. B. Smejkal and A. Lazarev (Eds.), *Separation Methods in Proteomics*, CRC/Taylor & Francis, Boca Raton, FL, 2006.

23. S. R. Y. Wong, M. W. Linder, and R. Valdes (Eds.), *Pharmacogenomics and Proteomics*, American Association for Clinical Chemistry, Washington, DC, 2006.

24. T. D. Veenstra and J. R. Yates, *Proteomics for Biological Discovery*, Wiley, Hoboken, NJ, 2006.

25. E. Bertrand and M. Faspel (Eds.), *Subcellular Proteomics: From Cell Deconstruction to System Reconstruction*, Springer, Dordrecht, The Netherlands, 2007.

26. I. Eidhammer, K. Flikka, L. L. Martens, and S. O. Mikalsen, *Computational Methods for Mass Spectrometry Proteomics*, Wiley-Interscience, Chichester, UK, 2007.

27. S. Sechi (Ed.), *Quantitative Proteomics by Mass Spectrometry*, Methods in Molecular Biology Series, Vol. 359, Humana Press, Totawa, NJ, 2007.

28. T. Wehr, Quantitative proteomics—tools and results, *LC-GC* **25**(10), 1030–1040 (2007).

29. R. Westermeier, T. Naven, and H. Hans-Rudolf, *Proteomics in Practice*, 2nd ed., Wiley-VCH, Weinheim, Germany, 2008.

30. J. Lederberg and A. T. McCray, "Ome Sweet" omics—a genealogical treasury of words, *Scientist* **15**(7), 8 (2001).

31. C. W. Gehrke, R. L. Wixom, and E. Bayer (Eds.), *Chromatography: A Century of Discovery (1900–2000)—the Bridge to the Sciences/Technology*, Vol. 64, Elsevier, Amsterdam, 2001, Chap. 4.

32. C. W. Gehrke, R. L. Wixom, and E. Bayer (Eds.), *Chromatography: A New Discipline of Science (1900–2000)*; see Supplement on Internet at Chem. Web Preprint Server (http://www.chemweb.com/preprint); see Chaps. S9, S12, and S15.

AUTHOR/SCIENTIST INDEX

SUBJECT INDEX

Chromatography: A Science of Discovery. Edited by Robert L. Wixom and Charles W. Gehrke
Copyright © 2010 John Wiley & Sons, Inc.

COVER PHOTOGRAPHS KEY

Upper panel

1	Marie-Claire Henion	13	Volker Schurig
2	Mario Vega	14	Philip Marriott
3	Edward L. Cussler	15	David Neville
4	Richard Van Breeman	16	William Stein
5	Stanford Moore	17	Terry Butters
6	Patrick J. Sandra	18	Terry Berger
7	Peter C. Uden	19	Ben Burger
8	Ernst Bayer, Emanuel Gil-Av, & Volker Schurig	20	Arne W.K. Tiselius
9	Stephen Wren	21	Klaus Gerhardt, Roy Rice, & Charles W. Gehrke
10	Charles W. Gehrke	22	Phyllis Brown
11	Linus Pauling & Charles Gehrke	23	Richard L.M. Synge
12	Paul Haddad	24	Yoshio Okamoto

Lower panel

25	Thomas L. Chester	37	Todd Maloney
26	Apryll Stalcup	38	Christopher Lowe
27	Del A. Koch	39	Mitch Jacoby
28	Patrick Nolan	40	David S. Hage
29	Robert W. Zumwalt, Charles W. Gehrke, & David L. Stalling	41	Walter G. Jennings
		42	Ron E. Majors
30	Frantisek Svec	43	Kenneth C. Kuo, Robert W. Zumwalt, & Charles W. Gehrke
31	Yong Guo		
32	Roberto Saelzer	44	Michael Dong
33	Vadim Davankov	45	Janusz Pawliszyn
34	Robert L. Wixom	46	Raymond Naxing Xu
35	William S. Hancock	47	Archer J.P. Martin
36	Edward Yeung		

Printed in the United States
By Bookmasters